American Eugenics

American Eugenics

Race, Queer Anatomy, and the Science of Nationalism

Nancy Ordover

University of Minnesota Press
Minneapolis
London

Excerpts from the Margaret Sanger Papers are reprinted by permission of the Sophia Smith Collection, Smith College, Northampton, Massachusetts, and from Alexander C. Sanger, literary executor of the Margaret Sanger estate.

Published by the University of Minnesota Press
111 Third Avenue South, Suite 290
Minneapolis, MN 55401-2520
http://www.upress.umn.edu

Library of Congress Cataloging-in-Publication Data

Ordover, Nancy.
   American eugenics : race, queer anatomy, and the science of
nationalism / Nancy Ordover.
      p.   cm.
   Includes bibliographical references (p.   ) and index.
   ISBN 0-8166-3558-7 — ISBN 0-8166-3559-5
   1. Eugenics—United States—History—20th century.   I. Title.
   HQ755.5.U5 073 2003
   363.9'2—dc21   2002011518

Printed in the United States of America on acid-free paper

The University of Minnesota is an equal-opportunity educator and employer.

12 11 10 09 08 07 06 05 04 03   10 9 8 7 6 5 4 3 2 1

For my parents, Rochelle and Jordan, who made everything possible for me, and in memory of my grandparents, Avrum, Yitta, Mindl, and Wolf

"my lifetime listens to yours"

Contents

Acknowledgments   ix

Introduction   xi

I

National Hygiene: Twentieth-Century Immigration and the
Eugenics Lobby

   ImagiNation   3
   Calculating Hysteria   9
   The Immigrant Within   32
   The Pioneer Fund: Scientific Racism and the Eugenic
      Endowment   45
   "Indiscriminate Kindness" and "Maudlin Sentimentalism":
      Fighting the "Philanthropic" Impulse   51
   The Abiding Panic   54

II

Queer Anatomy: One Hundred Years of Diagnosis, Dissection, and
Political Strategy

   Science as Savior   59
   Delineating Deviance: Moral Imperatives, Hereditarian Hypotheses,
      and the Letter of the Law   70
   Biological Apologists: Appeals and Miscalculations   83
   Gender, Race, and the Strategy of Metaphor   88
   Homosexuality and the Bio/Psych Merge: An Additive Model of
      Causation Theories   102
   AIDS, Backlash, and the Myth of Liberatory Biologism   119

III

Sterilization and Beyond: The Liberal Appeal of the Technofix

Liberal Loopholes   127

Buck v. Bell and Before   133

Margaret Sanger and the Eugenic Compact   137

Physical Fallout: Racism, Eugenics, and Liberal Accomplices
after World War II   159

New Technologies, Old Politics: Norplant and Beyond   179

Disability and Eugenics: The Constant Consensus   195

Quinacrine, the Next Wave   202

Conclusion   206

Notes   217

Index   275

Acknowledgments

These thank-yous are what I've most looked forward to writing. Yet if I were to enumerate all the ways I've been fed and inspired and stupefied by other people's patience and generosity and work and commitment, this volume would be twice as long.

That this work was undertaken and completed at all has to do more with the merits of my friends and family than my own. They gave me more support than I knew how to ask for: intellectual, emotional, technical, postal, and nutritional. Many thanks to my comrades, whose own labors in the world kept mine going: Thomas Kane Landry, Henry Serrano, Eileen Ordover, Chana Pollack, Jill Pierce, Shawn O'Toole, Rachel Pfeffer, Annie Kastor, Melanie Kaye/Kantrowitz, Gail Kielson, Amarpal Dhaliwal, Iris Garcia, Natasha Banta McDermott, Amalia Cabezas, Heidi Dorow, Herb Green, and René Poitevin. Amalia Aboitiz and Judith Romero caught me more times than I can count.

In addition to being good and true friends, Rachel Levitsky, Dana Greene, and Rachel Rosenbloom remained on call as sensitive readers and emergency researchers. Seb Katz, Francis "Bugs" Dooling, and Aaron Ze'ev Ordover Sary were just plain fun to be with.

Most of these individuals shared my sense of rage and urgency and irony, many laughed at my jokes, and a select few could be depended on for physical comedy. These are the things that keep a person going.

I am extremely grateful to Michael Omi, Caren Kaplan, and Evelyn Nakano Glenn for years of encouragement, guidance, and unflagging belief in the importance of this project. Bill Walker, erstwhile archivist of the Gay and Lesbian Historical Society of Northern California and all-around *mentsh*, enabled much of my research. Thanks too to Barbara Quan, Carla Atkins, and Althea Grannum-Cummings.

To say that George Lipsitz's comments were invaluable would be a gross understatement. For his insight, I thank not only him but also the University of Minnesota Press for sending him my manuscript. One phone call with William Murphy, my first editor at Minnesota, made it impossible for me to even consider another publisher. Working with Richard Morrison and Pieter Martin has only cemented that decision.

There aren't words for the love, respect, and gratitude I feel for my parents, Jordan and Rochelle Ordover. They gave me two great gifts: complete support and reason to take them for granted.

Introduction

> You talk of your breed of cattle
> And plan for a higher strain
> You double the food of the pasture,
> You heap up the measure of grain;
> You draw on the wits of the nation
> To better the barn and the pen,
> But what are you doing, my brother,
> To better the breed of men?
> —Rose Trumball, "To the Men of America"

On October 16, 1994, the cover of the *New York Times Book Review* sported a full-page color graphic of a DNA double helix alongside the headline, "How Much of Us Is in the Genes?" No less than five books on the subject were covered that Sunday, the most prominent being *The Bell Curve* by Charles Murray and Richard Herrnstein: "The articulation of issues touching on group intelligence and ethnicity has been neither fashionable nor safe for the last three decades, but these scholars argue that the time has come to grasp the nettle of political heresy, to discard social myths and come to grips with statistical evidence."[1] With this plaudit, the *Times* reviewer valorized a recent incarnation of a less than novel ideology: scientific racism. *The Bell Curve* merely restated old claims, chief among them being that intelligence can be quantified (by IQ tests), that African-Americans score an average of fifteen points lower than white Americans on these tests, and that genes are accountable for this rift.

A few pages later was a review of Dean Hamer and Peter Copeland's *The Science of Desire*, which also dealt with heredity, honing in on Hamer's quest for a "gay gene."[2] Hamer was, at that time, the latest in a long succession of theorists intent on establishing a biological or physiological basis of homosexuality. It is no accident that these books emerged at the same time or that each in turn was greeted by an on-air media blitz and splashed across the cover of every major magazine in this country.[3] A renewed respectability was being lavished on eugenics, a movement perhaps more overt in its principles and purpose in the early twentieth century, but still very much with us.

Early eugenics proponents, drawn from the ranks of scientists, politicians, doctors, sexologists, policy makers, reactionaries, and reformers, held that through selective breeding humans could and should direct their own evolution. Most believed in the supremacy of Nordic and Anglo-Saxon peoples, and to this end agitated for immigration restriction and supported antimiscegenation laws. Eugenicists advocated compulsory sterilization of the poor and the disabled and the "immoral." The legislation they drafted, the interventions they backed, the medical regimens they prescribed stemmed from a belief that everything from intellect to sexuality to poverty to crime was attributable to heredity.

This premise, and its attendant policy recommendations, have remained largely unchanged since Francis Galton first coined the term "eugenics" in 1883. Positive responses to *The Bell Curve* and *The Science of Desire* exemplified the way in which a retread of eugenic invective can be resuscitated, repackaged for public consumption, and hailed as brave, groundbreaking, and legitimate. There was nothing new here, save perhaps for a slightly abridged and terrifying vocabulary. "Eugenics" was avoided, but phrases such as "cognitive elite" and "cognitive disadvantage" began to gain currency in some circles, along with the resurrected "dysgenesis" (referring to the deterioration of allegedly heritable qualities from one generation to the next—in this instance the outbreeding of high IQ testers by those with lower scores).

In the interest of full disclosure, let me pause here and place myself in this tangle of social Darwinist affronts. I am Jewish and I am queer. I see my peoples among those dubiously honored as eugenic castoffs at both the entrance and exit of the twentieth century. Declarations of the limited cognitive abilities of Jewish immigrants—indeed of all immigrants from southern and eastern Europe, Latin America, and Asia—garnered the men who made them a respectability that drew strength from and reinforced their offensive against the native poor and the racialized in the early decades of this century. Likewise, escalating attacks on lesbians, gay men, bisexuals, and transgendered people seem to have coincided nicely with a flurry of scientific data proclaiming the genetic basis of homosexuality as the 1990s drew to an end. I cannot deny that a sense of my own history and my own future impelled me to embark on this project.

I began my research in the mid-nineties when eugenics enterprises were receiving rousing support from some political quarters and tepid validation from others. I was living in California at the time, where a white judge sentenced an African-American woman to Norplant; where Proposition 187, denying vital and heretofore legally guaranteed services to anyone suspected of being an undocumented immigrant, passed overwhelmingly with the help of a substantial donation from a foundation that backs race-based intelligence research. I lived then in a city where each new biology-based theory on homosexuality made the front page. I did not then, nor do I now, believe that this disinterred zeal to pin everything on biology is anything but a product of the historical moment that gave rise to it. It is no coincidence and it is not going away.

The long-lasting appeal of eugenics has rested on its protection of the status quo, on its emphasis on individual and group "failings" over analyses of systemic culprits, and on its bedrock insistence on scientific/technological remedies over fundamental social and institutional change. It has thrived in times of mainstream anxiety over genuine or perceived gains of marginalized groups, making it an attractive tool for conservatives. And so, decades after litigants and activists, doctors and attorneys proved that African-American, Latina, and Native American women and girls were being singled out for coerced, eugenically informed sterilization procedures, Norplant began to be foisted on these same communities with the full force of the judiciary and the medical establishment and with the blessing of both conservative politicians and liberal organizations. After generations of queers resisting pathologization, exactly twenty-five years after the Stonewall uprising, at a time of increased visibility in the political, social, and cultural realms, *The Science of Desire* appeared on the scene to cast us as genetically distinct from the rest of humankind. Eugenics is, once again, making a very public ascent.

To grasp the resiliency of this often discredited but never dormant philosophy is to understand the consolidation of race, gender, class, sexuality, and nation—not only as categories but also as ideological weapons of a state committed to eugenic curatives.[4] The first section of this book, "National Hygiene," examines eugenics as it served, and was served by, nationalism. Anointed guardian of national health and character, eugenics served the restrictionist lobby well for over a hundred years, predicting dire consequences for the country's bloodline if immigration of the "unfit" was not curtailed. The eugenic verdict on what constituted sound and unsound bodies was imposed on legislative debate, constructing immigrants as both contaminated and contaminators of the body politic. Eugenicists and their fellow travelers singled out everyone whose origins could not be traced to northwestern Europe, including southern Blacks making their way north during the Great Migration. As Charles Mills wrote of the era of de jure white supremacy, "[T]his period had the great virtue of social transparency.... One didn't have to look for *sub*text, because it was in the text itself."[5]

Science, and at its zenith eugenics was considered science, was not summoned solely by exclusionists and supremacists. It was, and continues to be, used simultaneously to persecute and to vindicate. The second section, "Queer Anatomy," examines the use of scientific pronouncements on homosexuality by a medical establishment hostile to all gender transgressors, including lesbians, gays, bisexuals, and transgendered people (LGBTs), and by queers and our allies who have long relied on eugenic, psychiatric, hormonal, and now genetic theories of homosexuality in a bid to win acceptance. In both instances, science has become savior, a cure for either homosexuality or homophobia, depending on your standpoint. And in both instances, the results have been catastrophic.

Finally, the third section, "Sterilization and Beyond," examines liberalism's propulsion of eugenics. An Enlightenment faith in progress has meant an unwavering belief in science as apolitical. This has, at different historical moments, extended to adherence to biological determinism.[6] Coupled with liberalism's investment in quick-fix individual solutions to basal inequities, this easily translated into support for race-, class-, and disability-based sterilization policies, from compulsory tubal ligations to mandated Depo-Provera acceptance. These measures were portrayed as a means to "help the poor help themselves." But wherever biologism and public policy have intersected, they have extracted a terrible price from the poor, physically and politically.

No treatise on U.S. immigration can stand without a discussion of nationalism and racism. Nationalism, George Mosse noted, has "attempted to co-opt most of the important movements of the age."[7] Eugenics, far from being exempt from this process of absorption, has played a pivotal role in nationalist and racist enterprises, as practice and as theoretical prism, an ideology in the service of other ideologies. As Étienne Balibar has written, "for the nation to be itself, it has to be racially or culturally pure."[8]

> This is an obsessional imperative which is directly responsible for the racialization of social groups whose collectivizing features will be set up as stigmata of exteriority and impurity, whether these relate to style of life, beliefs, or ethnic origins.[9]

Central to this imperative, Balibar contends, is a "racism of *extermination* or elimination (an 'exclusive' racism) and a racism of *oppression* or exploitation (an 'inclusive' racism), the one aiming to purify the social body of the stain or danger the inferior races may represent, the other seeking, by contrast, to hierarchize and partition society."[10] Eugenics employed and rationalized both "inclusive" and "exclusive" racism—the former, most notably, though not solely, through eighty-five years of IQ tests (an exam whose initial purpose in this country was to manufacture and expose the alleged intellectual inferiority of Jews, Italians, and Slavs), the latter by means of antimiscegenation laws, coerced sterilization, and most central to the first section of this book, immigration restriction.

"National Hygiene: Twentieth-Century Immigration and the Eugenics Lobby" examines eugenicists' contribution to the construction and monitoring of the nation, beginning with the confluence of anti-immigrant and pro-eugenics rhetoric that informed the Immigration Acts of 1917 and 1924 and continued on, through the decades, in an only slightly mutated form. It is an exploration not only of the blatant white supremacy and xenophobia that licensed these enactments, but also of the racial cataloging of immigrants and the ways in which national character was biologized and quantified—both in the pages of eugenicist journals and on the floor of the U.S. Congress.

Eugenicists' "scientific" validation of racism, their dexterity in exploiting labor strife, environmentalism, urban poverty, world war, crime, disease, and white anxiety over interracial social contact, is revealed in their own documentation of their movement. The *Journal of Heredity* (originally the *American Breeders Journal*) and *Eugenical News* proved essential to my investigation, both in terms of reconstructing eugenicists' campaigns and in grasping the meaning they themselves ascribed to their crusade. In this, they offered a unique portal to eugenicists' restrictionist agenda. More important, a survey of the publications' eclectic contributor pool reveals the broad base of their support. Eugenicists' own annals tended toward self-aggrandizement. They had everything to gain by exaggerating their political pull.[11] Their claims to political relevance might easily be dismissed as an exercise in self-importance if not for evidence provided by congressional records. Transcripts of various immigration debates and hearings attest to the fact that eugenicists were well received, even courted, by lawmakers. Indeed, some legislators counted themselves among their ranks.

Eugenicists owed their influence over twentieth-century immigration policy in large part to nineteenth-century legislation: the Page Law of 1875 and the 1882 Chinese Exclusion Act.[12] By 1917, decades of anti-Asian policy had already racialized the debate, though "debate," implying more dissent than was actually voiced by lawmakers, is certainly too active a term. The primacy accorded the 1917 and 1924 statutes here is not intended to downplay either the singularity or precedent-setting nature of these earlier bans, but rather to highlight the evolving and increasingly technical matrix of charts, test results, surveys, and computations that eugenics brought to the fore in immigration legislation in the decades before World War II.

Doctors wrote with alarm about America's disastrous attempt to "assimilate bad germ plasm."[13] Laymen, the most notorious of whom may be Madison Grant, warned of Jewish characteristics being "engrafted upon the stock of the nation."[14] Harry Laughlin, of the Eugenics Research Association, was invited to give expert testimony before the House Committee on Immigration and Naturalization's 1920 hearings on the *Biological Aspects of Immigration*. The language he employed was the language of chemists and anatomists and statisticians. Jews and Italians, he indicated on a chart submitted to the committee, composed a disproportionately high percentage of the foreign-born in U.S. asylums.[15] Citing the alleged immorality, innate propensities to crime and vice, smaller crania, illiteracy, compromised heredity, and diminished mental capacity of non-Nordic peoples, anti-immigrant agitators promised national calamity if Congress did not act fast. Immigrants, racialized and demonized, were posited as a threat to what Harvard professor Robert DeCourcey Ward called the "American race."[16]

During this period, the menacing "immigrant" was coded as male, with the important exceptions of the prostitute (corrupter of men) and the mother of future citizens (corrupter of bloodlines). It was this latter categorization that enabled eugenicists (although not only eugenicists) to advance and exploit white hysteria over bi- and multiracial children, actual and anticipated. Eugenicists were, among other things, believers in and tabulators of blood quantum. The Great Migration, along with the entry of large numbers of southern and eastern European immigrants, would, Grant wrote in 1916, "produce many racial hybrids and some ethnic horrors that will be beyond the power of future anthropologists to unravel."[17]

Eugenics used whatever was at hand to advance its legislative and social goals. It dovetailed nicely with virulent state animosity toward socialists, anarchists, and communists, as radicalism was, in the view of restrictionists, the domain of Italian and Jewish immigrants. During and after World War I, eugenicists warned of unwanted refugees flooding the United States—bad enough, they warned, in and of themselves, even worse because of the leftist "provocateurs" who would, in the chaos, be able to slip through existing immigration bans and spread propaganda. At the same time, eugenicists appealed to American-born, white workers. Opposition to immigration restriction was said to be a ploy by industrialists to undermine workers' gains and the health of the nation by importing cheap labor and defective germ plasm—all in the same bodies.

Most inquiries into the eugenics lobby and its impact on U.S. immigration end in 1924, when passage of the National Origins Act effectively slammed the door on any significant nonwestern/nonnorthern European immigration. Eugenicists and their allies, however, saw the legislation as a temporary compromise at best, and reinvigorated earlier warnings of white "amalgamation" with Puerto Ricans, Japanese (in Hawaii), Filipinos, and Mexicans (who, Laughlin exclaimed, were threatening to retake the Southwest).[18] Other accounts date the beginning of the end to the early 1930s, when the Third Reich made eugenics an ugly word, even in the United States. An examination of eugenicist publications, congressional records, and other supporting documents tells a different story. Neither the rise of fascism nor the reading of its record at Nuremberg spelled the end for eugenicists. Their influence was not as overt as in prewar years, but they remained a political presence. The Pioneer Fund, for example, has been one of the torchbearers for a eugenically informed immigration policy since 1937. The Fund's original charter called for "racial betterment" and increased procreation by descendants of white settlers of the original thirteen colonies. Francis Walter, cosponsor of the 1952 McCarran-Walter Act, served on its board in the 1950s.[19] When the 1965 Immigration Act went before the House Committee on the Judiciary, the Fund's codirector testified in favor of preserving the national origins quota system.

An examination of immigration "reform," past and present, provides an opportunity to expose the intertwining strands of eugenicist thought, if for no other reason than the fact that individual eugenicists have seldom confined their attacks to a single demographic group or social policy. The final chapter of *The Bell Curve* advocated both IQ-based entry for immigrants and the termination of welfare (which, according to the authors, fosters reproduction among women with low intelligence quotients).[20] Like their pronouncements on race and intelligence, Herrnstein and Murray's views on immigration echoed the work of early twentieth-century eugenicists: U.S.-born whites were situated at the top of the intellectual heap, and African-Americans were eternally fixed in last place:

> White immigrants have scores that put them above the mean for the native-born American population (though somewhat lower than the mean for native-born American whites). Foreign-born blacks score about five I.Q. points higher than native-born blacks.[21]

Herrnstein and Murray blamed the low mean IQ of immigrants (likely, they wrote, to be less than 100) on members of ethnic groups who score "significantly below the white average," thereby deflating the median figure of all immigrants.[22] "Immigrant" has always been a racialized term in the United States. Today, and certainly not for the first time, it is frequently used as a club against Latinos and Asians. The following is a not so thinly veiled attack on the 1965 Immigration Act, which allowed entry for previously excludable Asian immigrants provided they had family in the United States:

> This is not the place, nor are we the people, to try to rewrite immigration law. But we believe the main point of immigration law should be to serve America's interests. It should be among the goals of public policy to shift the flow of immigrants away from those admitted under the *nepotistic rules* (which broadly encourage the reunification of relatives) and toward those admitted under competency rules, already established in immigration law.[23]

Elsewhere in their eight-hundred-page tome, Herrnstein and Murray lamented the passing of turn-of-the-century immigrants who were "brave, hard-working, imaginative, self-starting—and probably smart," attributes, they commented, lacking in current immigrants to the United States. Of course, few Americans around 1900 recognized any of these qualities in the newcomers, and scientists and scholars went to great lengths to document the intellectual, spiritual, and physical shortcomings of the perceived interlopers.

Herrnstein and Murray's unapologetic nativism would have been just another episode in academic immigrant bashing if not for the historical moment in which it appeared: shortly before the vote on Proposition 187.[24] The suggestion here is not that California voters went to the polls with copies of *The Bell Curve* tucked squarely under their arms, but rather that the book and, more important, the media attention it generated were products of—and contributors to—a climate of heightened scapegoating and xenophobia of national proportions. It was, and remains, a hysteria reinforced and made corporeal by a federal crackdown: in the year following the passage of Proposition 187, the United States deported a record 51,000 undocumented immigrants.[25] While current anti-immigration offensives can in no way be reduced solely or even primarily to manifestations of scientific racism or paranoid concerns over the national gene pool, it would be a mistake to minimize the resemblance between much of the current rhetoric and the previous eugenics crusades for restrictive immigration laws it recalls.

In certain instances it seems that Mother Nature hesitates over her
sexual determinations. It almost amounts to biological stuttering. In
such cases anything may happen.[26]

A few years ago, UCLA biologist Dr. Lauren Allen told attendees at the
American Association for the Advancement of Science that in the not
too distant future a brain scan could conceivably alert parents to their
child's sexual orientation.

[T]here are some unknown percentages of people that are not going to
grow up heterosexual. . . . If these techniques could eventually be used to
identify these children, [then this knowledge] could be used to raise
those children in a way that they would feel more comfortable.[27]

Allen's remarks reflect a genuine faith in the implicit humanitarianism
of science, especially when applied to social issues. Leaving aside the
question of why any child should be brought up with an eye toward any
sexual orientation, it is highly unlikely that a brain scan would dismantle homophobia inside or outside the family. "Early detection" thus far
(i.e., parental guesswork) has moved too many people to disown, beat,
and even institutionalize their children.

But predictions such as Allen's, like Hamer's undertaking, continue
to receive warm welcomes, in part because of a liberal belief in progress
and knowledge, in part because social categories, rather than substantive research questions, are driving scientific inquiry. This is true not
only of Hamer's work at the National Institutes of Health, but of Simon
LeVay's measurements of cell-group size in the "gay hypothalamus" and
of Bailey and Pillard's lesbian twin studies. Due to their veneer of plau-
sibility and rationality, these are far more perilous than cause-and-effect
claims perhaps more easily laughed off—such as the recent lawsuit filed
by a woman who insists she was heterosexual until an electronic bingo
board fell on her.[28] Yet, there are obvious biases implicit in their research, the most pronounced being the fundamental premise that homosexuality is not normative. This fresh interest in the roots of homosexuality is cause for extreme concern, especially given the rise in hate
crimes and homophobia in the streets and the Senate that accompanied
its onset. Scientific exploration is neither funded nor pursued for politically neutral ends. At this level it is conducted for only one of two reasons: to either replicate or eradicate what is being studied. Given the
backlash to queer visibility and the victim-blaming apparent in legis-

lative and cultural responses to AIDS, it is difficult to believe that time and money were being doled out to ensure a steady population of gays and lesbians.[29] With the losses to AIDS, breast cancer, and youth suicide, and with the escalating antigay violence faced by our community, the more relevant and pressing question is not "why are we here?" or "how did we get this way?" but rather "why is it so important to find out?"

The second section, "Queer Anatomy: One Hundred Years of Diagnosis, Dissection, and Political Strategy," documents a century of medical models and interventions imposed on (and sometimes embraced by) lesbians, gays, transgendered people, and bisexuals in the United States.[30] It highlights the danger posed by the intersection of medical and judicial discourse—a melding that enabled punishment to be enacted under the guise of treatment—and the cumulative nature of all derivation hypotheses. Expositions of moral turpitude, anatomical flaws, eugenic mishaps, psychiatric disorders, and hormonal imbalances—none of these theories eclipsed the others, though each was hailed as signaling a new day for either treatment or liberation. Causation theories were additive, each supplying new (and not so new) ammunition to an arsenal of antigay rhetoric and punishment.

Because of the medical profession's dedication to publishing, there is no shortage of documentation. Diagnoses and treatments were assiduously chronicled in medical journals and related texts, both by practitioners antagonistic to queers as well as by those who thought they were working on our behalf. Articles in the *Journal of Orificial Surgery*, the *British Medical Journal, Sexology*, and numerous medical texts, not to mention contemporary news accounts, record over one hundred years of physicians' boasts and disappointments. Particularly in the nineteenth century, as Siobhan Somerville notes, this literature was one of only a few venues available for explicit discussions of sexuality. Further, it "held substantial definitional power within a culture that sanctioned science to discover and tell the truth about bodies."[31]

This "truth" was bound up with other truths. It would be impossible to excavate the roots and the reach of the medicalization of queers without an understanding of scientific racism and other eugenics endeavors. Not only were they ideologically tethered, but they had the same champions and often the same victims. In 1904, G. Frank Lydston, a professor at the Chicago College of Physicians and Surgeons who lectured on the dangers of "sexual perversion," also wrote, "Physical and moral degeneracy...

with a distinct reversion to type, is evident in the Southern negro . . . especially manifest in the direction of sexual proclivities."[32] Dr. F. E. Daniel of Texas advocated castration for "all sexual crimes or misdemeanors" as well as mandatory sterilization for the purposes of "race improvement."[33] African-American lesbians and gays were declared predatory; their anatomy, gender deviation, and sexual propensities sensationalized and exaggerated by researchers seeking to substantiate their own biases. Eugenics enterprises emerge as inextricably linked. Treatises on homosexuality enlarged and benefited from existing pronouncements of scientific racism, as well as eugenicist justification of the economic status quo and the relentless monitoring of female sexuality and compliance with gender norms.

When physicians and theorists connected nonprocreative sexuality among women to ancestry, insanity, and anatomy, lesbians and other gender nonconformists became obvious recipients of their attention. The simultaneous ascription of lesbianism to insanity and inheritance foreshadowed the more formal consolidation of psychiatric and biological approaches to homosexuality that has circulated for the better part of a century. "Orificial" surgery gave way to lobotomies and hormone regimens—both of which remained legitimate responses to homosexuality even in the midst of a lesbian and gay civil rights movement. Aversion therapy is still an accepted procedure among many practitioners.

There is, to borrow from Foucault, no doubt that the medical and juridical discourse on homosexuality that emerged over a hundred years ago enabled all manner of inventive social control over queers. But, he further noted, the emerging discourses "also made possible the formation of a 'reverse' discourse: homosexuality began to speak in its own behalf, to demand that its legitimacy or 'naturality' be acknowledged, often in the same vocabulary, using the same categories by which it was medically disqualified."[34] Queers were not only subjects of the new research, not only its consumers, but also its exploiters, if failed exploiters. As a political strategy, this has wrought grave consequences. The emerging heredity-based models resulted in further jeopardy for gays (not the least of which was inclusion in many states' mandatory sterilization statutes). Medical doctors' forays into the legal system had forged the ideological links between eugenics, "perversion," and crime. Inattentive to the punitive nature of the eugenic discourse that circulated around them, Havelock Ellis and Magnus Hirschfeld advanced biological theories of "inversion" and the "third sex" in the late nineteenth and early twentieth centuries. Their efforts to decriminalize homosexuality, resting on medical models, bolstered claims of inborn deviance. Work citing science as exonerator—if not emancipator—of a despised "sexual minority"—from Hirschfeld to the gay press of the past thirty years—document the painful truth that, in desperation, queers have frequently cooperated with our own oppression.

In a time marked by sheep cloning and the Human Genome Project, eugenics often operates under the guise of a sort of liberatory biologism. When Colorado's antigay Amendment 2 went to court in the fall of 1993, Donna Minkowitz of the *Village Voice* reported that "it was not the state's attorneys but the pro-gay ones whose arguments could set back the clock on gay liberation." In their attempt to prove the unconstitutionality of the amendment, attorneys called on "expert" witnesses to offer testimony on the "masculinization" of genitals of future lesbians in utero and "temporal lobe pathology" among transsexuals.[35] The amendment, which would have banned antidiscrimination protections for lesbians, gays, and bisexuals, was struck down, but predicating civil rights on anatomy remains a risky strategy and one generally used against marginalized peoples.[36]

Hamer, who also took the stand in Colorado, has reflected somewhat on the political uses of his work. Scientists, he has written, can take measures to ensure that their research is not abused, such as asserting intellectual property rights, opposing genetic tests and treatments for sexual orientation, and helping to set guidelines for the ethical use of genetic research.[37] Yet, one need only look to the physicists who split and "harnessed" the atom to realize that scientists cannot control the intellectual or cultural currency of their work. Conceding that today's research could lead to a range of prejudicial tactics, from employment and insurance discrimination (based on DNA scans for genes implicated in sexuality) to homophobia-induced decisions to abort, Hamer warned:

> Although such hypothetical dangers are frightening, I believe there is a far greater danger. A danger that is present now, not in the future. A danger that is real, not hypothetical. . . . The real danger is not to study sex at all.[38]

But such concerns are not so easily reduced to the realm of the hypothetical. Employers and insurers do discriminate.[39] Parents do discard and dismantle their queer children.

In response to concerns about a hypothetical "genetic sexual orientation test," Hamer wrote, "biology is neutral. There is no reason that
gays . . . could not use such a test to abort fetuses that had the 'straight
gene,' or refuse to hire people who didn't have a 'gay' genetic profile."
This, however, assumes a power equilibrium that simply does not exist.
Between 1992 and 1997, twenty-eight antigay initiatives passed in this
country and nine gay rights laws were repealed.[40] Laurie Andrews, genetic ethicist and author of *The Clone Age: Adventures in the New World
of Reproductive Technology,* cites a recent survey of potential parents that
found that 33 percent of respondents want control over their offspring's
sexual orientation.[41] All this leaves little doubt as to who would screen
for a genetic propensity toward homosexuality and toward what ends,
and yet, as of this writing, no major LGBT organization is monitoring
the ethics or implementation of genetic research on queers.[42] The question remains: if queers were truly safe, would the scientific community
be having this discussion?

> While the United States shuts her gates to foreigners, and is less hos-
> pitable than other countries in welcoming visitors to this land no
> attempt whatever is made to discourage the rapid multiplication of
> undesirable aliens—*and natives*—within our own borders. On the
> contrary: the Government of the United States deliberately encourages
> and even makes necessary by its laws the breeding—with breakneck
> rapidity—of idiots, defectives, diseased, feeble-minded and criminal
> classes.[43]

After World War II, eugenic ideology was largely considered to be the
property of foreign fascists, racists, and nationalists. The United States
was able to ignore its own history of scientific racism, supposedly negligible in comparison with the brutality of the Third Reich. Far from admitting our participation in the production and practice of eugenics,
we have cast our nation as savior of Europe's eugenics survivors, and
thus we can, to paraphrase Jonathan Boyarin, eulogize someone else's
victims while denying our own.[44] This is not hard to do, given the enor-
mity of the Nazi crusade. But the truth is more complex. Science claims
for itself a certain internationalism, and there was a tremendous exchange
of ideas in the eugenics world. Likewise, eugenics was not solely the
purview of the extreme right wing. It crossed not only international waters, but political ones as well. While there can be no doubt that U.S. eugenics has been staked and sustained by all manner of reactionaries and

conservatives, congressional xenophobes, racist policy makers, paternalistic medical practitioners, and a judiciary hostile to the poor and to the disabled, it has also been promoted by many who would recoil at being included in these ranks.

The third section, "Sterilization and Beyond: The Liberal Appeal of the Technofix," does not seek to exonerate or minimize the role of eugenicists on the political right. They were, and remain, the backbone of scientific racism. However, they were not alone. Liberal voices have, for the better part of this century, been among the loudest praising eugenicist undertakings, including mandatory sterilization policies. These champions of eugenic sterilization were driven by liberalism's elevation of the individual and by a persistent reliance on what I refer to as the "technofix" (a term collapsed from Robert Blank's assessment of "technological fixes" that "divert attention from the need for more basic social changes").[45]

The tribute Margaret Sanger is paid by mainstream feminism offers an excellent illustration of liberal acquiescence to and participation in the eugenics movement. Her activism in the birth control movement frequently overshadows her advocacy for income-based population control measures, yet there can be no mistaking her allegiance to the prominent eugenicists of her day. In 1919 Sanger wrote, "Like the advocates of Birth Control, the eugenicists... are seeking to assist the race toward the elimination of the unfit. Both are seeking a single end but they lay emphasis on different methods." Eugenics without birth control, she declared, "seems to us a house [built] upon the sands."[46] So inevitably entwined were these two movements in Sanger's mind that in 1925 she polled readers of her *Birth Control Review* on whether or not the journal should merge with a "Eugenics magazine." Though no such joint venture materialized, her publication, both during and after her tenure as editor, remained a de facto amalgamation of the two. Sanger, and the unproblematized homage she continues to be paid, embody the quintessential liberal-eugenic alliance, an alliance that has yet to be completely severed.

While Angela Davis, Linda Gordon, and Betsy Hartmann have exposed Sanger's adherence to eugenics doctrine, she continues to be vindicated by biographers and others as a woman who merely forged alliances where she could, rather than a true believer in eugenics. Ellen Chesler, author of *Woman of Valor: Margaret Sanger and the Birth Control Movement in America,* and Carole McCann, author of *Birth Control Politics in the*

United States, 1916–1945, characterize her as a product of her age, and on this point they are certainly correct. Her peers in the eugenics movement were preoccupied with concepts of "race hygiene" and "feeble-mindedness" and with mandatory sterilization statutes. Sanger shared with them their desire to punish those deemed genetically unworthy, their paternalism toward the poor, and their preoccupation with the morality and sexuality of people of color. Though McCann and Chesler deny Sanger's racism and minimize the extent to which her views on the disabled and the poor factored into her pro-eugenics stance, Sanger's articles, books, and personal correspondence all disclose an unabashed courtship of eugenicists. A discussion of the ties between eugenics and birth control is in no way intended to discredit the latter. At the same time, a feminist commitment to reproductive rights and freedoms must lay bare the ongoing reverberation of that early compact, particularly in welfare policies that encroached more and more into the bodies of poor women—especially women of color—as time wore on.

Sanger and her colleague Clarence Gamble focused their efforts on eliminating not poverty but the poor. Sanger endorsed and enabled Gamble's efforts to establish a direct link between welfare and sterilization in the southern states in the years after World War II. Gamble's vision was not his alone, and in the decades that followed, doctors, social workers, and government agencies took up the cause. Sterilization in the South was referred to as the "Mississippi appendectomy," not only because it was so common but because medical staff relied on deception to obtain "consent." Similar "protocol" was followed in clinics across the United States as hysterectomies and tubal ligations were performed on Chicanas, Puerto Ricans, African-Americans, and Native Americans without patient (or, in the case of young girls, parental) consent. As this campaign was commonly rationalized as a panacea to poverty, an attendance to class is essential to any examination of eugenic sterilization practices. Class dictated who would be viewed as unworthy of procreation and who would easily fall to economic coercion. What's more, after the Holocaust, overtly race-based sterilization proposals were a hard sell, necessitating a linguistic re-encoding of sorts. Similarly, today we hear not of eugenics, but of "gene therapy" and "health care rationing." This is not to understate or sublimate the importance of race as a central determinant in targeting entire groups of women for sterilization in the 1950s, 1960s, and 1970s. Sterilization abuse of African-American women, for example, was an extension of a long history of medical atrocities perpetrated by eugenicists eager to "put the Negro Question into Science and Science into the Negro Question."[47]

For the poor and racialized and criminalized, tubal ligation, hysterectomies, and vasectomies were never value-free medical procedures. Rather, they were technological fixes imposed on individual bodies in lieu of meaningful correctives to economic inequity. The latest round of reproductive technologies—quinacrine (tested on women in the Middle East, Latin America, and Asia), Norplant, and Depo-Provera—have not altered the technofix paradigm, nor have they shifted its demographics. All physician controlled, they stand in for substantive challenges to a grossly inadequate welfare system and a booming criminal justice system. A critique of liberal assent to such "solutions" as the targeting of teenage girls as Norplant accepters, the imposition of implants and other chemical contraceptives on AFDC (Aid to Families with Dependent Children, now Temporary Assistance to Needy Families) recipients, and Norplant-for-probation plea bargains concludes the third section.[48] The history of eugenic sterilization is also the history of the steadfast refusal of liberal organizations, including some liberal feminist organizations, to oppose practices that constitute population control policies and state-sponsored assaults on poor women and girls of color, even as they espouse reproductive "choice." But "choice" is not a static or immovable construct. It does not stand outside of history or politics or society. It is bound by race and class, particularly where technology is concerned.

Eugenics has always been an extremely nimble ideology. It cannot be isolated from the movements it bolstered and was conscripted by: nationalism, "reform-oriented" liberalism, out-and-out homophobia, white supremacy, misogyny, and racism. Its longevity relies on these confederacies for the simple reason that even as one falls into relative disrepute, others remain intact. Moreover, its appeals to biases of every stripe has meant that most individuals singled out for exclusion or genetic extrapolation or surgical intervention have multiple identities and belong to more than one target group. If for no other reason than this, "[t]he current round of genetic fetishism,"[49] as Adolph Reed calls it, demands an integrated rebuttal.

But if eugenics is agile, it is also reductionist and therefore highly amenable to sound-bite reportage. It promises simple, indisputable, predetermined core divisions—divisions that naturalize disfranchisement and disparity. Eugenics takes externally imposed social categories, with all their insufficiencies and hazards (and often violent origins), and from these postulates eternal medical truths. But these will ultimately prove insufficient. The world, as Muriel Rukeyser reminded us, is made of stories, not of atoms.

I

National Hygiene: Twentieth-Century Immigration and the Eugenics Lobby

> The character of the nation is determined primarily by its racial qualities; that is, by the hereditary physical, mental, and moral or temperamental traits of its people.... It is now high time that the eugenical element, that is, the factor of natural hereditary qualities which will determine our future characteristics and safety, receive due consideration.
>
> —Harry Laughlin, testimony before the U.S. House Committee on Immigration and Naturalization, "Biological Aspects of Immigration"

> The fact of the matter is that nationalism thinks in terms of historical destinies, while racism dreams of eternal contaminations, transmitted from the origins of time through an endless sequence of loathsome copulations: outside history.
>
> —Benedict Anderson, *Imagined Communities*

ImagiNation

Proposition 187 was voted into law in the state of California on November 8, 1994. Challenged in court and ultimately superseded by federal enactments, it sought to bar undocumented immigrants and their children from a host of services, including health care and public education. Service providers would have become mandated reporters, demanding verification of legal residency from potential clients, patients, and students, and turning over the names of any "suspect" individuals to the Immigration and Naturalization Service. Just two weeks before the election, National Public Radio reported that a key backer of Proposition 187, the Federation for American Immigration Reform (FAIR), had received some of its financing from the Pioneer Fund. Founded in 1937, the Pioneer Fund has provided millions of dollars to researchers working for "racial betterment," particularly in the area of race-based intelligence and similar eugenics enterprises. The ballot initiative (aka Save Our State or S.O.S.) received an additional twenty-thousand-dollar donation from Florida state senator Don Rogers, who is affiliated with the Christian Identity movement.[1] This confluence of anti-immigrant, pro-eugenics, and white supremacist support—unnamed and unexamined by the mainstream press—underscores the very quality that has endowed eugenics with long life: its capacity to both absorb and be absorbed by other ideologies and political agendas.

Certainly, neither the passage of Proposition 187 nor the subsequent introduction of other anti-immigration offensives can be reduced solely, or even primarily, to manifestations of scientific racism or paranoid concerns about the national gene pool. The image of immigrants (documented or otherwise) and their children as threats to the nation's well-being was ensured largely by constant references to "tax-payer burdens" and outright appeals to racism and nativism throughout the campaign.[2] The proposition was part of a larger backlash that grew to include the abandonment of affirmative action, the gutting of welfare at both the federal and state levels, and a record-breaking number of deportation proceedings. Yet, along with this has come a rash of biological determinist conjecture and suggestions that immigration policy, like education, criminal justice, and civil rights strategies, should bend to genetic conjecture. In such a climate, it would be a mistake to minimize the resemblance between much of this rhetoric and the previous eugenics crusades for immigration restriction it recalls. These earlier campaigns were also constituted by, and constitutive of, racism, xenophobia, and nationalism. This dynamic interplay, together with the opportunism of the eugenics movement, asserts itself repeatedly in U.S. immigration history.

Any examination of eugenic agitation underscores the inseverability of racism, xenophobia, and profound class bias. Eugenics was, after all, a response to both internal and external "perils." The Great Migration, the growing immigrant population, hysterical fears about "miscegenation"—all propelled eugenics among reactionaries and reformers alike. Biology would explain and ultimately solve the problems of the northern cities, obviating the need for social analysis. It was neither the first nor the last time science was called on to provide a quick fix for systemic inequities. Likewise, class cannot be separated out of the equation, affected as it is by racism and given the fact that "immigrant" was a buzzword for a specific stratum of workers in so many legislative bills governing entry and naturalization. Most importantly, class divisions were, and continue to be, validated by eugenic explanations. The poor have borne the brunt of eugenic enactments.

The "National Hygiene" section follows the trail of this legislation, beginning with eugenicists' efforts to solidify the fusion of race and nationality in the years before the passage of the 1917 Immigration Act and ending with recent anti-immigrant harangues. It is a history complicated by the pliant usage of "race" during the late nineteenth and early twentieth centuries to refer, at any given time, to religion, color, class, and/or national origins. Eugenicists' contributions to this legislation included the designation of moral, physical, and mental "inadequacies" as the immutable property of racialized immigrants and U.S.-born outsiders.

Bolstered by statistics garnered from IQ tests, they sounded panicked alarms over the entry of Slavs, Italians, Jews, Poles, and others with a "dysgenic" bent toward "feeble-mindedness." After the 1917 enactment, they deftly exploited fears over a post–World War I "overflow" of refugees and an atmosphere of anticommunist, antianarchist, and antilabor persecution. Like eugenicists involved in campaigns to regulate sexuality and procreation among other marginalized groups, they began with social categories and then sought to legitimize them through statistical and biological "evidence."

Exclusionist bids eventually culminated in the 1924 National Origins Act. Accepting neither the 1917 legislation nor the 1921 Three Percent Restrictive Act as fully satisfactory, eugenicists redoubled their efforts. No fringe element, they exerted direct influence on immigration debates: culling new test results, drafting legislation, testifying before Congress, and offering "evidence" of the disproportionate number of "insane" among Bulgarians, Chinese, Jews, Irish, Italians, Mexicans, Poles, Russians, and Turks. Eugenicists and eugenics sympathizers could be found in the House of Representatives, the Senate, and the White House.

After the National Origins Act, eugenicists continued to push for inspection at ports of departure (a component of their model eugenic legislation not realized in the 1924 law) as well as an extension of the quota system that would target Latin American and Caribbean immigrants. Prominent eugenicists called for a multipronged approach to "dry up . . . the streams that feed the torrent of defective and degenerate protoplasm":[3] immigration restriction, sterilization of the nation's "manifestly unfit," and a heightened assault on miscegenation. Theirs was a comprehensive plan, and, as evidenced by statements before congressional subcommittees and in their own journals, they were fully cognizant of the links they forged between nation building, white supremacy, xenophobia, and the eugenic dream.

Two early eugenicist publications, the *Journal of Heredity,* which began as the *American Breeders Magazine* in 1910, and *Eugenical News,* which hit its stride more than a decade later, offer a record of eugenicists' own perceptions of their impact on immigration legislation, particularly the enactments of 1917 and 1924. While pro-eugenics pieces appeared in other publications (such as Margaret Sanger's *Birth Control Review*), the *Journal* and the *News* were eugenicists' very own. Certainly, journal contributors, desirous of self-promotion, tended to exaggerate their sphere of political influence (their manipulation of statistics is particularly striking). Their words alone are not sufficient to judge their impact on the legislative process. At the same time, their bent toward self-aggrandizement made them diligent chroniclers of both the eugenics movement and the major anti-immigrant campaigns of their day.

For the better part of this century, eugenics, like the immigration bans it supported, has played a pivotal role in the "exclusive racism" Étienne Balibar has identified as essential to "purifying" the nation.[4] But the physical and ideological construction of the nation is a project that can never be completed, requiring constant monitoring and patrolling of literal and figurative borders. Eugenics has always been primarily concerned with what the nation would look like: surveying which women were bearing which men's children, tabulating (and frequently fabricating) hereditary disabilities, asserting biological determinants of sexual and political behavior, and zealously guarding the entryways to America's bloodlines. By warning against Latin American, Asian, eastern and southern European, and North African immigration, twentieth-century eugenics was a significant tool in the hands of those seeking to construct and preserve an Anglo-Saxon nation. These "interlopers," along with American Blacks, were viewed as both contaminated bodies and contaminators of the body politic.

All this is not to say that the nationalism and racism that found expression in—and contributed to—restrictive immigration policies *needed* eugenics. Long before eugenicists fully undertook to insinuate themselves into the immigration debate, nation building was already a racialized project. In 1882, during debate on the Chinese Exclusion Law, Senator John T. Morgan of Alabama decried Chinese immigration as an impediment to the complete (white) settlement of the West: "the young man with his wife upon his arm who chooses to go to the West and settle himself upon a farm and toil for a livelihood with some expectation of greater growth in later years, will not go to the West because the Chinaman is there."[5] Nevertheless, eugenics gave racism and nationalism substance by bringing to bear the rationalizing technologies of the day. In turn, these ideologies were prime movers of eugenic thought and its accompanying legislation.

Nationalism, because it relies so heavily on other ideologies for coherence and practical application, is impossible to isolate. As Balibar has written, it cannot be disentangled from racism[6] or from xenophobia—both of which it absorbed, both of which marked the same immigrant groups as inadmissible. The nationalism cultivated by early twentieth-century eugenicists relied largely upon a default mechanism. In their attempts to re-create a nation that never was and to distill an American phenotype, they engaged in what Balibar has called "the obsessional quest for a 'core' authenticity that cannot be found."[7] Thus, eugenicists constantly lamented the "watering down of our nation's life-blood" by the admission and procreation of "alien defectives."[8]

In constructing entire racialized categories of demonized others, eugenicists put forth an ideologically purified America—purged of past sins and guarded against future menace. The eugenics project revolved around imagining the nation: what it was (now threatened) and what it might be (with and without government and medical intervention). It was the sort of creative visualization that demanded both historical revisionism and ominous prophesy. Few were as adept at mapping this to a legislative agenda as Robert DeCourcey Ward.

Ward, a professor of climatology at Harvard and a member of that university's administrative board, holds a prominent place in the history of the U.S. eugenics movement. His 1931 eulogy read: "For years, every patriotic legislator struggling against odds to obtain the needed legislation, and every conscientious official working for proper enforcement against misrepresentation and malign political influence, felt the result of his labors."[9] In 1912 Ward wrote in the *American Breeders Magazine*, "Our country was founded and developed by picked men and picked women."[10] Measured against this counterfeit history was his account of impending national degradation, published the following year:

> The days of a dominant Anglo-Saxon immigration are over, forever.... Southern and Eastern European immigration has increased until it now numbers about 70 per cent of the total.... Asia is contributing more each year. British India has begun to send the advance-guard of its coming millions....
> ... There can be absolutely no doubt that the recent change in the races of our immigrants will profoundly affect the character of the future American race. What the resulting physical and mental changes will probably be, various authorities have told us. The ethnic composition of an "average immigrant" (whatever that may be!) has radically changed during the past few years, the Baltic and the Alpine stocks giving way to the Mediterranean.[11]

In warning against the encroaching huddled masses, Ward conjured what Balibar has termed a "mythical national type,"[12] dating back to the country's origins (but not too far back, lest indigenous peoples be exalted). Nationalism, says Balibar, is "the product of a fictive ethnicity," an "imagined unity." It is "a force for uniformity and rationalization and it also nurtures the fetishes of a national identity which derives from the origins of the nations and has, allegedly, to be preserved from any form of dispersal."[13] Ward cautioned readers about the dilution of what he called "energetic blood." He spoke of the "conservation and preservation of the American race," typifying the conflation of race and nationality that marked this era.[14]

The critical question for Ward and others was, "who will sire and bear American children?" "National eugenics for us," he explained, "means the prevention of the breeding of the unfit native, as well as the prevention of the immigration, and of the breeding after admission of the unfit alien." Ward's article "Our Immigration Laws from the Viewpoint of Eugenics" chastised readers for "legalizing the begetting of criminal children." He denounced parenthood on the part of "degenerates," be those propagators of poor blood native or alien. However, Ward deemed the latter the greater threat. "To admit to this country the feeble-minded, the insane, the epileptic, the habitual criminal," he continued, "is . . . a crime against the future."[15]

> [I]t is actually true that we are today taking more pains to see that a Hereford bull, or a Southdown ewe, imported for the improvement of our cattle, is sound and free from disease than we take in the admission of an alien man or woman who will be a parent of American children.[16]

Neither Ward nor the journal's editorial staff were concerned with *actual* children. Their interest lay in *potential* offspring and on the impact of "negative eugenics" on once "race proud families."[17] They called on "the sub-races of the aryo-germans . . . [to] assure themselves of the continuance of their dominance in the world's affairs, and of the permanence and even brilliant expansion of the splendid civilization they have created, by scientifically directing their evolution."[18] The fit must take care to neither bear nor sire children of the racially unfit, be they natives or newcomers.

Calculating Hysteria

Writing on Europe between the two world wars, George Mosse has suggested that racism is a visually centered ideology—stereotyping physical and mental characteristics of outsiders and insisting on recognizable, undeniable, immutable differences between "inferior" and "superior" peoples.[1] American eugenicists, armed with charts, photographs, and even human skulls, were there to provide the visual and mathematical support that rendered racism scientifically valid and politically viable. As national and ethnic and racial identities merged in public and political discourse, eugenic rhetoric acted as court empiricist, justifying, sustaining, and often initiating anti-immigrant attacks in the name of "bettering and protecting" the white race.

The litany of moral, intellectual, and physical deficiencies of racialized outsiders was no longer mere conjecture, but backed by the creative number crunching of exclusionists and others. Immigrants and their progeny joined the native poor, the physically and developmentally disabled, the sexual outcasts, African-Americans, prostitutes, alcoholics, addicts, convicts, and others as subjects of eugenic and statistical scrutiny. This was inevitable, not only because of a stalwart faith in science and progress, but because eugenicists were motivated by a desire to substantiate and sustain existing social hierarchies and not by legitimate research imperatives. Biologism, in short, served not only to preserve the status quo, but also to evade analyses of socioeconomically generated inequities.

The scientific expertise of the day intruded more and more on judicial and legislative renderings. It had to. As George Stocking explains, eugenicists believed that in the "civilized" world, natural selection had all but ceased and could no longer move forward without hands-on human intervention.[2] Group demerits (biological and otherwise), quantified and given a veneer of rationality, equipped eugenicists with a persuasive argument they could take to Congress. The campaign for inclusion of a literacy test in immigration legislation is a good case in point. Meaningful in its own right, it foreshadowed the political uses of later standardized tests.

Presidents Cleveland, Taft, and Wilson all vetoed the literacy entry requirement that Warren G. Harding eventually signed into law. Between the vetoes and the executive signature were years of relentless restrictionist lobbying. The Grand Jury of King's County, New York, urged Congress to "prohibit the immigration into this country of all who cannot read and write English," as they "do not possess an intelligent understanding of the fundamental ideas of human liberty."[3]

> The stream of our national life cannot rise higher than its source. To permit any further pollution of this stream is to intensify both our foreign as well as our domestic problems. It will foster disunion, instead of promoting union. Instead of continuing as a nation of high ideals, we shall degenerate into a mere medley of races, a hodgepodge of nationalities.[4]

Literacy was a convenient benchmark of alleged unassimilability. In the hands of eugenicists, it became a mark of inborn mental and intellectual health, a conviction that continued to hold strong currency for decades to follow. But literacy's real benefit was that it was quantifiable. In 1894, Robert DeCourcey Ward cofounded the Immigration Restriction League (IRL), an early force for a literacy test, for the express purpose of keeping the Irish out of Boston. Italians, whose inferiority the IRL constantly proclaimed, became their secondary target.[5] Proponents took pains to document exactly which groups held the preponderance of illiterate immigrants. In *The Tide of Immigration*, Frank Julian Warne classified over one-third of Bosnian, Italian, Rumanian, Lithuanian, and Turkish arrivals illiterate. In 1914 alone, he claimed, illiteracy among Turkish immigrants was 64 percent, while for the English, Welsh, Scandinavian, and Finnish, it was under 1 percent. After calling readers' attention to what he cast as racial and national discrepancies (and, by extension, their failings), Warne conceded that a literacy test would be an exam "of opportunity, rather than character." Nevertheless, he wrote, it would be valuable as volume control, "solely for the purpose of bringing [immigration] within a reasonable degree of our ability to absorb and assimilate its elements."[6] Others, like Sidney Gulick of the New York–based Commission of Relations with Japan, worried that a literacy test was not *enough* of a race-conscious device, as such a test "could not wisely be applied to Asiatics, for it would admit millions."[7] Fortunately for eugenicists, there was an ever-growing pool of statistical evidence being accumulated and fabricated that far exceeded the limitations of a literacy requirement. Standardized tests were fast becoming cutting-edge methodology.

By the time Henry Herbert Goddard entered the scene as translator of Alfred Binet's articles (Binet developed the first intelligence tests and administered them in France) and as proctor of the first IQ tests on American soil, there was strong precedent for linking race to a string of physical and behavioral traits deemed dangerous to the republic.[8] It was Charles Davenport of the American Breeders Association (and later co-founder of the American Eugenics Society) who first suggested that Goddard, a psychologist, use the Binet test to document the hereditary shortcomings of immigrants to the United States.[9] In 1912 immigrants disembarking at Ellis Island became the first group in the country to whom the IQ tests were administered. Like so many other eugenic undertakings, Goddard's had a built-in class bias: only those who came steerage were subject to examination. According to his results, over 80 percent of all Jewish, Polish, Italian, Hungarian, and Russian immigrants were "feeble-minded defectives."[10]

A year after his initial foray into the gray matter of new male arrivals, Goddard sent two women to Ellis Island to test female immigrants. Goddard believed women had keen intuition and that his testers would be able to pick out intellectually and mentally "deficient" women by sight. They did not disappoint him, proclaiming that of the 152 women they tested, 83 percent of the Jews, 80 percent of the Hungarians, 79 percent of the Italians, and 87 percent of the Russians were "feeble-minded."[11] Though in later years he would significantly alter his views, at the time Goddard was thrilled by the implications of his work. He perceived it as directly impacting policy, crediting the adoption of "mental tests" with an increase in deportations for "feeble-mindedness." The doctor praised "the untiring efforts of the physicians who were inspired by the belief that mental tests could be used for the detection of feeble-minded aliens," and called on the public to demand congressional support for testing facilities at all ports of entry.[12]

There was no legally binding codification for the term "feeble-minded," and thus, no limit to its misappropriation and resulting abuse. Howard Knox, assistant surgeon with the U.S. Public Health Service at Ellis Island, celebrated the ambiguity of the designation in a 1914 article in the *Journal of Heredity.* "Fortunately," he wrote, "the term 'feeble-mindedness' is regarded by most alienists as a sort of waste basket for many forms and degrees of weakmindedness, and since it is incorporated in the law as a mandatorily excludable defect, it is especially suited to the needs of [Ellis Island] examiners."[13]

By means of a battery of tests, officers at Ellis Island began screening not only for "feeble-minded defectives," but for "potential defectives"— those who *seemed* healthy enough, but might succumb at a later date. According to Knox, parents of "defectives" were burdened with "temperamental peculiarities . . . precursors of actual deficiency in the immediate or future descendants." These "peculiarities" included "nerve storms, poorly controlled grief, . . . untimely mirth, sensual morbidity and perversion, sullenness, facial tic . . . sick headaches . . . signs of genius in certain lines . . . the formation of strong habits, mannerisms, speech defects and other noticeable qualities that brand the possessor as 'queer.' . . . [T]hese ill-defined entities come in one generation, while in the next two or three, definite psychoses and mental deficiency appear." In 1914 Knox wrote that "the number of alien insane in the country is much larger than is generally supposed. . . . there are alien insane who never find their way to hospitals, their relatives knowing that if they do they are liable to be deported." Their countrymen, he maintained, "because of race pride or for other reasons do not wish the amount of insanity and disease they harbor to become known."[14] Given the unqualified tabulations being offered in those years, it is difficult to take Knox's claim of an undercount too seriously. The reviewer of Warne's book, citing the 1910 U.S. census, reported that while foreign-born whites constituted only 14.5 percent of the total population, they composed 28.8 percent of asylum inmates.[15] This was not, of course, a comment on which groups were the most likely to get ensnared in the asylum system and why. In 1915, Assistant Attorney General L. E. Cofer claimed that in the previous fiscal year precisely 41,250 arriving immigrants out of 1,485,957 were certified for all causes, including 1,360 for "mental deficiency of various kinds." Cofer actually named these deportations "eugenics." According to his calculations, "1,000,000 children [are] destined to be born from the immigrant women arriving last year."[16] "Immigrant" usually implied male in eugenic tirades, with the exception of marked references to prostitutes and mothers (or potential mothers) of "defectives." In any case, deportation was considered eugenic self-defense. And it was about to be codified in H.R. 10384.

By 1917, the United States had a store of immigration restrictions that were ideologically, if not explicitly, aligned with eugenics goals—a fact not lost on eugenicists. Bemoaning Europe's alleged use of the United States "as a dumping ground for its convicts, paupers and insane," Cofer declared that immigration enactments, "which purport to exclude some twenty-one classes of mentally, physically, morally and economically undesirable persons, were originally intended to protect the country from the dumping process above described."[17]

Congress debated one immigration bill after another between 1875 and 1924. Each one, as Daniel Kelves has noted, widened the circle of inadmissibility.[18] Race and class considerations were ever at the fore. The 1875 Page Law banning Asian prostitutes, real or imagined, virtually halted the immigration of Chinese women to the United States. In 1882 the Chinese Exclusion Law was passed as well as the first federal immigration ban on "lunatics, idiots, and persons likely to be a public charge." The exclusion of Chinese laborers and the insane was upheld in 1891 by yet another law that also singled out "persons convicted of crimes of moral turpitude." In 1903, the base of unwanteds was again broadened as a new law declared epileptics inadmissible.[19] Four years later, anarchists, paupers, "professional beggars," and those with tuberculosis joined the rolls.[20] That same year, the Gentlemen's Agreement barred Japanese laborers with the help of the Japanese government, which stopped issuing passports to the would-be emigrants. 1907 also saw the presidential appointment of the Immigration Commission. In its 1910 report to Congress, the commission urged the exclusion of all unskilled laborers arriving without families or wives, limits on the number of immigrants arriving yearly at any port, and an increase in the amount of money an immigrant must have at the port of entry.[21]

Then, in February 1917, Congress overrode a presidential veto and passed H.R. 10384, "An Act to regulate the immigration of aliens to, and the residence of aliens in, the United States."[22] The act revealed as much about America's feelings toward its native marginalized as it did about the fear of incoming unwanteds. It emerged in the heyday of American eugenics, and should be viewed as part of the political and cultural backlash that accompanied not only the influx of eastern and southern European immigrants (and what remained of Asian immigration after decades of restrictive enactments) but also the large-scale movement of African-Americans to the Northern cities during the Great Migration. It reinscribed the ethnic/racial scapegoating cultivated by eugenic treatises on criminals and the mentally ill, and dovetailed nicely with half-hearted gestures to labor and to popular antianarchist sentiment.

The new law either barred from entry or strengthened old bans against prostitutes, anarchists, and convicted or admitted felons. Prostitution in urban centers was already being ascribed almost exclusively to immigrant women. Likewise, a host of illegal activities was laid at the door of entire immigrant communities. Criminal anthropologists, led by Cesare Lombroso (1835–1909), had already propagated the notion of the "born criminal."[23] Issues of crime, health, insanity, and morality all fell to eugenic inspection, and human failings were meted out with scientific precision to the ethnic, often immigrant, target of the day. This was by no means a new endeavor. Half a century earlier, in the 1860s, *Harper's Weekly* declared that 75 percent of violent crimes were the work of Irish immigrants.[24] In 1908, New York City Police Commissioner Theodore Bingham stated that "Hebrews, mostly Russian" composed a full 50 percent of the city's criminal element.[25] Cofer insisted that the preponderance of juvenile delinquents could be found in the north Atlantic states, "where immigrants form a larger proportion of the population than in any other section of the country."[26]

More than once, eugenics has posed as pan-class solidarity. In the ongoing rhetoric surrounding immigration restriction, the livelihood of American-born workers was (and is) often invoked. The 1917 act also strengthened the ban on contract laborers. Shortly before the bill's passage, an unnamed book reviewer for the *Journal of Heredity* stated:

> It is pretty well recognized now that the low birth rate among the most useful and enlightened classes is principally economic in origin. Any successful eugenic propaganda must, therefore, be preceded by such economic and social changes as will make it economically and socially possible for young married people to have children; and it seems

probable that a restriction of the volume of unskilled labor arriving in this country would be one of those changes.[27]

In other words, the jobs of eugenically fit American workers had to be protected from unfit foreigners so that the former would have the financial wherewithal to raise children. This paternal attitude toward poor workers represented yet another inconsistency on the part of eugenicists who elsewhere categorized poverty as a manifestation of innate "dysgenesis" (thereby scientifically rationalizing the inequitable distribution of wealth under capitalism).

Robert Ward approached the labor-eugenics bridge from a slightly different angle. "We constantly speak of the need of more 'hands' to do our labor. We forget that we are importing, not 'hands' alone, but bodies also" (a sentiment he reiterated in nearly every one of his articles).[28] While some of his contemporaries called for dispersal of immigrant arrivals throughout the nation's agricultural areas, Ward worried that such a plan would mean the "replacement of the native by mentally and physically inferior foreign stock, which is already going on in the North." Instead, Ward called on the southern states to stand firm on eugenic immigration. "The people of the South at present hold the key to the immigration problem, eugenically considered." They must, he said, "insist on having none but honest, industrious, healthy and fit immigrants" for laborers.[29]

Eugenicists had been able to make common anti-immigrant cause with large segments of organized labor by raising the specter of cheap, imported labor.[30] World War I expanded this rhetorical opportunity. Gulick outlined the following scenario:

> Now, while war suspends the tide of new-comers to our shores, is the time for enacting new laws to regulate the coming of fresh aliens.
>
> No one can foretell how large or how small will be the immigration from the war ravaged countries of Europe. One factor in the problem that is generally overlooked is this: Wages in America will be high after the war and demand for cheap labor will be urgent. Immigration companies and steamship lines will seek for fresh sources of cheap labor to bring to America.

"What," asked Gulick, "is to prevent them from securing hundreds of thousands from West and North Africa, Egypt, Syria and Asia Minor?"[31] On the surface, this invocation of an external threat to domestic labor might have registered as concern for the plight of U.S.-born workers (albeit a xenophobic one) were it not for Gulick's emphasis on possible recruitment from Africa and the Middle East (why wasn't he alarmed over potential refugees from, for example, France or Germany?). Instead, it provides a good example of what Balibar has called "crisis racism," wherein there exists a "social 'consensus' based on exclusion and tacit complicity in hostility towards [foreigners]. . . . [It] becomes a determining factor of the consensus which makes the difference between classes seem only relative."[32]

Ward, however, was anxious about European refugees, though his use of the war was slightly more veiled than Gulick's. By admitting the inevitable "flood" of postwar refugees, the United States would hamper political, social, and religious reform abroad.

> [O]ur duty as Americans, interested in the world-wide progress of education, of religious liberty, of democratic institutions, is to help the discontented millions of Europe and Asia to stay in their own countries, and work out there, for themselves, what our own forefathers worked out here for us.[33]

Nevertheless, Ward acknowledged, "Our immediate, paramount interest is eugenic."[34] The previous year, Ward, along with the rest of the Committee on Immigration of the American Genetics Association, had called on lawmakers to take advantage of the "breathing space" provided by the war and pass new immigration legislation. He claimed that a decrease in immigrant arrivals since the war's outset enabled more rigorous medical examinations at ports of entry. "This . . . has resulted in a marked increase in the numbers of aliens certified as having physical or mental defects" and the number of immigrants debarred or deported.[35] The *Journal of Heredity* also called on Congress to seize the lull as an opportunity to adopt "a rational policy for future guidelines."[36] By early 1916, Ward was already worrying about a postwar influx of refugees. European immigration would swell, while the "physical and mental quality of the immigrants is likely to show a decrease."[37] Europeans in their prime, with initiative and courage, the Committee on Immigration maintained, either would be dead or would remain in their home countries to rebuild, while the "least fit" would emigrate. In fact, they predicted, the nations of western Asia and southern and eastern Europe, owing to "differences in race, political institutions, education and social habits,"

probably wouldn't do much rebuilding at all.[38] Their desperate popu-
lace would turn to the United States.

> From the war-ravaged Balkans hundreds of people have already come.
> The flotsam and jetsam from the disturbed conditions in Europe and
> Asia Minor—the backwash of the war—has begun to find its way to our
> shores. From Syria, from Turkey, from Egypt, from Greece, from Siberia,
> have come refugees—the first trickles of the vast stream that will flow
> here when the war is over.[39]

Madison Grant's concerns were more long-term. "When the Bolshevists
in Russia are overthrown, which is only a matter of time, there will be a
great massacre of Jews and I suppose we will get the overflow unless we
can stop it."[40]

Just how large a factor the war was in the framing of the 1917 legislation is unclear. To be sure, the scope of the legislation was broad enough
to encompass "the mental and physical derelicts of the war" and many
others.[41] Legislators targeted the disabled and the racialized, as backers
and eugenicists worked to make them synonymous. Section 3 of H.R.
10384 denied entry to

> All idiots, imbeciles, feeble-minded persons, epileptics, insane persons;
> persons who have had one or more attacks of insanity at any time
> previously; persons of constitutional psychopathic inferiority; persons
> with chronic alcoholism; paupers; professional beggars; vagrants; per-
> sons afflicted with tuberculosis in any form or with a loathsome or
> dangerous contagious disease; persons . . . certified by examining sur-
> geons as being mentally or physically defective, such physical defect being
> of a nature which may affect the ability of such alien to earn a living.[42]

A deeper appreciation of this clause rests on an understanding of
three critical factors. The first involves the aforementioned elasticity of
the term "feeble-minded" and the exploitation of that leeway by immigration inspectors. The second concerns the construction of American
cities as catalyst for manifest dysgenics. The third is the extended risk
period for deportation. Though reinscribed by the 1917 Immigration Act,
"feeble-mindedness," along with many of the other listed conditions,
was already on the books as grounds for exclusion. What was new in
this legislation, however, was the five-year period allotted for deportation of immigrants found, after their arrival, to belong to one or more
of the excludable classes. Prior to 1917, officials had three years from the

date of entry to deport these individuals. Eugenicists had complained that even inborn "feeble-mindedness" or "imbecility," or an innate propensity toward "pauperism," might not manifest itself until the harsh realities of urban life had sunk in and the deportation deadline had lapsed. In a less than ringing endorsement of life in the United States, Howard Knox explained that "a great number of alien insane would not have become so had they not come to America. Such people are not fitted to become citizens of this country, although they may get along very well at home."[43] Ward, too, warned of this "constitutional psychopathic inferiority" (an excludable condition that made its legislative debut in the 1917 act), declaring that East Coast hospitals were full of immigrants who could not cope with the stress of their new environment.[44] Eugenicists may have fixed immediate blame on environmental factors, but there is little doubt as to which groups they believed the least capable of warding off madness. Lest anyone get the impression that it was the English or German immigrants who were suffering breakdowns in the streets of New York, one writer offered this clarification: "The mental weakness appears only after [the immigrant] has been here some years, perhaps inevitable [innate] or perhaps because he finds his environment in, say, lower Manhattan Island much more taxing to the brain than the simple surroundings of his farm overlooking the bay of Naples."[45] Officials now had a full five years to watch new arrivals unravel and to file deportation papers under that most malleable of terms, "feeble-mindedness."[46]

Eugenicists did see their handiwork in the bill's fine print, but they did not see everything they wanted. For example, while responding to paranoia about a less than pristine gene pool, the Immigration Act of 1917 did not shut the door on Jewish immigration, as Madison Grant had hoped (indeed, it made some provisions for those fleeing religious persecution). That would come seven years later. It did, however, effectively exclude South Asians and others by banning

persons who are natives of islands not possessed by the United States adjacent to the Continent of Asia, situate south of the twentieth parallel latitude north, west of the one hundred and sixtieth meridian of longitude east from Greenwich, and north of the tenth parallel of latitude south, or who are natives of any country, province, or dependency situate on the Continent of Asia, west of the one hundred and tenth meridian of longitude east from Greenwich and east of the fiftieth meridian of longitude east from Greenwich and south of the fiftieth

parallel of latitude north, except that portion of said territory situate
between the fiftieth and sixty-fourth meridians of longitude east from
Greenwich and the twenty-fourth and thirty-eighth parallels of latitude
north, and no alien now in any way excluded from, or prevented from
entering the United States shall be admitted to the United States.[47]

These lengthy specifications from the congressional debate emphasize
just how exacting lawmakers were in their efforts to circumvent judicial
rulings declaring South Asians to be Caucasian.[48] As Charles Mills explains in *The Racial Contract,* "membership requirements for Whiteness
are rewritten over time with shifting criteria."[49] The 1917 act was an opportunity for Congress to affirm its own parameters of whiteness, as Mills
describes, to enshrine white personhood simultaneously with nonwhite
personhood, and to further reify race as an immigration consideration.

On the day of the final vote, Senator Reed of Missouri noted that the
original bill, unlike the version about to be passed, allowed that "white
persons" living in the excluded territory could enter the United States,
but that the stipulation was cut when the bill was in committee. It was,
he said, the first time people were excluded by "lines of latitude and
longitude, not by races, not by intellectual qualifications, not by moral
attributes," and he lamented that in its present form "it absolutely ex-
cludes every person within that mighty portion of the earth's surface,
and this regardless of the intellectual or moral qualification of the immigrant even though he be of the purest of pure white blood" (i.e.,
British). Georgia Senator Hardwick responded,

> The Senator knows that the Hindus have been held by our courts to be
> white people. Now, even if they are technically white people, they are not
> the sort of white people the Senator wants to come in, and if we had left
> in the bill the words "white people" we would not have gotten rid of this
> immigration that we wanted to avoid.[50]

Barred zone aside, thousands of South Asians were denied entry to
the United States between 1908 and 1924 based on clauses that excluded
potential public charges (another useful and highly malleable term).[51]
Much of the 1917 bill was couched in economic terms, banning anyone
who might conceivably become dependent on the state as well as workers recruited as contract labor. Invocations of questionable health and
crime statistics and incantations of fiscal woe betrayed the law's class
bias, augmenting rather than muting the openly racist intent of the law.
"You want to preserve the purity of our race; you want to prevent the

influx of great hordes of undesirable people," Reed exclaimed, "and starting from that laudable object... the Senate absolutely refuses to consider the bad propositions that are loaded into the bill." Those "bad propositions" were the ones that allowed entry to the "Moors from Africa" and the "head-hunters and cannibals of Patagonia and the Fiji Islands."[52] Reed and others voted against the bill not because it banned entry to those from India, parts of Russia, Persia, and what is now Turkey (to name a few), but because it was not restrictive enough. Neither was Reed pleased with the clause that allowed refuge from political persecution. Here he was in full accord with President Woodrow Wilson, whose veto message he quoted:

> Such a provision, so applied and administered, would oblige the [immigration] officer concerned in effect to pass judgment upon the laws and practices of a foreign government and declare that they did or did not constitute religious persecution. . . . [I]t is not only possible but probable that very serious questions of international justice and comity would arise between this Government and the Government or Governments thus officially condemned should its exercise be attempted.[53]

Wilson was, ostensibly, still trying to keep the United States neutral, and he did not want that stance threatened by the acknowledgment of religious persecution overseas, even if the position of Europe's religious minorities was made more precarious by the heightened nationalism of war.

Reed called for the drafting of a new bill that would "protect the citizenship of the United States from an influx of improper foreigners, by which I mean those who, by blood or race, are incapable of amalgamation into the body and life and spirit of the American people. That is not to be done by arbitrary lines on a map, but by blood."[54] His legislative day was coming.

For a short while, eugenicists basked in their legislative victory. Of course, it was not only a coup for eugenicists, but also for less scientifically inclined racists, certain segments of organized labor, anti-Bolsheviks, and antianarchists. Nevertheless, eugenicists called it their own. Long before the bill's passage into law, the Committee on Immigration of the American Genetic Association (AGA) was celebrating its 308 to 87 victory in the House, claiming that it "embodies several important eugenic provisions." The committee "heartily" endorsed the bill's provisions and called on the officers and members of the AGA to "make every possible effort to secure favorable action by the Senate upon this bill."[55] After the 1917 Immigration Act became law, Robert Ward declared it to be "in its essentials, a eugenic measure—perhaps the most comprehensive and satisfactory ever passed by Congress." He continued in his praise:

> The American Genetic Association, as well as every one who has at heart the eugenic welfare of the United States, has good reason for satisfaction in the final enactment into law of a measure which cannot fail to result in a marked improvement in the mental and physical qualities of future alien immigrants.[56]

The committee credited the enactment with 23 percent of all 1918 rejections of immigrants by the authorities. Still, it was not enough. Members called for a temporary suspension of "alien immigration," claiming that such a measure would have "highly desirable eugenic results."[57] The passage of the 1917 law did little to curb the tide of hysterical outpourings by restrictionists. "All records are going to be shattered from January on," exclaimed one New York official in late 1920; "[w]hole races of Europe are preparing to remove to the United States."[58]

Such panic stemmed, in part, from fear of a large-scale postwar refugee immigration. A heightened animosity toward communists, anarchists, socialists, and other radicals in the wake of the Bolshevik victory in 1917 further augmented more general anti-immigrant attitudes. Eugenicists like Ward were quick to take advantage. Included in his 1920 article, "The Immigration Problem Today," was this lengthy excerpt from a cablegram sent by the *Philadelphia Public Ledger*'s foreign correspondent in Warsaw:

> The most extraordinary, hopeless, destitute and pathetic emigration which the world has known is making its way to America, the promised land, through Poland from as far away as Kief *[sic]*, and from the Russian territory north and east of the Black Sea. Even from Georgia, masses of poor, disease-laden people are making their way to America.
> Within three weeks, 150,000 have reached the Warsaw territory. This is only a beginning. Unfortunately, Bolshevist agitators and Communists are with the majority of the hordes and are confident that in the general confusion they will be able to get into America, where they propose to spread propaganda.[59]

Such rhetoric was typical of an era marred by the Palmer Raids (during the course of which thousands of real and suspected radicals, most of whom were eastern and southern European immigrants, were violently rounded up in nationwide sweeps by federal agents and local police), deportations of immigrant communists, anarchists, labor leaders, and socialists, and the execution of Sacco and Vanzetti. Anti-anarchist and pro-eugenics appeals were often wedded, perhaps because eugenicists identified radicals and carriers of "deficient germ plasm" with the same demographic groups. More likely, it was due to eugenicists' position vis-à-vis various political ideologies as both consumer and consumed. In this instance, they were able to exploit an antileftist nationalism and to trade on xenophobia by pronouncing immigrant radicals eugenically dangerous. Bolstered by reactionary sentiments, eugenicists professed to be guarding the nation's biological *and* political future, while simultaneously validating the aims and efforts of restrictionists and reactionaries outside their immediate circle.[60] Although the 1917 Act had fortified bans against anarchists (frequently a veiled reference to all Jewish and Italian leftists), and despite the campaign of harassment undertaken by state and federal authorities, antired vitriol remained a powerful device for eugenicists for years to come.[61] In 1919, the *Journal of Heredity* published "The Racial Limitation of Bolshevism," by Frederick Adams Woods, a biologist at the Massachusetts Institute of Technology. "Is the Anglo-Saxon temper by nature averse to Bolshevism?" Woods queried. His answer was a resounding "Yes." For Woods, anarchism and communism were synonymous with chaos and violence. He explained the Russian revolution and warned of what would inevitably follow:

> The racial elements in the make-up of Russia are mainly Slavic. The true Russians constitute nearly three-fourths of the population of Russia, the rest are chiefly Letto-Lithuanians, Poles, Jews, Finns, Turco-Tartars, and Mongols. They have indulged in much anarchy in the past. Historical evidence strongly suggests that there is something inherent in the temperament of the Slav causing him to yield much more easily than his Nordic neighbor to the temperament of mob violence. Let us hope that now is another time when nature will assert itself as stronger than nurture and that Bolshevism will find itself delimited on the Anglo-Saxon frontiers.[62]

Eugenicists have always been poised to use whatever hysteria-inflated doctrine was at hand. In this instance, the preservation of a "pure" Anglo-Teutonic bloodline was the last defense against the Red menace. In a more recent eugenics-as-bulwark-against-Bolshevism theory, Ben J. Wattenberg of the American Enterprise Institute warned in 1987 that the

United States would find it "difficult to promote and defend liberty" if American women did not catch up to women in communist countries, who were outbreeding them 2.3 to 1.8 children per mother. Wattenberg recommended cash bonuses to encourage U.S. births and fend off economic, military, and ideological losses on the world stage. His supporters included Jack Kemp and Pat Robertson (who stated that the nation could be "committing genetic suicide").[63]

Eugenicists were nothing if not dexterous, and therefore partook not only of xenophobia and red-baiting, but again appealed to native labor's concerns. The finger was pointed at greedy industrialists. Ward cautioned, "if we do not bestir ourselves, the steamship companies, and the large employers of 'cheap labor,' and the societies of foreign-born hyphenates will carry the day, as they have so often done in the past."[64] Thus, though promoting elitist concepts of the "higher" and "lower" born, proponents simultaneously invoked a sort of eugenic populism, wherein big business was posited as a menace to both the native workforce and the national gene pool.[65] Consider the comments of John C. Box of Texas, member of the House Committee on Immigration and Naturalization, before the House of Representatives in early January 1921. According to Box, "powerful influences oppose restriction": immigrants, desirous of entry for their relatives; and "corporate greed which disregards the present and future welfare of the mass of Americans and their children, because it wants money and power over labor." Box contended that the "Inter-Racial Council" was the mouthpiece for both groups and that subscribers included General Electric, the Standard Oil Company of New Jersey, Colt's Patent Firearms Manufacturing Company, Bethlehem Steel, General Motors, textile interests, sugar refineries, and steamship companies, to name a few.[66] What is of interest here is not that manufacturing interests would have a vested interest in a continuing influx of immigrants whose labor could be easily exploited; rather, the significance lies in the fact that Box's remarks had nothing to do with genetics, eugenics, or any related field, yet were cited at length in the *Journal of Heredity.* There was not even the pretense of a scientific claim in his text, despite the venue. By this time, having already established the alien-eugenic link as a given, the journal was carrying articles on immigration that made no mention of heredity whatsoever.

The occasion for Box's remarks was debate on an immigration suspension bill introduced to Congress by Representative Albert Johnson.

An ardent adherent of eugenics, Johnson was a member of the Eugenics Record Association, the Galton Society,[67] and later the Eugenics Committee of the United States. In 1919, with the help of the Immigration Restriction League's lobbyist in Washington, he was appointed chair of the House Immigration Committee.[68] Barry Mehler has documented the anti-Semitic rhetoric Johnson employed in promoting his bill, particularly the congressman's reliance on a State Department report that derided Jewish immigrants as "filthy, un-American, and often dangerous in their habits." Immigrants on the whole were "wasted by disease" and "abnormally twisted," and Polish Jews were singled out as being "of the usual ghetto type."[69]

Johnson's goal was complete suspension. The Three Per Cent Restrictive Act, signed into law by President Harding in 1921, fell short. It decreed that "the number of aliens of any nationality who may be admitted...to the United States in any fiscal year shall be limited to 3 per centum of the number of foreign-born persons of such nationality resident in the United States as determined by the United States census of 1910." This national origins act was to remain in effect through June 1922.[70] It was a culmination of various anti-immigrant harangues—some begun by eugenicists, others annexed by them. This act, however, like its predecessors, was merely a short-term landing spot. The real victory was three years away.

During debate on the 1917 Immigration Act, Senator Reed complained that immigrants were excluded not by race, by "character and blood, or even by countries," but "in accord with parallels of latitude and degrees of longitude."[71] He could take no such issue with the 1924 National Origins Act (aka the Johnson-Reed Act), which set quotas at 2 percent of the 1890 census. The 1924 law owed its success in large measure to years of anti-immigrant agitation on the part of eugenicists. Their ideological fingerprints were to be seen everywhere on the text of the law. Certainly, the 1924 legislation should be considered the consummate payoff for eugenicists and others who felt previous bills did not go far enough. It was much closer to the legislation they had lobbied for even prior to the 1917 act, and thus eugenicist input into the National Origins Act should be viewed as spanning a much longer period than the mere seven years that separated the two laws.[72]

In large part, the 1924 act and the lobbying efforts that precipitated it constituted a reprise of its 1917 predecessor. Arguments for both bills relied on a steady stream of flawed statistics, in particular calculations of race-based "feeble-mindedness" and the much-celebrated IQ tests. The difference lies in the immediacy of eugenicists' involvement in the passage of each bill. In 1917, eugenicists were instrumental in establishing an ideological and rhetorical backdrop for exclusionary legislation. In 1924, their impact was more tangible: they provided critical testimony before Congress and helped frame the actual bill. Harry Laughlin ("eugenics expert" of the U.S. Congress and of the Chicago Municipal Court, director of the Eugenics Record Association and cofounder of a handful of like organizations) and Lothrop Stoddard (author of several volumes, including *The Rising Tide of Color against White-World-Supremacy* [1922], whose later testimony before Congress on immigration was lauded by President Hoover) held regular sessions with the bill's cosponsor, Representative Johnson.[73] Madison Grant, who wrote in 1916 that he was convinced that once "the true bearing and impact of the facts are appreciated by lawmakers, a complete change in our political structure will inevitably occur," was among the bill's framers.[74] Finally, Johnson himself was a member of the American Eugenics Society.[75]

The tests and tabulations generated by eugenicists during the seven-year stretch between 1917 and 1924 had an enormous impact on the upcoming bill, in particular the work of Robert M. Yerkes, Carl C. Brigham, and Harry Laughlin. Just months after the passage of the 1917 Immigration Act, Yerkes assembled a group of hereditary determinists, including Goddard, to devise what would become known as the Army Mental Tests. Yerkes's undertaking, as well as its legislative impact, has been well documented. Briefly, the Harvard professor tested 1.75 million U.S. Army recruits.[76] The results spelled disaster for true believers. The mental age of white Americans was a shockingly low thirteen years. The blame for this was placed squarely on interracial unions between whites and African-Americans, the alleged fecundity of the poor and "feeble-minded," and the inundation of native stock by southern and eastern European immigrants. According to test results, these immigrants were less intelligent than the "fair stock" of northern and western Europe. Russians had a mental age of 11.34, Italians 11.01, and Poles 10.74. Bringing up the rear were Black recruits, who scored an average mental age of 10.41. Examiners split African-Americans into three test groups based on "intensity of color," claiming that the lighter skin groups scored higher.[77] Yerkes also tested prostitutes working around the camps. He concluded that 53 percent (44 percent of white prostitutes and 68 percent of Black prostitutes) had a mental age of 10 or under. Conscientious objectors and other "disloyals" were also tested, but, to the disappointment of their detractors, they scored above average.[78] Testing during this era was not, of course, confined to IQ exams. In 1921, M. G. Wilson and D. J. Edwards announced that, according to their measurements, Black children could not exhale as much air within a given time frame as their white counterparts and therefore had a difficult time equaling "western standards."[79]

Carl Brigham, assistant professor of psychology at Princeton, picked up where Yerkes left off, undertaking a broad application of the army test data. In later years, he would use the exams as a basis for the Scholastic Aptitude Test, but in 1922 he had other measurements in mind.[80] In *A Study of American Intelligence,* he praised his mentor's work as a validating force, and one that rationalized a wariness toward incoming immigrants.

> For our purposes in this country, the army mental tests give us an opportunity for a national inventory of our own mental capacity, and the mental capacity of those we have invited to live with us.

He continued:

> These army data constitute the first really significant contribution to the study of race differences in mental traits. They give us a scientific basis for our conclusions.[81]

Extrapolating from this data, Brigham estimated that "over 2,000,000 immigrants below the average negro" in intellectual ability had been permitted to enter and remain in the United States between 1901 and 1921.[82] Here was a recurring tactic in the eugenicists' arsenal: using one despised group to denigrate another. The violent myth of Black intellectual inferiority, so long enshrined as fact in the white imagination, was a handy yardstick with which to measure all comers. Eugenicists were extremely agile in the strategic use of metaphor. Analogy has surfaced again and again in analysis and treatment of women, queers, and the disabled.

Brigham classified northern European Jews as "Alpine Slavs," but he was caught in a conflation of race and nationality of eugenicists' own making. "It is unfortunate," Brigham opined, "that our army data classify foreign born individuals only by country of origin, so that we have no separate intelligence distributions for the Jews." This oversight on the part of army examiners did not seriously hinder Brigham, as he was nevertheless able to assert the intellectual shortcomings of Jews and caution that "every indication would point to a lowering of the average intelligence of the Nordic if crossed with the Alpine Slav."[83] Seeking to dispel any notion of mental prowess among Jews (who he noted were highly overrepresented among Russian immigrants in the United States), he wrote that extreme group variability, not group intelligence, explained the presence of Jewish thinkers:

> Our figures . . . would rather tend to disprove the popular belief that the Jew is highly intelligent. Immigrants examined in the army, who report their birthplace as Russia, had an average intelligence below those from all other countries except Poland and Italy. It is perhaps significant to note, however, that the sample from Russia has a higher standard deviation . . . than that of any other immigrant group sampled. . . . If we assume that the Jewish immigrants have a low average intelligence, but a higher variability than other nativity groups, this would reconcile our figures with popular belief, and at the same time with the fact that investigators searching for talent in the New York City and Californian schools find a frequent occurrence of talent among Jewish children. The able Jew is popularly recognized not only because of his ability, but because he is able and a Jew.[84]

Yerkes's data was a jumping-off point for Brigham, who sought additional measurements. To truly determine the contribution and cost of each "European race," Brigham suggested, "even approximate estimates of the percentage of Nordic, Alpine and Mediterranean blood in each of the European nations sending immigrants to this country" would be more helpful than the "Northern and Western, and Eastern and Southern classification." To that end, he furnished his readers with a table designated "Tentative estimates of the proportion of Nordic, Alpine and Mediterranean blood in each of the European countries." The figures therein were a distillation of statistics from the 1920 *Statistical Abstract of the United States* and Carl Brigham's own bias. Along with the "findings" on distribution of test scores presented in chart form, the table points to a conclusion that, by 1922, was already preordained: Nordics were intellectually superior to Alpines and Mediterraneans, and the countries with the highest "Nordic blood quantum" were Sweden, Norway, Denmark, the Netherlands, Scotland, and England—in that order. Countries with the lowest percentage of Nordic blood, and ergo with the highest percentage of intellectual inferiors, were Turkey, Romania, Greece, Italy, Russia (including Poland), and Portugal.[85] Brigham concluded his volume with a call for immigration barriers, though, like many of his contemporaries, he viewed future immigration as but one of many eugenic hazards. The immigration issue was a politically expedient foot in the door for those seeking more exhaustive measures, including eugenic regulation of reproduction among and across domestic demographics.

> The steps that should be taken to preserve or increase our present intellectual capacity must of course be dictated by science and not by political expediency. Immigration should not only be restrictive, but highly selective. And the revision of immigration and naturalization laws will only afford a slight relief from our present difficulty. The really important steps are those looking toward the prevention of the continued propagation of defective strains in the present population. If all immigration were stopped now, the decline of American intelligence would still be inevitable. This is the problem which must be met, and our manner of meeting it will determine the future course of our national life.[86]

Yerkes's army tests and Brigham's wide-scale application of their results were seized upon by eugenicists and restrictionists alike. Sociologist Paul Popenoe, an editor at the *Journal of Heredity* and an advocate of eugenic sterilization, wrote that the army tests clearly proved the claim that "the immigration of recent decades into the United States, coming of late years largely from the Mediterranean region and the Near East, is eugenically below the average of the older immigration, and of the American stock of native parentage, both of them largely Nordic in composition."[87] Yerkes himself was fully cognizant of the potential bearing his work might have on policy. In a foreword to Carl Brigham's *A Study of American Intelligence,* he wrote, "no one of us as a citizen can afford to ignore the menace of race deterioration or the evident relations of immigration to national progress and welfare."[88] The army test data was constantly cited in congressional debate over the 1924 Immigration Restriction Act, fueling eugenicists' efforts to junk previous quotas and virtually end immigration of all "inferior" nationals.[89] After the bill's passage, Lewis Terman, a psychologist who worked with Yerkes on the army tests, bragged that the exam enabled psychology to "become the beacon of light of the eugenics movement... [and] is appealed to by Congressmen in the reshaping of national policy on immigration."[90] Terman was also a supporter of research undertaken by Robert Latou Dickinson's Committee for the Study of Sex Variants between 1935 and 1941, namely, the inspection of pelvic structure, genital shape and size, skin complexion, hair texture, nipple erection, and so on, of gays and lesbians in New York City.[91]

Congressmen were equipped with more than the findings provided by Terman and company, however. In April 1920, they called Harry Laughlin, whose *Eugenical Sterilization in the United States* had not yet been published, to testify before the House Committee on Immigration and Naturalization. Laughlin's statement, entitled the "Biological Aspects of Immigration," was solicited by the committee's chair, Representative Johnson. It was Johnson, whose suspension bill had failed the year before, who secured Laughlin's official expert status with Congress.[92]

In that first appearance before the committee, Laughlin warned that "our failure to sort immigrants on the basis of natural worth is a very serious national menace."[93] He did not limit his testimony to criticism of the current policy, but came prepared with an alternate plan. His Model Eugenical Sterilization Law identified ten classes of "socially inadequate persons:"

> (1) Feeble-minded; (2) insane (including the nervous and psychopathic); (3) criminalistic (including the delinquent and wayward); (4) epileptic; (5) inebriate (including drug habitués); (6) diseased (including the tuberculous, the syphilitic, the leprous, and others with chronic infectious segregated diseases); (7) blind (including those with greatly impaired vision); (8) deaf (including those with greatly impaired hearing); (9) deformed (including the crippled); and (10) dependent (including children and old folks in "homes," ne'er-do-wells, tramps, and paupers.[94]

Laughlin insisted that his concern was "not with inferior nationalities, but inferior individual family stocks,"[95] an assertion gainsaid by his further testimony: he claimed that Mexicans and Mexican-Americans were overrepresented in southwestern "schools for delinquents." Laughlin presented the committee with a "comparative statement of the nativity of the foreign born insane in New York state"—a table which offered "conclusive" evidence that "Italians, Russians, [and] Austrians (largely Jews) constitute[d] a large portion of the insane."[96]

Laughlin's plan was twofold: compulsory sterilization for America's own "unfit" and eugenic-based immigration policy. He recommended that "an examination of immigrants be made in their hometowns, and that 'immigration passports' be issued by the American consuls to individual would-be immigrants who pass the required physical, moral, sanitary, mental, and family-stock tests." If an undesirable immigrant somehow managed to slip through this net and gain entry to the United States, Laughlin had a backup plan. He called for a national registry of aliens. "As good Americans, concerned in the conservation of our country, we must follow-up the immigrants process of naturalization and Americanization. If, because of insanity, feeble-mindedness, moral turpitude, or shiftlessness, the immigrant does not make good, he should be deported."[97]

While Laughlin's prototype did not become law in the United States, it was the partial blueprint for Germany's Hereditary Health Law, under which two million people were sterilized during the twelve years of the Third Reich.[98] A year after that law was enacted, Laughlin wrote that "Hitler should be made an honorary member of the E.R.A. [Eugenics Record Association]."[99] Sometime later the sentiment was reciprocated, and Laughlin was given a citation by the Nazis. This is not to imply that Harry Laughlin went unappreciated in his own country. At the time, he was assistant director of the Department of Eugenics at the Carnegie Institution.[100]

Laughlin was called to testify again in 1922. This time around, his statement before the committee, "An Analysis of America's Modern Melting Pot," employed a quota system. The ratio of an immigrant group's prison or asylum population to the total U.S. prison or asylum population should, according to Laughlin, correspond to said group's percentage of the total U.S. population. According to his calculations, Turkish immigrants exceeded their quota of the segregated "insane" by over 200 percent; Russians, Finns, and Poles by 265 percent; Bulgarians and Irish by over 300 percent; and "Serbians" by an alarming 400 percent. As for crime, he reported that all the "Nordic" countries fell well below their allotment, but that Italians and African-Americans, once again classified as more alien than native, exceeded twice their allowance. Likewise, immigrants from Turkey, "All Asia," "All Balkans," and Greece fell into the 240 to 294 percent range. "West Indians," Chinese, and Bulgarians overfilled their quotas by 318.14 to 366.26 percent, Mexicans by 549.6 percent, Spaniards by 660.21 percent, and Serbs by 1,400 percent.[101] Needless to say, Laughlin conceded neither the machinations of cultural bias on the part of doctors nor the greater likelihood of the poor and racialized becoming caught up in various institutional nets. Likewise, no mention of selective enforcement of criminal statutes or judicial bias was made by Laughlin or by the House committee, nor were the differences in economic standing that might impact individuals' and groups' standings with the authorities.

The importance of Laughlin's appearances before Congress cannot be overstated, skewed statistics notwithstanding. Shortly after the 1924 Immigration Act was signed by Calvin Coolidge, Johnson wrote that Laughlin's "investigations" were "of the greatest value to the House Committee on Immigration and Naturalization in the preparation of laws affecting these two important subjects."[102] Congress could not have been more sympathetic to his platform. In addition to the regular planning meetings Laughlin and other eugenicists had with Johnson, the "expert eugenics agent" conducted many of those "investigations" using congressional letterheads and mailing privileges.[103]

Though the 1924 act was the most successful of eugenicists' bids for immigration restriction during this era, it was not their last. Eugenicists considered the national origins legislation of the early 1920s to be a compromise at best. What they really wanted, agents and testers screening for the unassimilable in ports of entry *and* departure, was not feasible.[104] Even as eugenicists praised their own role in the passage of the Johnson-Reed Act, they continued to lament its limitations and to push for more exhaustive race-based enactments.

The Immigrant Within

As quick as eugenicists were to declare victory and emphasize their pivotal role in the passage of the Johnson-Reed Act, they were slow to relinquish their hold on immigration issues. The second great eugenicist victory of the 1920s, the *Buck v. Bell* Supreme Court decision declaring compulsory sterilization constitutional, was three years away, and neither Laughlin nor his cohorts, devoted as they were to the concept of eugenic sterilization, were ready to completely abandon their work on immigration restriction. In the years ahead, eugenicists would remain active in the push for further restriction, intensifying their scrutiny of Mexicans and Mexican-Americans (regardless of their immigration status or country of birth), watchful and laudatory of what was unfolding in Germany, and ready to launch new organizations in the United States.

Far from being an organization whose interest in immigration and related issues of "national fertility" was waning, the Eugenics Record Association sought to broaden its base through its sponsorship of essay contests. These competitions were an attempt to get the eugenics word out, to popularize the eugenics mindset. In 1928 the ERA offered $1,000 for the best entry on the topic "A comparison of both the crude birth-rate, the birth-rate per 1,000 females 15 to 45 years of age, and the 'vital index' (or 100 births/deaths ratio) of the Nordic peoples and non-Nordic peoples in the Americas." A simultaneous competition asked entrants to contrast "the Nordic peoples and non-Nordic peoples of Europe."[1] According to the judges, the information found in contest submissions pointed to a drop in "the net fecundity... for the last forty years in different European countries." In response, the ERA more than tripled the prize for the following year's contests, awarding $3,500 for "the best essay upon the causes of this fall in birth-rate, with special reference to Europeans and persons of European stock." According to the guidelines, "The treatment should be historical, should include an analysis of studies already made upon the subject, and should lay stress upon the phenomenon in peoples of Nordic, or chiefly Nordic origin in all parts of the world. Preference will be given to essays which are based upon objective studies rather than expressions of opinion."[2] In the midst of the Depression, the ERA had the financial wherewithal to offer a total of $4,000 ($3,000 for first prize, $1,000 for second) for research on the "probability of commitment for a mental disorder of any kind, based on the individual's family history."[3]

A second organization, the American Eugenics Society, was also sponsoring contests during this period. In addition to their "Fitter Families Contests," the society's Committee on Cooperation with Clergymen offered $500 for the best sermon on eugenics. Participants were instructed to submit their sermons, along with a study of their respective parishes. "The study will reveal the birth rate in the 'church population' of each church, will throw some light on heredity in such matters as the holding of church office and will give valuable material in regard to the size of families according to groups based on profession and also on church relationships."[4] This was, in many ways, a logical expression of eugenicists' aims. Eugenics was, and remains, an additive ideology. While invoking the new science, eugenicists were able to utilize religious rhetoric, just as they had used nationalism and antired invective.

Originally known as the Eugenics Committee of the United States of America, the American Eugenics Society (AES) was founded in 1922 by some of the most prominent and prolific eugenicists in the country.[5] From 1924 to 1936, its Committee on Immigration, working side by side with both Albert Johnson and the House Immigration Committee, generated a steady stream of anti-immigration literature in book and article form.[6]

On July 3, 1925, the organization's Sub-Committee on Selective Immigration, headed by Laughlin, Grant, and Ward, issued its third report, calling for examinations of prospective immigrants prior to their departure for the United States. In addition, the committee suggested that it would be "entirely proper" for an individual applying for a visa to

be required to "produce reliable witnesses to support his own statements, even to the extent of demanding medical and other expert testimony to the effect that he is mentally and physically up to the standard
required by our laws and that he belongs to sound family stock."[7] In
1928, the subcommittee called for three addenda to current legislation:
First, relatives of potential immigrants had to display a healthy "American stock," as assurance that their kinsmen would be an asset to the
country. Second, no immigrant who ranked below a C on the Army
Mental Test intelligence scale would be admitted. Third, future immigrants would only be allowed entry if all of their forebears were of Caucasian descent.[8]

With this third goal in mind, eugenicists turned their attention southward, relying once again on the tropes that had served them so well in
1917 and 1924: claims of intellectual inferiority and a race-specific threat
to white labor. On 4 January 1927, Grant, Ward, Laughlin, Charles
Davenport, H. F. Osborn, sociologist Edward Ross (whose most enduring contribution to eugenics remains the coinage of "race suicide"—a
term enthusiastically embraced by Theodore Roosevelt), and others submitted a "Memorial on Immigration Quotas: To the President, the Senate, and the House of Representatives":

> We urge the extension of the quota system of North and South America
> from which we have substantial immigration and in which the popula-
> tion is not predominantly of the white race.... During each of the last
> two fiscal years we have been admitting upwards of 75,000 immigrants
> from Mexico, the West Indies, Brazil and elsewhere, who are for the most
> part not of the white race and who, because of their lower standards of
> living, are able to compete at an advantage with American workers
> engaged in various forms of agricultural and unskilled labor.[9]

This same line of argument had been invoked against each successive wave
of immigration, with a particularly exhaustive and long-lasting impact
on the Chinese decades earlier. As in 1882, there were sympathetic ears
in Congress. Representative John Box of Texas railed against Mexicans,
whom he cast first and foremost as temporary workers, entering the
country as laborers and then abandoning their employers. "What will
prevent them from going where they please and staying if they please?
Hosts of the most undesirable kinds have entered and stayed in this
manner, causing this element of our population to increase rapidly. Let
the citizen of Texas, the Southwest and the Middle West look about him

and see." Box then went on to offer a litany of charities whose funds, he claimed, were being depleted by Mexicans. These "formed communities of misery constituting a heavy charge upon the taxpayers" were cropping up in Dallas, Galveston, Denver, Fort Worth, and Houston.

> In Los Angeles . . . the Outdoor Relief Division states that 27.44 per cent. of its cases are Mexican. The Bureau of Catholic Charities reports that . . . Mexicans . . . consume at least 50 per cent. of the budget. . . . The City Maternity Service reports that 62½ per cent. of its cases are Mexicans, using 73 per cent. of its budget. The Division of Child Welfare . . . state[s] that 40 per cent. of their clients are Mexican.[10]

Mexicans' alleged overreliance on relief agencies simply underscored what Harry Laughlin had maintained for years. He had already declared Mexicans and Mexican-Americans to be mentally and socially inadequate during his 1920 appearance before Congress:

> [O]ne notices, by the names of the individuals found in institutions, that the lower or less progressive races furnish more than their quota. In the schools for delinquents at Whittier, California, and at Gainesville, Texas, about half of those names were American and the other half were Mexican or foreign sounding.[11]

In 1928, abandoning his earlier assertion that he was concerned not with the eugenically sound races but family stocks, Laughlin called on the House Committee on Immigration and Naturalization to limit immigration and naturalization to "white races." The *Eugenical News* relayed Laughlin's breakdown of "the epochs of American history in which race problems were uppermost":

> (1) the struggle of the white colonists against destruction by amalgamations with the Indians; (2) the conflict between British colonists, on the one hand, and those of French, Dutch and Spanish origin, on the other; (3) the introduction of negro slaves; (4) oriental migration; (5) the great immigration following the rapid expansion of industry in the 1880's; (6) the entrance of colored races into the southwestern part of the United States since 1920.[12]

Laughlin claimed that the 1924 Act allowed greater entry for Mexicans and that they were threatening to retake Texas, California, and Arizona.[13] In 1928, the *Eugenical News* listed three objectives for future immigration legislation in its February issue. Panic over Mexican immigration topped the list.

1. Inclusion of the Western Hemisphere in the Quota. The alarming influx of Mexican peons tends to inject another serious color problem into American life.... With all the Western Hemisphere under the quota, control of bootlegged aliens will also become more effective.

2. Registration of aliens is inevitable. The clamor against it is mostly by "hyphenates," in whose native lands there exist systems far worse than any mild registration as suggested for America.

3. Deportations of all aliens illegally entered is but justice. The man whose introduction to American life comes through breaking the Quota Act is *prima facie* an undesirable.[14]

Expansion of the quota system and a crackdown on "the border-jumper and the smuggler alien" were key components of eugenicists' post-1924 agenda. Secretary of Labor Davis favored an "alien" registration whereby immigrants would be required to enroll annually or face fines. This plan garnered the hearty endorsement of Robert Ward who added that, in addition to curtailing illegal entry, the process would protect immigrants from "those of his own nationality, already here, who are always ready to take advantage of his ignorance to exploit him.... Registration would thus be a simple act of Christian charity."[15]

Like earlier pronouncements on the intellectual ineptitude of Asians, Jews, Slavs, and especially African-Americans, eugenic "scientific" validation of attacks on Mexicans and Mexican-Americans continued to strike a responsive cord. Despite attestations from the same quarters that the Depression had stemmed the tide of Mexican immigration, echoes of Laughlin's 1920 testimony could be heard well into the 1930s.[16] The Princeton League of Women Voters warned readers in its 1935 publication *Heredity and Twelve Social Problems* that

Tests in the schools in southern California show that the average Mexican child is as far below the average Negro child in abstract intelligence as the average Negro is below the average white child. Every one of these children born in the United States becomes an American citizen by birth. The 1930 Federal census showed 1½ million Mexicans in the United States.[17]

Once again, Black inferiority was a given—so much so that its mere invocation served as an anti-Mexican rallying cry. Reminiscent of the anti-Irish and anti-Italian agitation of the IRL decades earlier, the treatise decried the entry of "over a million illiterate Mexicans" subsidizing the ranks of cheap labor previously filled by southern and eastern European immigrants prior to 1924.[18] Osborn wrote that intelligence tests, having proven the intellectual underdevelopment of immigrants, were testament enough to the need to extend the quota principle "to North America (especially south of the Rio Grande), South America, and the Atlantic and Pacific Islands."[19] Drawing on the familiar tropes that had served them so well in their assaults on other groups—immigrant as diseased, feebleminded, job-stealing, taxpayer-gouging pariah—eugenicists continued to push for western hemisphere quota legislation. In his presidential address to the twenty-fourth annual meeting of the Eugenics Record Association, C. M. Goethe berated Congress for its failure to pass what he considered vital enactments, enactments that would proclaim,

> America, still overwhelmingly Nordic, proposes so to remain! ... The fate, in Congress, of successive Western Hemisphere Quota Acts is another example of hindrance of progress. These represent the next most important improvement, eugenically, in immigration restriction. Against them is always the smoke screen: "Latin-American Trade." Would all profits made south of the Rio Grande warrant an epidemic of the Black Death? Dr. S. J. Holmes wrote of the Los Angeles outbreak: "32 cases of pneumonic plague with 30 deaths, 7 cases of bubonic plague with 5 deaths were confined to the Mexican quarter. 145,000 rats were exterminated, 2,473 buildings were demolished, 7,499 buildings were rat proofed. Total cost was $2,777,000.[20]

Goethe's characterization of the "Mexican quarter" is critical, pointing as it does to the dual image of immigrants as both *containing* foreign bodies, in this instance disease, and *constituting* foreign bodies within the national entity. It was emblematic of eugenicists' larger classification of the allegedly unassimilable as foreign bodies in the body politic. In this paradigm, the epidemic that ravaged the Mexican community in Los Angeles was but a microcosm of the threat Mexicans themselves posed to the eugenic health of the United States.

Mexicans did not bear the brunt of eugenicists' contempt alone. *Eugenical News* cited Filipino immigration to California and Hawaii as a "major problem of Oriental migration-control."[21] A 1935 treatise labeled young Filipinos on the Pacific Coast a "racially threatening element." White mainlanders were warned that "[e]very Japanese child born on the [Hawaiian] Islands is an American citizen with full right of entry" to the continental United States.[22] Likewise, in 1935, congressional debate on Puerto Rican statehood triggered the wrath of eugenicists. "Apparently," a writer for *Eugenical News* fumed, "the desires, or the best interests, of the 125 million of the present population of the States have not been considered. Nothing seems to have been said about the nature of the population of Puerto Rico and whether its fusion with that of the States is the best thing for the future population of the States."[23]

Eugenicists' basic precept of Black deficiency once more played a central role, as the Sacramento Church Federation declared that the Puerto Ricans they worked with were "largely morons, overwhelmingly negroid, [and] had a jungle fecundity." The Eugenics Society of Northern California reported low intelligence and a "willingness" to work for one to three dollars a month. The Society further cautioned that "the introduction into our body politic of a new fecund negro group means further racial clashes, with *eventual* absorption into our white stock, or absorption of our white stock into the negroid."[24] It was a warning reminiscent of Grant's 1916 lament that "the coast region of our Gulf states and . . . the lower Mississippi Valley must be abandoned to the negroes."[25] Moreover, Latin and Caribbean Blacks provided the perfect prototype for the eugenic imagination: a biracial threat—both internal and immigrant—all in the same body. "If sentimental views engendered by the Civil War can be lived down, we will check this mulatto menace. Otherwise the absorption of Negroes, Mexicans, Filipinos, and Japanese . . . will produce a racial chaos such as ruined the Roman Empire."[26]

Anxiety over race mixture was long-standing. In 1856, Paul Broca, the founder of modern physical anthropology, declared that interracial unions resulted in progeny that were weak, with attenuated lifespans. Mentally and morally, these children could not measure up to either "parent race" and were frequently infertile. Furthermore, while the union of a white man and a Black woman might result in as many progeny as an *intra*racial union, that of a Black man and a white woman would typically produce no offspring. Without "continuous replenishment" by the former coupling, he declared, the mixed race group could even die out.[27]

Broca's pronouncements of infertility among the interracial might have soothed eugenicists' nerves, but seventy years later, a fearful constituency looked southward and railed against "race mixture." Eugenicists were fixated not only on Mexicans as Mexicans, but on Mexicans as an embodiment of interracial unions. "If we do not uphold the integrity of our immigration restriction laws against the low grade Europeans and against the mixed Indian and Negro population to the south of us," Grant wrote in 1928, "we white Americans will forfeit the world's leadership and then we shall have to face the yellow and the black peril in reality."[28] Back in 1916, Grant's *The Passing of the Great Race* invoked rapturous dystopia with its demographic projections. His preoccupation with intermarriage continued until his death in 1937. His colleagues shared his concern. Carl Brigham worried about "a cross between the Nordic in this country with the Alpine Slav, with the degenerated hybrid Mediterranean, or with the negro."[29] This notion of the hybrid body was not confined to racialized subjects. Siobhan Somerville has suggested that sexologists and others borrowed the model of a mixed body to understand and theorize the homosexual "invert."[30]

Eugenicists routinely drew on antimiscegenation sympathies to rally against the entry of all "undesirable stocks." The editors of the *American Breeders Magazine* complained:

> We are allowing race mixture to proceed without . . . having a knowledge of whether certain undesirable physical and psychic characters of other races are not so strongly dominant as to breed out some of the most desirable and distinctive characteristics of the aryo-german race; whether we are breeding to build up or undermine our own civilization; whether the aryo-german race, which took possession of this continent, assumed dominion over it and planted in it its civilization, institutions, and ideals, is not in danger of being obliterated if the present influx of immigration from southern and southeastern Europe and from Asia Minor is allowed to go unchecked.[31]

For his part, Madison Grant was convinced that "colored people have an elaborate program to defend mixed marriages." The American Eugenics Society, which he cofounded, supported antimiscegenation legislation in Michigan, Texas, Washington, D.C., and in Virginia where Grant worried that "many mulattoes are claiming to be Indian." Unlike African-Americans, Native Americans did not constitute a grave threat, to Grant's way of thinking, because he assumed they would, inevitably, "disappear."[32]

According to Mehler, Grant was dissatisfied with the level of AES involvement in antimiscegenation agitation.[33] *Eugenical News*, however, did not disappoint. In 1931 the journal printed a favorable review of *Crime, Degeneracy, and Immigration*, by Chicago attorney D. A. Orebaugh. Orebaugh's main thesis was that crime in America was "due to a degeneration of the national stock resulting from its intermixture with disharmonic and inferior races." The "growing degeneration in journalism, in the theater, [and] in respect to law," could be traced to the same source.[34] In 1943, at a point when U.S. eugenicists must have been largely, if not fully, cognizant of the consequences of eugenic doctrine overseas, Eugenics Record Association president C. G. Campbell outlined his own position on "race mixture":

> One or two states have laws against miscegenetic marriage, i.e., between those of white and negro blood. While democratic society is still reluctant to make racial distinctions, all eugenic study, and the science of genetics as well, lead to the conclusion that all race mixture goes to attenuate, or to lose, valuable human traits and survival values that have been built up in pure bred races by a long process of selection. Indeed there is a well supported thesis for the theory that race mixture is the main cause for racial degeneration.[35]

What loomed large for eugenicists was not only the specter of crime, disease, and "feeble-mindedness" gnawing away at the national stock as a direct result of immigration and interracial unions. As eugenicists, as tabulators of blood quantum, as statisticians of gene-fueled crime waves and "racial degeneracy," they were facing the possibility that the future citizenry of the nation would no longer be readily classifiable. They had defended the fiction of an ethnically homogeneous American past and the importance of such a future.[36] If the lines they drew blurred, what hope could they have of establishing—in myth or in reality—a white nation? Their continuing quest for racial purity led to an affinity for eugenic policies taking shape in Europe.

While some scholars have noted the challenge posed to biological determinism in the 1930s by liberal social science,[37] the latter's ascent fell short of relegating scientific racism to "the trashbin of history." The outright dismissal of such theories was far more prevalent inside the academy than outside.

An examination of the American eugenics press reveals that, far from abandoning their position, many U.S. eugenicists applauded and often took credit for *Rassenhygiene* measures being adopted by the Third Reich. From the earliest enactments of Third Reich race laws to the full disclosures of Nazi eugenics, *Eugenical News* in particular could barely contain its excitement. In the early days, it made a token nod at distancing itself from some of the more extreme advocates of Aryan purity. An editor's note in *Eugenical News* advised readers of "The Nordic Movement in Germany" that "[w]ithout race tolerance, and paradoxical as it may seem, without race pride, practical eugenics cannot be very success-ful. . . . It is understood that the signed papers published by *Eugenical News* represent the work and conclusions of their authors, who are there-fore solely responsible for their statements."[38] Disclaimer aside, the con-clusions of these authors were increasingly and overtly embraced by *Eu-genical News,* the official organ of the ERA, the Galton Society, the Third International Congress of Eugenics, and the International Federation of Eugenic Organizations.

It is not surprising that *Eugenical News* would be an early and con-stant champion of Third Reich policies. Laughlin, as previously noted, made no secret of his admiration for Hitler. He received his honorary degree of doctor of medicine from the University of Heidelberg in 1936 in acknowledgment of the impact his model sterilization legislation had on German policy.[39] Two other members of the *Eugenical News* editorial committee were also so honored. When the magazine published the text of Germany's 1933 Hereditary Health Law, an unsigned preface boasted of the influence U.S. eugenicists had had. In much the same manner as eugenicists had claimed victory with the passage of the Johnson-Reed Act, the writer declared:

> Doubtless the legislative and court history of the experimental
> sterilization laws in 27 states of the American union provided the
> experience which Germany used in writing her national sterilization
> statute. To one versed in the history of eugenical sterilization in
> America, the text of the German statute reads almost like the "American
> model sterilization law."[40]

This self-awareness on the part of U.S. eugenicists is further evidenced by the comments of C. M. Goethe at the 1936 annual meeting of the ERA. "The last twenty years witnessed two stupendous forward move-ments, one in our United States, the other in Germany. These first in-cluded the unique Immigration Quota Acts of 1921–24. . . . These acts began a gigantic eugenical experiment in population control."[41] There is no doubt that American eugenicists saw themselves as aligned with and ideologically supportive of Nazi eugenics. They proudly anointed their own efforts as the predecessor for what was unfolding in Germany.

The most explicit links between these "forward-thinking movements" were in the realm of "intellectual" currency among individual eugeni-cists. The Nazis were avid students of U.S. eugenics doctrine. In addition to Laughlin's contribution, the works of Madison Grant and Lothrop Stoddard were translated into German and found a wide audience.[42] In 1941, Yerkes remarked, "Germany has a long lead in the development of military psychology.... What has happened in Germany is the logical sequel to the psychological and personnel services in our own army in 1917–1918."[43] Nazi doctors named American eugenicists their ideological mentors at the Nuremberg trials. Historians have also cited a shared affinity for sterilizing the "unfit," along with Jim Crow laws. Less noted is the sympathy American eugenicists expressed for a Germany grappling with its own "foreigners" (whether they were German-born or not). While the preponderance of *Eugenical News* dispatches were concerned with the "wisdom" of mass sterilization programs, the journal did, over the years, heap praise on other eugenic measures more closely aligned with its own immigration position.

> One may condemn the Nazi policy generally, but specifically it remained for Germany in 1933 to lead the great nations of the world in the recognition of the biological foundations of national character....
> ... Just as nations have codified their sanitary laws, one of the legal and scientific tasks for each of the nations in the near future is to codify its own eugenical statutes—particularly those relating to immigration, deportation, marriage, and sterilization.[44]

Throughout the duration of the Holocaust, the journal published laudatory articles on the progress of the Nazi project. "Such race-purification may be accompanied by hardship to the individual," wrote a reviewer of *Erblehre und Rassenhygiene im volischen Staat* in a 1934 issue of *Eugenical News*, "but society follows nature's method in regarding the progress of the race as more important than that of the individual."[45] *Eugenical News* ran translations of works by German eugenicists—including Dr. Frick, Reichsminister of the Interior—and printed decontextualized comments gleaned from other journals so that their authors appeared sympathetic to Nazi goals, when in fact they were not. Nicholas Fairweather's essay, which initially ran in the April 1932 issue of the *Atlantic Monthly*, was decidedly not pro-Nazi. The excerpt published in *Eugenical News*, however, seemed to imply that Fairweather was offering a rational and supportive explanation for the Nazi platform and, by extension, for that of American eugenicists:

> The state is not an end in itself; it is only a means to an end. That end is to protect, preserve, and promote the German race. The state is the

vessel and the race the contents. If non-Germans are absorbed, they merely lower the German racial level. Foreigners can be taught to speak German, but they cannot be made German.[46]

Because of its own xenophobic efforts, as well as its support for seg-regation and antimiscegenation laws, *Eugenical News* and the men be-hind it were highly sympathetic to the initial anti-Jewish statutes that Germany enacted in the 1930s. They matter-of-factly reported on mar-riage bans, "cancellation of naturalizations," and expulsions of German Jews. Not only did they fail to condemn these enactments—indeed how could they, having lobbied for and supported similar laws in the United States—they saw the measures as a learning opportunity.

> [T]he student of race history will be interested to find out what effect, in the long run—several generations hence—the present policy will have on German nationality and on the character and quality of its leadership, and what biological credits and debits—physical, mental and spiritual— these refugees will constitute for their respective receiving countries.[47]

Eugenical News wavered between stalwart defender of and apologist for Nazi eugenics. They dutifully printed a letter from Alfred Ploetz, or-ganizer of the German Society for Race Hygiene (Ploetz objected to the journal's use of the word "expelled" in describing Jewish refugees). They published dispatches from the fatherland defending measures against Jew-ish physicians: "after all ... the large German cities were literally swamped by these physicians.... The city of Berlin quite logically is trying to re-duce the number of its Jewish physicians, which is not in keeping with the racial composition of the general population."[48] Furthermore, de-spite—or maybe because of—Germany's invocation of "race hygiene," the term remained on the journal's masthead: "Eugenical News: Current Record of Human Genetics and Race Hygiene." Even as critics of Nazism began to surface, the journal stood firm.[49] Its March–April 1936 edition led off with a lengthy ode to German *Rassenhygiene*, penned by C. G. Campbell, doctor and honorary president of the ERA:

> It is unfortunate that the anti-Nazi propaganda with which all countries have been flooded has gone far to obscure the correct understanding and great importance of the German racial policy....
> This policy is not the creation of political opportunists designed to flatter national vanity or to engender racial antagonisms, but it is the integration of the well-considered conclusions of its anthropologists, its biologists, and its sociologists, the latter of whom, in contradistinction

to many in other countries, take full cognizance of the biological basis of collective life. No earnest eugenicist can fail to give approbation to such a national policy. Indeed it goes to realize the hopes that eugenicists have entertained for many years, but have despaired of ever seeing adopted in the present generation.[50]

It would have been disingenuous for *Eugenical News* or the ERA to find any serious fault with Nazis when the latter were living what U.S. eugenicists had only dreamed. Germany was, for Campbell and others, an "encouraging example of a nation that is intelligent enough to see that its first necessity is the biological one of improving its racial quality." Germany, he wrote, had the requisite patriotism, resolution, self-discipline to "augment its survival values." Other nations, Campbell warned, could ill afford to deceive themselves.[51]

Popular wisdom has it that once the tide of Nazism was fully unleashed and its intent known, eugenicists checked their enthusiasm. Not so. In 1941, eight years into the Nazi regime, six years after the passage of the Nuremberg Laws, three years after the first mass deportations, a writer for *Sexology* could still dismiss these disclosures: "There is much silly anti-eugenic criticism abroad nowadays much of which will not stand logical examination."[52] To be sure, far from being deterred by Nazism, U.S. eugenicists were encouraged. The late 1930s saw the founding of a new eugenics enterprise, an organization with enormous staying power: the Pioneer Fund.

The Pioneer Fund

Scientific Racism and the Eugenic Endowment

For over sixty years, the Pioneer Fund has bankrolled academics with explicitly racist research imperatives. In one way or another, its officers have remained active in anti-immigration efforts—including testifying before Congress during debate on the 1965 Immigration Act—throughout the organization's life.

The Pioneer Fund was incorporated in 1937, courtesy of funds provided by Wycliffe Draper, textile magnate and advocate of Black "repatriation." It was founded and maintained by some of the same men who led eugenicists to victory in 1924, including Frederick Osborn and Harry Laughlin, its first president.[1] The original charter mandated the pursuit of "racial betterment" via research in eugenics and heredity. The fund was also to be used to encourage procreation by white descendants of white settlers of the original thirteen colonies.[2] In 1985 the charter was amended, apparently to clarify original goals and to establish guidelines for financial aid. Paragraph 2A reads:

> The children selected for such aid shall be children of parents who are citizens of the United States, and in selecting such children, unless the directors deem it inadvisable, consideration shall be especially given to children who are deemed to be descended predominantly from persons who settled in the original thirteen states prior to the adoption of the Constitution of the United States and/or from related stocks, or to classes of children, the majority of whom are so descended.[3]

Like other racist crusades of this century, eugenics was interested in recruiting the young. One of Laughlin's first acts as president of the

Pioneer Fund was to import *Applied Eugenics in Present-Day Germany*, a Nazi propaganda film in which eugenically healthy and physically and mentally sound young Aryans live in poverty while people with disabilities lived it up in a posh sanitarium. Laughlin added subtitles and offered screenings to three thousand high schools in the United States. Twenty-eight schools were interested.[4] Another early funding priority was the disbursement of cash awards to junior flying officers to enable them to pass on their highly valued genes to children they otherwise could not have afforded.[5]

After World War II, the fund continued to underwrite the translation of eugenics texts from German to English. It financed the AES through the 1950s.[6] During this period and on into the 1960s, Fund grantees produced tracts for the Ku Klux Klan and the Southern White Citizens Councils. The Pioneer Fund itself supported Joseph McCarthy and opposed civil rights legislation.[7] Though its eugenic undertakings were diversified, the fund and the men who ran it persevered in their anti-immigrant platform.

Twenty-eight years after the passage of the Johnson-Reed Act, and with the continued revelations of Nazi eugenics still coming to light, eugenicists were not in the position they had once been vis-à-vis overt congressional support. It would be difficult to make a case for explicit and exhaustive eugenicist impact on the drafting and passage of the McCarran-Walter Act of 1952. Nevertheless, there was a critical link. Francis Walter, representative from Pennsylvania who would chair the House Un-American Activities Committee (HUAC) for five terms, sat on the Pioneer Fund's leadership board in the 1950s. While the legislation that bears his name did remove some obstacles for some immigrants (a small number of immigrants from the "Asian-Pacific Triangle" were permitted, the Japanese were given a meager immigration quota of 185 per year, and Issei were granted naturalization rights), the law tightened existing limitations and significantly reduced immigration from "nonwhite" countries. It provided for the deportation and denaturalization of immigrants deemed "subversive." Resident aliens became deportable for engaging in political activity, and deportation cases were moved to specialized boards, not bound by due process of the courts. Eight years later, Francis Walter was found to be involved with the Draper Project, which worked to establish the genetic inferiority of African-Americans.[8]

Over the years, members and affiliates of the Pioneer Fund continued to press for exclusionist immigration legislation. John Trevor, who at various points served as the endowment's codirector, treasurer, and secretary, testified against the passage of less restrictive immigration laws in 1965, railing against "unlimited diversity" in immigration and in the population at large. Trevor's father, John Trevor Sr., was credited by Representative Michael A. Feighan of Ohio with "providing the basic thinking out of which evolved a national origins quota system." Trevor was chairman of the American Coalition of Patriotic Societies (ACPS), which claimed a membership of three million and was founded by his father in 1929, when he testified before a subcommittee of the House Committee on the Judiciary.[9]

Like his exclusionist forebears, Trevor summoned an array of worst-case scenarios for the congressmen: a decline in "our moral standards," "our standard of living," "our standards of health," and in our effectiveness as world leader in "the common struggle against communism." Trevor denied that the national origins system was discriminatory, arguing circuitously in front of the committee that it was "like a mirror held up before the American people and reflecting the proportions of their various foreign nationals. . . . National origins simply attempts to have immigration into the United States conform to our own people." Despite his claims of racial and national neutrality—hardly a credible assertion given his in-depth involvement with the Pioneer Fund and the pro-apartheid ACPS—Trevor took care to read into the record statistics citing a 3 percent annual increase in Latin American population, along with a predicted rise of 400 million over the next thirty-five years.[10] More telling, however, was his lengthy citation of a 1959 article by John M. Radzinski that first appeared in the *American Journal of Psychiatry*. Radzinski indulged in his own brand of paranoia when it came to racial "mingling," casting American society as a body in decline:

A review of ethnic mingling among the peoples of Europe demonstrates that the process has been associated with cultural dislocations and decline. Eventually, after centuries of incubation and biological fusion, new cultural patterns became established, often superior, but at times inferior, to the ones they displaced. The immigration of many millions of people into the U.S.A., particularly during the last 80 years, has brought together here the greatest assortment of ethnic stocks in the world and probably in history.

... such a conglomeration of racial and ethnic elements renders a serious cultural decline inevitable. Symptoms of the decline are already apparent in the deteriorating state of some aspects of our culture, ... and in the virulence of frank antisocial behavior among our people far in excess of that encountered in West European countries, Canada, and Australia.

In short, we must face the unpalatable truth that the present American society is sick.[11]

Representative Arch A. Moore of West Virginia called Trevor's testimony and his work with the ACPS "refreshing." Frank Chelf, representative from Kentucky, took the opportunity to praise the "good stock" of "those who settled this country... and banded themselves together in 13 small colonies."[12] Despite the apparent longevity of the "good stock" concept in congressional quarters and the warm reception given to Trevor by some (but by no means all) of the subcommittee members, the 1965 Immigration Act abolished the national origins quota system. The mere fact that Trevor was invited to testify at all, no less during the height of the Civil Rights movement, demonstrates the ongoing validity accorded to eugenics by the legislative branch of the U.S. government (and its power as an agent of backlash) even if eugenicists no longer possessed the pull or persuasion of their glory years.

If eugenicists lost some ground on the floor of Congress, they gained it back in the halls of academe. Mehler estimates that the Pioneer Fund dispensed in excess of $10 million between 1971 and 1992. Other assessments are higher. In 1994, Adam Miller found that between the fund's $5-million investment portfolio, trusts, bequests, donations, and other sources, the Pioneer Fund generated an annual income of approximately $1 million, most of which is divided among twenty recipients per year. These include anti-immigrant groups and academic eugenicists.[13] Garrett Harding, professor emeritus of biological sciences at the University of California at Santa Barbara, sat on the boards of two Pioneer Fund–backed groups, both of which have singled out Latino immigration as a eugenic threat and both dedicated to its curtailment.[14] An outgrowth of this financial validation is the way in which pseudoscience reinforces pseudoscience. Herrnstein and Murray cited the work of at least thirteen Pioneer Fund grantees in the pages of *The Bell Curve*.[15]

More disturbing than this monetary support, however, is the fund's backing of the Federation for American Immigration Reform (FAIR).

Founded in 1979, FAIR is an offshoot of Zero Population Growth. In 1982, it secured its first grant from the Pioneer Fund and by 1993 had received over $1 million from the endowment.[16] FAIR operatives Harold Ezzell and Alan Nelson, a former INS commissioner, cowrote California's Proposition 187. Both were on the federation's payroll as lobbyists at the time.[17] As such, they were indirect beneficiaries of the Pioneer Fund's goodwill. This connection comes as no surprise. Eugenics has always had as its goal the singling out of the already marginalized for institutional bias, excluding them from what is available to the rest of the population (in this instance, basic services such as health care and education are seen as a right of the "general public" from which undocumented immigrants are now isolated). Like their ideological predecessors eighty years before, FAIR, with 30,000 members in the state of California alone, has been able to make common cause with other interests in an effort to undermine immigrants' standing and inflame xenophobia in its ultimate aim to cut all undocumented immigration and slash legal immigration by half.[18]

As always, exclusionists have been drawing from the ranks of other movements, including the environmental movement. Key figures in the Sierra Club were instrumental in establishing FAIR.[19] The common cause here is a shared investment in the neo-Malthusian belief that "overpopulation" bears primary responsibility for poverty, inequity, and environmental degradation.[20] Uninterrupted, the logic goes, population growth will lead to the total ruination of both civilization and habitat. Neo-Malthusianism has cropped up in a variety of progressive campaigns. Margaret Sanger and her associates frequently made recourse to the jargon of "population pressure" in their birth control advocacy. By ignoring issues of comparative per capita consumption and the distribution of resources, population control rhetoric has been used to promote draconian measures against the most vulnerable groups and individuals in society (including compulsory sterilization in the United States and abroad).[21] Today, right-wing groups such as Negative Population Growth (NPG) are blaming immigrants for "the population growth that is propelling our nation down the road to environmental ruin."[22] In 1996, NPG ran ads in *Harpers, Atlantic Monthly,* and *E Magazine* predicting a "demographic and environmental disaster" if the United States did not deport five hundred thousand people annually.[23] The Coalition for United States Population Stabilization, founded in 1995, lobbies for a near moratorium on immigration, an end to "welfare incentives" (also blamed for encouraging fecundity), and a decrease in national fertility. Among the Coalition's organizational founders was the Massachusetts Audubon Society.[24] In recent years, the Sierra Club has again found itself mulling over immigration. Many, including those serving on the San Francisco chapter's "population committee," concur with NPG's assessment of immigrants as an environmental hazard. In early 1994, that committee recommended severe cuts in permissible immigration.[25] Sierra Club members were pressured in 1996 by the conservative Population-Environment Balance (PEB) to adopt a pro-restriction stance as club policy.[26]

This is not intended as an indictment of the environmental movement. The Sierra Club is not NPG or PEB. There were splits in the organization over immigration and many members resisted the xenophobic diatribes coming at them from both inside and outside the organization. In the end, the Sierra Club decided to remain "neutral" on immigration issues.[27] However, the position of the Sierra Club (and, to a lesser extent, the Massachusetts Audubon Society) is indicative of pseudoscientific assaults that appeal to, and cannibalize, other movements.

That an organization presumably dedicated to saving wetlands and wildlife should be debating immigration issues is not a scenario entirely without precedent. Madison Grant, along with other eugenicists of his generation, was deeply involved in conservation efforts. Elk reserves and giant redwoods in California bear his name in testimony to his work on their behalf. For Grant and others, environmentalism and eugenics were points along the same continuum. There is a certain historical consistency in the perception among some strands of contemporary environmentalism that efforts to save the planet from toxins and corporate greed are intimately connected to population control and, by extension, to immigration restriction. It is an associative link that is especially frightening because it lends a veneer of not only rationality but also progressiveness to reactionary, racist, and inherently coercive legislative and medical tactics.

"Indiscriminate Kindness" and "Maudlin Sentimentalism"
Fighting the "Philanthropic" Impulse

In 1994, Charles Murray told a reporter for the *New York Times Magazine* that people were no longer poor because of social barriers, but rather because of their inherent lack of intellectual prowess.[1] This argument, appealing to some precisely because it obviates any societal obligation toward the disfranchised, has strong precedent within the eugenics movement. In 1929, *Birth Control Review* ran an article by E. J. Lidbetter explaining the persistence of "a race of chronic pauper stocks."[2]

> The problems of population are no longer regarded to be essentially as economic, but mainly as biological. In this respect, the teaching of the economists of the 19th century needs to be modified.... so also must the social theories that arose out of it, particularly in the matter of social inadequacy, in its various forms, for these, we now know, depend to a very large extent upon heredity and not, as was formerly supposed, upon economic conditions.[3]

Lidbetter, too, had ideological predecessors, most notably Grant and Ward who urged resistance to "sentimental" impulses that might weaken eugenicists' resolve and compromise the outcome of the immigration debate. Grant closed *The Passing of the Great Race* with this cautionary paragraph:

> We Americans must realize that the altruistic ideals which have controlled our social development during the past century, and the maudlin sentimentalism that has made America "an asylum for the oppressed," are sweeping the nation toward a racial abyss. If the Melting Pot is allowed to boil without control, and we continue to follow our national

model and deliberately blind ourselves to all "distinctions of race, creed, or color," the type of native American of colonial descent will become as extinct as the Athenian age of Pericles, and the Viking of the days of Rollo.[4]

Having thus reduced a sense of socioeconomic justice, more imagined than real, to mawkish emotion, Grant heard his words echoed the following year in a 1917 piece by Robert Ward:

> I do not believe that sentiment can solve grave national problems. I do not believe that the indiscriminate kindness we may seem to be able to show to some thousands, or hundreds of thousands, or millions, of Europeans and Asiatic immigrants can in any conceivable way counterbalance the harm that these people may do our race if large numbers of them are mentally and physically unfit.

Ward went so far as to describe this "indiscriminate hospitality" as a "short-sighted, selfish, ungenerous, un-American policy." He wrote,

> It is in the highest degree ungenerous for us, who are the custodians of the future heritage of our race, to permit to land on our shores mental and physical defectives who, themselves and through their descendants, will lower the mental and physical standards of our own people, and will tremendously increase all our future problems of public and private philanthropy. We have no right to saddle any additional burdens upon the already overburdened coming generations of Americans.[5]

In Ward's scenario, it was not immigrants or the poor who were vulnerable, but the American taxpayer. Later that year, another of the journal's contributors, Maurice Fishberg, declared that Jewish philanthropy supported the survival and proliferation of "defectives" and was therefore "an important dysgenic factor... undoubtedly responsible in a large measure for much of the degeneracy observed among the Jews of today."[6] This line of reasoning reasserted itself again and again in "cycle of dependency" theories, "culture of poverty" arguments, and, most recently, in debates about workfare.

Passage of the 1917 Immigration Act had already been secured by the time Ward and Fishberg issued their warnings. Similarly, in 1921—the year Harding signed the Three Per Cent Restrictive Act—then Vice President Calvin Coolidge, who would later sign the 1924 Immigration Act, chided Americans for what he perceived as the nation's warmheartedness. His article "Whose Country Is This" appeared in *Good Housekeeping*, cautioning, "The retroactive immigrant is a danger in our midst...."

There is no room for him here. . . . There are racial considerations too grave to be brushed aside for any sentimental reasons. . . . Quality of mind and body suggests that observance of ethnic law is as great a necessity to a nation as immigration law."[7]

Eugenics meant, to its proponents, the victory of rationality over short-sighted altruism, reason over "un-American" (yet "distinctively American") hospitality, science over sentimentalism.[8] What better way to fight for "racially sound" citizens than with a rational immigration policy? "Scientifically" informed immigration restriction would not only mark an era of "better-bred" Americans, but would signal the sound, legislative mettle of a healthy nation with its priorities intact.

The Abiding Panic

Race demonization, partially though not exclusively rooted in pseudo-biology's reemergence as a legitimate discourse, accompanied, benefited from, and propelled the round of anti-immigrant legislation that brought the twentieth century to a close. FAIR founder and board member John Tanton penned a confidential memo back in 1988, warning that the continuation of immigration from Latin America would result in the takeover of the United States by "a group that is simply more fertile."[1] Today, the preoccupation with immigrant fertility is couched in concerns over expenditures rather than in classic eugenicist worries about the depletion of the national gene pool. However, the foregrounding of immigrants' imagined overfecundity resonates with earlier eugenic treatises. Once again, it is the threat of immigrants' progeny that is summoned in current attacks that, by their very nature, single out women and children. In Virginia, state Senator Warren E. Barry introduced a bill that would require state officials to identify and report undocumented immigrants who use publicly funded social services. In addition to being reported to federal authorities, immigrants would be liable for "debt collection for the services they have illegally obtained." Said Barry, "We have to make an appeal to the federal government to either put an end to this tremendous influx of pregnant women and undocumented workers and children that are pouring into the commonwealth or to pay for it."[2]

The image of immigrants and their children as unassimilable foreign bodies, as polluters, as both infected and infectors, is still with us. In an editorial published in the *San Francisco Chronicle* not three months before the vote on Proposition 187, Alan Nelson wrote, "The best way to stop contagious disease is to stop illegal immigrants from entering the United States."[3] In a society terrified by disease and viruses it cannot control, this type of rhetoric is extremely loaded and best seen as part of a multipronged campaign bent on what Mosse has called the "medical legitimizing of the outsider."[4] Where Nelson fabricated a public health threat, California State Senator Rob Hurtt relied on metaphor to forge the same stranger-sickness link, going so far as to say that the state's immigration "problem" was reminiscent of the bubonic plague.[5] His comments harken back to Goethe's 1936 declaration that Mexican immigration meant "an epidemic of the Black Death."

Other anti-immigrant tirades also recall ominous titles of eugenic tracts of yesteryear, such as Stoddard's *The Rising Tide of Color against White World Supremacy* (1922) or Grant's *The Passing of the Great Race.* On the eve of the NAFTA vote in 1993, the newsletter of the California Coalition for Immigration Reform *(9-1-1)* trumpeted, "Nafta = Totally Open Borders (more citizens murdered) . . . the death of our nation and the death of our people." The American Immigration Control Foundation published a booklet entitled *The Path to National Suicide—An Essay on Immigration and Multiculturalism.*[6]

In the past, all invocations of a once-great-now-declining nation were predicated on the racialization and racist derision of Others (immigrants, northbound southern Blacks, Mexicanos on the "wrong" side of the border). They relied on and resorted to hysterical images of those Others and the threat posed to the metaphoric and literal national body/bodies. Little has changed. Immigration foes have neither abandoned nor outgrown the rhetorical and strategic devices employed by early twentieth-century eugenicists. There are, of course, important differences between those early decades and today. Eugenics was an esteemed partner in sustaining anti-immigrant sentiment and ensuring passage of exclusionist legislation and eugenicists wielded their influence openly. While it may not command the power it once knew, eugenics continues to find its niche in "respectable" political circles. Its ongoing assault on immigrants, predicated on its service to, and recruitment of, nationalist and racist dogma, has been but one of its trajectories.

George Mosse has written that nationalism, during its long career, reached out to conservatism, liberalism, and socialism.[7] Eugenics, too, has been a highly opportunistic and interactive ideology. In its role as nationalism's house biologist, eugenics has drawn supporters from across the political spectrum. At different historical moments, eugenicists have been able to join forces with less scientifically inclined racists, wartime patriots, environmentalists, progressives, red-baiters, and anarchist haters, all the while claiming victory and vindication for themselves with the passage of each new anti-immigrant measure. This was especially true with regard to the 1917 and 1924 Immigration Acts, which best epitomize the consolidation of anti-immigration forces and the way in which eugenic measurement acted as catalyst. United States eugenicists were at the height of their power during the debates on these bills, but the amalgamation of scientific racism, xenophobia, and nationalism, evident in the eugenics lobby, can lay claim to a certain continuity, even if the strength of that lobby has been less than constant.

Eugenics was never concerned exclusively with immigration matters— by the time H. H. Goddard brought the IQ test to Ellis Island in 1912, five states characterized their sterilization laws as eugenic and/or therapeutic, and many more would follow—targeting the "feeble-minded," the syphilitic, the epileptic, prostitutes, "hereditary criminals," "sexual perverts," and "moral degenerates," among others.[8] Eugenics served as an arbiter and defender of immutable lines and blood boundaries between peoples. Its insistence on group pathologies that predetermined social and economic arrangements provided a facile explanation for inequity without threatening its disruption. Within this paradigm, a society of equal access and reward and status was, as Richard Lewontin has written, biologically impossible.[9]

No matter what branch of science they came from (and many merely invoked the vocabulary of science), eugenicists were driven by a need to rationalize and shore up social categories. This imperative would have chilling effects on other outsiders who came under the microscope, for proponents of "racial betterment" were very clear about what they perceived to be complementary crusades. Eugenics is, after all, an inherently comparative enterprise. For eugenicists, immigration was the peril from without. To truly "purify" the nation, racially and culturally, the perils from within also demanded attention.

II
Queer Anatomy: One Hundred Years of Diagnosis, Dissection, and Political Strategy

The difference between homosexuals and heterosexuals should be in the brain, not in the heart.

—Dick Swaab, neuroscientist

Science as Savior

In February 1999, the *Atlantic Monthly* ran an article on the work of Paul Ewald and Gregory Cochran, a biologist-physicist team who have theorized that homosexuality, along with heart disease, cervical cancer, and mental illness, may be caused by a germ rather than a gene. This assertion is in direct opposition to a host of recent claims on the cause of homosexuality and gender "deviance": Simon LeVay's 1991 announcement that a cell group in one portion of the brain's hypothalamus is twice as large in straight men than in gay men; J. Michael Bailey and Richard Pillard's 1992 contention that lesbianism or bisexuality in the identical twins they examined points to genetic determinants in the brain in utero; Dean Hamer's 1993 declaration that a not yet isolated gene on the X chromosome influences male sexual orientation; and Dick Swaab's 1995 report that a region in the hypothalamus is 60 percent larger in men than in male-to-female transsexuals.

There's nothing new about all this conjecture. Speculation on the causes and symptoms of homosexuality has been both a professional and a popular preoccupation for over one hundred years: prenatal stress results in gay offspring; lesbians don't menstruate; lesbians have asymmetrical faces; gay men can't whistle; lesbians are proficient whistlers; fluoride in the drinking water causes homosexuality; and, ripped from the headlines of the *Oakland Tribune*, "Crash Made Him Gay, Jury Made Him Rich."[1]

The latest findings are not even unique in their genetic emphasis. Yet the media has treated them all as breakthrough science. Hamer and LeVay garnered most of the attention appearing on *20/20* and *Nightline* and in the pages of the *New York Times Book Review,* the *Washington Post,* the *San Francisco Examiner, USA Today,* and the gay press. The popular press, along with mainstream publishers, has been instrumental in creating and disseminating the story of genetic homosexuality. Not only have they given the public access to what might have remained dense and unreadable in scientific journals, they have worked with researchers to market their findings to a broader audience. Ewald and Cochran's work appeared in the *Atlantic Monthly* long before it faced the rigorous peer review that accompanies publication in professional journals. While Hamer took pains to explain that he had not isolated a gay gene per se, merely uncovered evidence that such a gene exists, he marketed the buzzwords very effectively. His book, *The Science of Desire,* bore the subtitle *The Search for the Gay Gene and the Biology of Behavior,* enabling him to simultaneously disavow and profit from a misleading turn of phrase.

The warm reception that greeted these hereditarian hypotheses (Ewald and Cochran's germ theory has not garnered the same kind of reception) raises two issues: What is it about causation theories that is so appealing to mainstream institutions and heterosexual America? What is it about the research that has so many in the queer community looking to it for deliverance? Mainstream media and its predominantly straight consumers look for a good story; if it holds an unspoken promise of curatives, so much the better. More than that, a focus on what causes queerness eclipses the larger question: who wants to know and why? Significant segments of the gay community, on the other hand, hold that causation theories can be honed into a strategic tool and integrated into a larger legal and political struggle. For many, there may also be personal attachment to biological explanations, a comfort in being able to tell straight family and friends that "we were born that way." The stakes are clearly different, but there is a commonality here. Genetic promises have been embraced without interrogation by a community and a larger society eager to accept any quick-fix explanations (and consequent solutions) that modern science had to offer. Whether the hope was for an antidote for homosexuality or homophobia, this embrace typifies the science-as-savior prism that has greeted so many determinist enterprises.

Of all the groups targeted by biological determinism, queers seem to be the only ones who have looked to eugenics to deliver us from marginalization. Why, in the face of so much evidence to the contrary—chemical and electric shock treatments, lobotomies, hysterectomies, castrations, vasectomies, and clitoridectomies all performed on lesbians and gay men in this country—do we continue to invest in medical models? A case in point is the legal challenge to Colorado's anti-gay Amendment 2, which sought to ban the enactment of antidiscrimination protections for lesbians, gays, and bisexuals.

On 14 October 1993, Hamer, a geneticist at the National Cancer Institute, took the stand in Denver's district court. He had been subpoenaed by lawyers for the Colorado Legal Initiatives Project (CLIP) who were challenging the recently passed amendment. Their legal argument held that homosexuality is a biological construct and that rights cannot be denied or abridged based on biology. According to this reasoning, homosexuality, if it is hereditary or congenital, endows lesbians and gays with an authentic, incontestable, protected minority status. It is not chosen, but doled out by nature; ergo, gays and lesbians should not suffer social ostracism or political disfranchisement.

This particular strategy was not without precedent, at least outside the courtroom. Lesbian and gay history is replete with champions who relied on evolutionary or biological arguments to agitate for our civil and human rights. Karl Heinrich Ulrichs, for example, pursued this line in the mid–nineteenth century. If homosexuality was recognized as inborn, he reasoned, gays could not be criminally prosecuted. Perhaps choice implied guilt, but the undeniable force of nature should not. Magnus Hirschfeld engaged in similar reasoning, but neither his Institute of Sex Research, which ultimately fell to Nazi violence, nor the Scientific-Humanitarian Committee, which he cofounded, could safeguard civil rights for gays and lesbians. More recently, studies on everything from hormone levels to maternal DNA have underscored a furious desire for morphological explanations of homosexuality. As evidenced by a quick succession of antigay ballot initiatives in Colorado, Oregon, Maine, and Idaho, such quests have contributed little to the abatement of homophobia. The best that can be said of such theories is that they promote not justice or equality or respect, but a grudging tolerance infused with pity. The worst that can be said of them takes up the bulk of the pages that follow.

Arguing the biological immutability of homosexuality, a tactic historically used against queers, was an acquiescence on CLIP's part to demands that the lesbian/gay/transgender/bisexual community explain or justify its very existence in order to secure full political and social enfranchisement—a troubling precedent for all civil rights movements. In choosing to invoke medical models, our legal advocates chose also to ignore not only a century of medical assaults on queers—indeed, no matter how well intentioned, CLIP's strategy is a product of this history—but also the current round of eugenics being unleashed on other marginalized groups. They put up witnesses such as Richard Green, a UCLA psychiatrist and author of *The Sissy Boy Syndrome*. Green stated in his affidavit that prenatal stress could turn a male fetus into a homosexual boy, and testified that large clitorises not treated by "corrective surgery" within the first year of life could turn little girls into lesbians. The court also heard from Judd Marmor, who claimed that some transsexuals have "temporal lobe pathology."[2]

Hamer was asked to testify because of his "discovery," some months earlier, that a small region on the X chromosome was the same in a disproportionately high number of gay brothers, indicating, to his mind, a genetic basis for homosexuality. Hamer later wrote, "More than once, sitting in that Colorado courtroom, I thought, what am I doing here? What does biology have to do with law, politics, and equal rights, anyway?"[3] This would be a sound and reasonable question if there were not a long history, in this country and abroad, of science and social policy reinforcing each other's agendas. It might be a valid question had the trial not been taking place a decade into the AIDS pandemic, the political and medical responses to which have so clearly demonstrated exactly what biology has to do with law, politics, and equal rights. Hamer's own book jacket displayed a quote from E. O. Wilson, standard-bearer of sociobiology: "The search for a genetic basis of homosexuality illustrates, in a profound way, the potentially turbulent relation between the biological sciences and public policy." Wilson called *The Science of Desire* "an authoritative and very readable explanation of how both to conduct good science and to manage its risky social implications." In truth, the book fails on both of these counts. While the primary endeavor of this chapter is to historicize these "risky social implications"—not the least of which is the specter of eugenics the book raised but never adequately dealt with—something needs to be said about just how good Hamer's science was. Adherence to faulty scientific data demonstrates both the right's commitment to homophobia *and* the political desperation of gay rights advocates. It also exposes the stalwart faith in the alleged neutrality of scientific theories, an enlightenment belief in progress that has marked most eugenics enterprises. For these reasons, the conditions of Hamer's research, and that of his like-minded contemporaries, require a brief discussion before proceeding further.

On 25 June 1995, the *Chicago Tribune* reported that Hamer was under investigation by the National Institute of Health's Office of Research Integrity. Just two years earlier the *New York Times,* the *Washington Post, Newsweek, USA Today,* the *San Francisco Examiner, Nightline,* and London's *Daily Examiner,* just to name a few, had all lavished favorable coverage on Hamer's "breakthrough." One headline blared, "Abortion Hope after 'Gay Genes Finding.'"[4] Even San Francisco's *Bay Area Reporter,* a queer weekly, was calling the geneticist a "trailblazer" and claiming that "the social, *medical* and other benefits of this could be enormous."[5] The *Tribune*'s report did not prompt serious critique in the national press. News that Hamer was now being accused of manipulating data during his search for a gay gene, specifically that he ignored data that contradicted his own theories, hardly made the same sort of media dent. When he released his second set of studies in October 1995, the news coverage made no mention of the investigation. The press greeted Hamer's latest report as if it were independent confirmation of his earlier work and not a scientist's reiteration of his own claims. More recently, researchers at the University of Western Ontario revealed that they were unable to find any evidence to support Hamer's claims—an announcement that also failed to garner much attention.[6] The fact is, most studies that contradict or fail to confirm a genetic basis for homosexuality go unpublished or underreported.

Long before Hamer came under the scrutiny of the *Chicago Tribune,* he was criticized by other scientists for failing to publish his original data and denying other researchers access to any of the genetic material he used during his study. No other research team has been able to duplicate, and thereby confirm, his work. Implications of his findings aside, Hamer's methodology was also questioned. He never examined the genes of heterosexual brothers, a control experiment that many in the field considered basic.[7] He eliminated potential subjects whom he deemed "atypical," including a family with two lesbian sisters and three gay brothers, for fear that "we might not find a typical gene," and because he had read that "homosexuality in males and females usually runs in separate families, so one family is not likely to have both gay men and lesbians." Likewise, Hamer's team removed from consideration families with gay sons and gay fathers because "this pattern also would not be consistent with X-chromosome linkage." He extrapolated about all gay men, despite an extremely homogeneous subject sample: 92 percent white with an average educational level of 15.5 years, average yearly income of over $40,000, average age 36. "The advantage of working with such a sophisticated group of subjects," Hamer rationalized, "was their openness about sexuality, a characteristic that would help rather than hinder our research." What was he implying, then, about working-class and poor gay men and gay men of color? Lastly, Hamer's working statistics placed the number of gays and lesbians in the staggeringly low range of 1 to 7 percent of the population (2.0 to 2.6 percent for the purposes of his study).[8] A numerical deflation of this magnitude has resounding political implications in and of itself.

Hamer's methodology is not the critical issue. Far more telling was the failure of most, though certainly not all, of the papers and magazines and news shows, which had lauded Hamer two short summers before, to come forward with follow-up stories. Hamer's study, and the packaging of that study, confirmed one hundred years of spoken and unspoken speculation, namely that queers are inherently deviant. The revelation that he may have rigged his results confirmed no one's hopes or fears, ergo it was not news. What happens when flawed, unproven theories and nonreplicable research enter the public discourse? If they support rampant bias, they become entrenched, regardless of later debunking. Such a scenario unfolded in early conjecture and reportage on AIDS, and the fallout is still being felt.

Hamer was only one player in the Great 1990s Medicalization of Queerness, a resurgence of a much longer, more entrenched biologism of homosexuality. His ideological cohort includes Sandra Witelson (who announced in 1994 that "the communication conduit between parts of the brain used for understanding speech and perceiving objects" is bigger in gay men than in straight men) and Dick Swaab.[9] Most significantly, Hamer acknowledged another scientist who, in his words, "ushered in the era of looking for biological differences in sexuality"—Simon LeVay.[10]

In 1991, two years before Hamer's star ascended, LeVay gained attention when *Science* published his findings regarding the hypothalamus in gay men. LeVay, then a researcher at the Salk Institute in San Diego,

wrote that a cell group in the interstitial nuclei of the anterior hypothalamus (INAH 3) was twice as large in heterosexual men than in gay
men. This conclusion was based on an analysis of brain tissue taken
during autopsies from forty-one subjects in New York and California
metropolitan hospitals: eighteen gay men with AIDS, one bisexual man
with AIDS, sixteen "presumed heterosexual" men (six of whom died
AIDS-related deaths), and one woman with AIDS (sexuality not stated).[11]
In early 1992, foreshadowing Hamer's search, LeVay told a reporter that
within five years researchers might be able to isolate a small number of
sex genes.[12]

LeVay's contention that the key to sexual orientation is in the brain,
and more specifically in the hypothalamus, was not new to the study of
sexuality and should be situated in its recent history. In 1972, two German neurologists, F. Roeder and D. Muller, reported on their efforts to
"cure" gays as "a matter of public health policy." Their "treatment" involved "the unilateral destruction of the sex behavior center located in
the central medial hypothalamic nucleus." They persuaded judges to release gay men serving prison sentences in exchange for their submission
to brain surgery. Needles—electronic probes—were inserted into the
men's brains, and the area then coagulated with the flip of a switch. In
other words, as one reporter noted at the time, they burned out the
brain cells they did not like.[13] Following the study, and with no regard
for medical ethics or the coercive tactics employed to secure subjects,
doctors in several countries began performing the psychosurgery, despite
possible side effects that included transitory diabetes, disturbance of
the appetite control center resulting in obesity, a diminished sex drive,
and asexuality.[14] Lancet, the official publication of the British Medical
Society, endorsed the surgery on the grounds that "castration is open to
criticism on ethical grounds while psychosurgery is not."[15] Not to be
outdone, scientists in the United States were also zeroing in on the brain.
Walter E. Stumpf, University of North Carolina, reported that nerve
cells that attract sex hormones are scattered in the frontal lobe, temporal lobes, the brain stem, and other parts of the central brain. This information, he declared, would open new doors in treating "emotional
and sexual disorders" through neurosurgery.[16]

At least LeVay's subjects were already dead, though the means of ac-
quiring those subjects deserves some attention. LeVay has acknowledged that samples of gay men's brain tissue were readily available to

him because of AIDS. For one of his test groups, Hamer recruited subjects from HIV-positive outpatients at the National Institutes of Health, the Whitman Walker Clinic in Washington, D.C., lesbian and gay twelve-step programs, and Emergence, an organization of lesbian and gay Christian Scientists.[17] He did not select a cross section of the gay male population. More disturbing is the fact that subjects were procurable because, as in other eugenics enterprises, they were enmeshed in an institutional net, in this instance the health care system.

Like Hamer, LeVay issued a disclaimer on his work, saying that it could not be determined that the difference in INAH 3 directly *caused* homosexuality; it might merely be the *consequence* of homosexuality. This hypothesis, too, has predecessors. Scientists have been arguing for over a hundred years that homosexuality, along with masturbation and "excessive" heterosexual sex, was either the manifestation of a biological deficiency or else the root cause of a wide array of anatomical "mishaps," from underbites to left-handedness.

History aside, contemporaries of LeVay have questioned his measurements. The structures themselves are difficult to see in tissue slices, and it is unclear what the most definitive measurement of size is. LeVay measured volume. Others say the actual cell count is more reliable.[18] He has also been criticized for his small sample size and for compiling inadequate sexual histories. Several of his colleagues have noted that the size of the nuclei could be impacted by AIDS, especially given the fact that INAH 3 is dependent on testosterone levels, which, studies have shown, may be reduced significantly during an end-stage HIV infection.[19] These criticisms, like those of Hamer's work, have been largely ignored relative to the fanfare that greeted the study's initial release. Regardless, the idea of a "gay brain" or a "gay gene" continues to be taken seriously.

The political climate in the country at the time these studies were unveiled goes a long way toward explaining their popularity. Their promise of liberatory biologism coincided with the spate of antigay initiatives that were on the ballot in towns, counties, and states across the country. Hate crimes against queers, including murder, were on the rise. Gays in the military became the seminal issue in gay politics, overshadowing larger issues of civil rights and AIDS. In short, the country was, once again, wondering what to do about and to queers. It should come as no surprise then, with all these pressing issues, that the focus would be displaced. The concern was and is not about rights, but rather about how we got to be the way we are. This is, of course, for those not being dissected, a much safer enterprise, requiring no change in the status quo and offering up the promise of eventual eradication not of marginalization, but of the marginalized.

Much of the commentary generated by genetic postulations is couched in terms of equal rights and an end to social and familial ostracism. A 1992 *Newsweek* article lauded these new theories as all-around good news for queers and their parents. "Theoretically, it could gain them the civil-rights protections accorded any 'natural' minority, in which the legal linchpin is the question of an 'immutable' characteristic. Moreover, it could lift the burden of self-blame from their parents." Parental guilt was a key flash point of the *Newsweek* piece. "For parents, a child's 'coming out' can lead to painful soul-searching." While this is too often true, scientific inquiry demands sounder inducements. Pillard, a psychiatrist who has worked with Bailey on lesbian twin studies, told an interviewer, "A genetic component in sexual orientation says, 'this is not a fault, and it's not your fault.'" Another researcher said, "There is a tendency for people, when told homosexuality is biological, to heave a sigh of relief. It relieves families and homosexuals of guilt." The *Newsweek* writer maintained that members of Parents and Friends of Lesbians and Gays (PFLAG) are "firmly behind any research that implicates biology as the source of gayness. It assuages the raging guilt some of them feel that they might be responsible."[20] Soothing anxiety has now been elevated to a research imperative for the hard sciences, a capitulation to the notion that there is indeed something to feel guilty about.

While the chief goal of much of the article seemed to be to temper doubt about parenting skills (the cover featured a close-up of a baby's face with the banner "Is This Child Gay?"), it also conveyed a simplistic and mislaid faith in the power of medical models. Biology here is not only destiny but liberation—liberation from homophobia but also from socialization exegeses. Socialization, as an explanatory argument, has, after all, also been used against LGBTs, routinely and systematically banning them from being teachers, foster parents, adoptive parents, and so on, for fear that impressionable minds will be corrupted and unsuspecting youth initiated into the queer fold. It is not surprising then that segments of the gay community would also embrace genetic interpretation, ignoring not only the histories of racism and eugenics, but also the medical profession's history of treating homosexuality. In this way, much of the community has fallen prey to the limiting, and ultimately destructive, binary exemplified by the plaintiff's counsel in the Amendment 2 case. But others, with longer memories, have contested a reliance on medical paradigms, hereditary or otherwise. As Doug Futuyma pointed out, such a reliance is perilous because new research could come along that suggests homosexuality has nothing to do with biology, and where would that leave an argument for gay rights that rests on genes, or hormones, or evolution?[21] In fact, this has already happened. The real question, as Wendell Ricketts wrote, is not whether human sexuality is influenced or dominated by either environmental or innate conditions, but "whether it will be the biologists or social scientists who prevail in demonstrating the conceptual barrenness of their theories, the arrogance of their research, and the ineluctable taint on 'objective' science of personal beliefs and cultural prejudices."[22] Yet it may not come down to such a choice. The reality is that causation theories have tended to be cooperative, not competitive.

There's a strong belief that there's something novel in these genetic theories, something that will dislodge older models of homosexuality that have been so destructive, something that will unseat socialization arguments or the psychiatric paradigm, with all its attendant "treatments" or charges of immorality fueled and sustained by Christian doctrine. But nothing in the long medicalization of homosexuality has ever displaced what came before. What we have seen instead is additive causation theories.

Causation theories, proposed and molded by both opponents and supporters of gay rights, have gone from moral turpitude to anatomical flaws (frequently race-based), to biological and eugenic culprits, to psychiatric disorders, to hormonal imbalances, and back to eugenics. These paradigms have not only reinforced one another, they have frequently made recourse to—and bolstered—eugenic pronouncements on race, crime, urbanization, and class. They are additive, and therefore none have dislodged their predecessors or posed any substantive challenge to homophobia. The "gay gene" will not displace psychiatric diagnoses any more than the latter displaced sin-based condemnations. The very need to locate a basis of homosexuality invests all causation endeavors with an antigay bias. Despite this book's focus on the bankruptcy and perils of biological models, what follows is in no way an endorsement of socialization arguments, but rather a call for queers to opt out of nature versus nurture arguments altogether.

Culturally and politically, queers have had to contend with the consolidated power of multiple explanations of homosexuality. This century has witnessed the emergence of extremely frightening syntheses. Biology has fused with psychiatry, disease with evolution, heredity with morality, psychiatry with genetics. "Curative" and punitive discourses have melded. Medical and legal assaults have merged.

Delineating Deviance

Moral Imperatives, Hereditarian Hypotheses, and the Letter of the Law

It almost goes without saying that seeking a cause or cure for homosexuality confers on it the mark of deviance and that the medical or psychiatric pathologization of homosexuality serves to normalize heterosexuality. The National Institutes of Health did not, after all, house the search for a straight gene, except, perhaps, by default. It is not surprising then that the term "homosexual," coined in 1869 by Hungarian physician Karoly Maria Benkert, gained currency before the term "heterosexual."

Any examination of eugenics and the construction of homosexuality is complicated by the shifting vocabulary of the last 130-odd years. Like the highly malleable designation of "race" employed by anti-immigrant eugenicists, the terminology and associative meanings ascribed to gender and sexuality by physicians, policy makers, allies, and apologists have been slippery. As George Chauncey notes, there was no "feminine" or "masculine" behavior that was consistently normative across class and ethnic lines at the turn of the century. Further, he states that middle-class culture insisted on exclusive heterosexuality as a measure of "normal" manhood long before African-American and immigrant European-American working-class cultures did.[1] Unfortunately, it was the first of these that set the medical and social standard.

To further confound the discussion, designations of homosexuality have been, for the better part of the last century, predicated on adherence to gender conventions—tricky when gender itself is a social construct. "Sexual inversion" referred solely to men who adopted women's gender norms and women who adopted men's. Thus, it was possible to engage in homosexuality without being an "invert" provided one maintained the physical and behavioral attributes deigned normative for their biological sex. Medical literature frequently, though not exclusively, referred to this group as "perverts."[2] "Urning," coined in 1898, was Karl Ulrich's term for a female mind in a male body, the result, the German lawyer said, of an undifferentiated human embryo.[3] Magnus Hirschfeld preferred the "third sex," describing homosexuality as a midpoint between total "femaleness" and total "maleness."

This idea of a continuum was very popular in the late nineteenth and early twentieth centuries. Xaviar Mayne, a crusader for homosexual rights, spoke of "intersexes...the half-steps, the between beings." British socialist Edward Carpenter also imagined an "intermediate sex." Both men, as Siobhan Somerville has noted, borrowed heavily from science's prevailing notions of race. Mayne declared that "[b]etween the whitest of men and the blackest of negro stretches out a vast line of intermediary races." So too was it with the intersexed. "Nature abhors the absolute, delights in the fractional." Carpenter wrote of women who were "one-quarter, or one-eighth male" as similar to "half-breeds," demonstrating once again how reliant one strand of eugenics was on another.[4]

Foucault noted that situating the discourse on sex (sexual relations, venereal diseases, matrimonial alliances, perversions) within broader hereditary paradigms invested it with urgency and granted it the power to "eliminate defective individuals, degenerate and bastardized populations." Its location linked it to other eugenic analyses across the board, and it became a supreme arbiter of the "moral cleanliness of the social body." Thus endowed with "biological responsibility" for the species, he wrote, "not only could sex be affected by its own diseases, it could also, if it was not controlled, transmit or create others that would afflict future generations."[5]

Hereditary notions that tied homosexuality, among other "propensities," to other manifestations of compromised bloodlines abounded during the late nineteenth century. In 1882, Charcot and Magnan determined "sexual inversion," along with kleptomania and dipsomania (an unsatiable craving for alcohol), to be part of a larger and more basic process of hereditary degeneration.[6] Though he later disavowed eugenic arguments, Richard von Krafft-Ebing also viewed homosexuality as but one of many manifestations of inherited weakness, stating that in "almost all cases where an examination of the physical and mental peculiarities of the ancestors and blood relations has been possible, neuroses, psychoses, degenerative signs, etc. have been found in the families."[7]

James Foster Scott applied similar eugenic reasoning to questions of sexual desire in his 1898 book, *The Sexual Instinct: Its Use and Dangers, as Affecting Heredity and Morals: Essentials to the Welfare of the Individual and the Future of the Race.* He was particularly concerned with the eugenic consequences of masturbation and onanism, catchall terms used in the nineteenth century to refer to all nonprocreative sex, including homosexuality. Masturbation, Scott wrote, was sometimes "a legacy inherited along with an unstable nervous system." According to Scott, an obstetrician at Columbia Hospital for Women, "almost all sexual perverts owe their anomalies of desire, inclination and fancy to . . . either their own or their ancestor's onanism."[8] Heredity was implicated in either instance, for these "anomalies," even if acquired, could still be passed along to the next generation.

> [T]he progeny of the impure are likely to suffer on account of the impaired and vitiated vigor of the parental reproductive functions; thus they are liable to have a proneness to sin—organic *fault,* or physical and moral damage—they inherit a neurasthenic sexual predisposition, . . . a constitutionally impaired physical and moral stamina.[9]

For Scott, the sexual instinct was "an inherent desire for the perpetuation of the species." Not surprisingly, he labeled every sexual act that did not involve propagation of the human race as perverse. Any man, and here Scott was writing exclusively about men, who engages in sexual acts not designed to further the race ultimately "bequeaths an undesirable legacy to his posterity, giving both his sons and daughters a proneness to psychoses and neuroses, especially in their sexual proclivities."[10]

"Men do not seem to realize," Scott opined, "the tremendous importance of heredity, and that their illegitimate pleasures and acquired preferences for impure courses are as likely to crop out in their daughters as in their sons, invariably in an evil way." In conjuring an image of generation upon generation destroyed by the masturbation and/ or homosexuality of their ancestors, Scott may have been urging restraint (which for gays and lesbians meant celibacy) as an alternative to legal punishment or medical treatment. However, his reference to an Oliver Wendell Holmes quote—"If you want to reform a man, begin with his grandfather"—suggests he was probably arguing for eugenic intervention.[11]

French writer Paul Moreau was much clearer about his intentions. Moreau's work, cited extensively in U.S. eugenics tracts of the late nineteenth and early twentieth centuries, exemplified the fusion of heredity- and morality-based approaches to homosexuality. According to Moreau, humans possessed a sixth sense, a genital sense susceptible to injury both psychic and physical. Environment, he wrote in 1887, could be a factor, but the existence of a "mixed class" of "[p]aederasts, sodomites, saphists," and others "can most frequently be explained only by one word: Heredity."[12]

> Not infrequently, under the influence of some vice or organism, generally of heredity, the moral faculties may undergo alterations, which, if they do not actually destroy the social relations of the individual, as happens in cases of declared insanity, yet modify them to a remarkable degree, and certainly demand to be taken into account, when we have to estimate the morality of these acts.[13]

Homosexuality per se might not be hereditary, but the propensity toward such moral decrepitude was. Heredity impacted morality and therefore eugenics did not undermine, but could actually bolster, sin-based condemnations.

It was based on his conviction of an innate pathology, passed from generation to generation, that Moreau called for *therapeutic* rather than *punitive* measures,[14] though given the conditions of asylums, the distinction blurs. He exhorted members of the medical faculty to sit on the bench during criminal trials, a strategy that moved gays—ideologically at first—from the prison to the hospital. The chief difference lay in who would determine the fate of homosexuals. Doctors, not judges, should evaluate those accused of sodomy, tribadism, or other "sexual excesses."

In *The Construction of Homosexuality*, David Greenberg has written extensively of doctors' and scientists' drive to claim jurisdiction over the realm of "social problems." Greenberg attributed the shift toward medicalization to a number of factors, among them the work of scientists seeking to "create a scientifically based and scientifically directed culture purged of metaphysics and religious superstition," which had been the guardians of antihomosexual rhetoric thus far.[15] But beyond Enlightenment reason, the changing economy of the country helped the medical profession stake their claim. Heredity-based explanations justified the existing distribution of wealth and power:

> As occupations became specialized and required technical skills, which often had to be acquired through costly advanced education, as work increasingly involved employment in bureaucratic organizations, and as larger amounts of capital were required to open a business, *moral* deficiencies no longer provided a plausible explanation for failure. Indeed, workers came increasingly to blame their difficulties on the class system, which severely limited their opportunities and threw them out of work at every downturn in the business cycle. This, however, was an uncomfortable thought to those who were relatively privileged; they found it far more attractive to explain failure in terms of innate intellectual deficiencies.[16]

Upper- and middle-class anxieties over urbanization, immigration, class conflict, and economics (there were those who blamed the depression of the 1890s on the biological deterioration of the American population) factored heavily into a belief in the hereditary nature of homosexuality, poverty, and crime.[17] The rhetoric around homosexuality was, in part, a product of apprehension. Homosexuality was viewed, perhaps contradictorily, as a pathological response to life in the modern world.[18] Many physicians diagnosed inversion as a strictly urban phenomenon, with the swelling of the homosexual ranks due, no doubt, to disembarking immigrants in port cities. As Chauncey observed, "The English tended to blame homosexuality on the French, and the French to blame it on the Italians, but Americans blamed it rather more indiscriminately on European immigration as a whole."[19]

Add to all this the biases brought to bear by doctors' own socioeconomic positions and their desire to advance those positions, and it is easy to see why they were so vested in medicalizing almost every aspect of human behavior. Yet, there was some resistance to locating homosexuality in eugenics gone awry. Then, as now, there were those who felt that any physical explanation or hereditary approach to homosexuality exonerated sinners. G. Frank Lydston, a professor at the Chicago College of Physicians and Surgeons and a lecturer on criminal anthropology at the Union Law School, reconciled this morality-eugenics split in "A Lecture on Sexual Perversions, Satyriases, and Nymphomania":

> Even to the moralist there should be much satisfaction in the thought that a large class of sexual perverts are physically abnormal rather than

morally leprous. It is often difficult to draw the line of de-marcation between physical and moral perversion. Indeed, *the one is so often dependent on the other that it is doubtful whether it were wise to attempt the distinction* in many instances. But this does not affect the cogency of the argument that the sexual pervert is generally a physical aberration.[20]

Lydston, a practicing surgeon, split "sexual perversion" into two major categories. "Congenital and perhaps sexual perversion," he wrote in 1889, consisted of one or more of the following:

a. sexual perversion without defect of structure or sexual organs
b. sexual perversion with defect of genital structure, e.g., hermaphroditism
c. sexual perversion with obvious defect of the cerebral development, e.g., idiocy

Acquired perversion, on the other hand, was composed of

a. sexual perversion from pregnancy, the menopause, ovarian disease, hysteria, etc.
b. sexual perversion from acquired cerebral disease, with or without recognized insanity
c. sexual perversion from vice
d. sexual perversion from over stimulation of the nerves of sexual sensibility and the receptive sexual centers, incidental to sexual excesses and masturbation[21]

Lydston's attention to the causes of "perversion" was a subset of his larger concern over the eugenic health of the nation. The "degenerate classes" (criminals, the poor, the insane) numbered, by his count, 215,000. "This is almost as large as the army organized by the United States during the war with Spain."[22] Lydston's confidence in the inheritability of acquired traits may have reflected not only a wide subject pool of "perverts" (encompassing but not limited to homosexuals) but also a certain belief in the fluidity of human sexuality that is largely absent in current discourse: his concern about the transmission of "sexual perversion" from one generation to the next stemmed from an acknowledgment that not all "sexual deviants" were solely homosexual in practice, but might have children from heterosexual contacts.[23]

The child of vice has within it, in many instances, the germ of vicious impulse, and no purifying influence can save it from following its own inherent inclinations. Men and women who seek, from mere satiety, variations of the normal method of sexual gratification, stamp their

nervous systems with a malign influence which in the next generation
may present itself as true sexual perversion. Acquired sexual perversion
in one generation may be a true constitutional irradicable vice in the
next, and this independently of gross physical aberrations.[24]

For Lydston, moral and physical manifestations of "sexual perversion"
were interdependent and multirooted. The benefit of connecting "vice"
to physical culprits was that it sanctioned physical curatives, already unleashed by nineteenth-century surgeons administering to the "deviant."
Cessation of dysgenesis, acquired or otherwise, demanded intervention.

Hysterectomy, vasectomy, castration, clitoridectomy, and blistering
of the vulva, prepuce, or thighs were all prescribed by physicians in cases
of perversion.[25] In 1859, the *Boston Medical and Surgical Journal* printed
a letter by a doctor who was treating a deinstitutionalized heterosexual
man for "satyriasis" (excessive desire) and "erotic mania." Would castration take care of the problem? The responding doctor confirmed that it
had been done repeatedly, notably at a hospital in Ohio "on quite an ex-
tensive scale." However, he wrote that he recoiled at the notion of removing the ovaries of a "nymphomaniac," stating, "I have found that
heavy doses of opium, long continued, do control the nymphomaniacal
disposition."[26] One contributor to the *Journal of Orificial Surgery,* in an
extensively illustrated piece on clitoridectomy and circumcision, concluded that "the hopes of the race for emancipation from sin as well as
from sickness will find their main reliance in orificial surgery."[27] Another wrote that children could inherit strong sexual passion from their
parents, grandparents, or great-grandparents. He supplemented this with
some nutritional advice: "Pork, ham, bacon, sausage, pickles, vinegar,
spices, alcoholic liquors, and tobacco and scores of other kindred articles should be banished from use, for such elements in the blood stimulate the passions and lessen the power of self-control."[28]

Whether curative or prophylactic, it was clear that submission to
treatment would not remain even nominally voluntary if Lydston and
his colleagues had their way. In his 1904 book, *The Diseases of Society,*
Lydston declared that "individuals whose physical and moral status is
such as to insure the unfitness of their prospective progeny should be
given the alternative of submitting to sterilization as the only condition
upon which matrimony is legally permissible."[29] His colleagues were
similarly preoccupied and eager to prescribe remedies, usually surgical

in nature. Dr. E. H. Pratt, for example, wrote that castration of "sexual monsters" who "startle the community with the boldness of their desperate propensities" would prevent them from siring a generation with like tendencies (though it would not "curb the propensities of the criminal or save him from the madhouse).["][30] The designation of homosexuality as disease or inheritance or congenital condition may have mandated "treatment," but it did not squash either environmental- or sin-based models. Most importantly, it did not preclude the possibility of imprisonment or other legislated punishment. Lydston and Pratt and others advised regulation by way of the consolidation of medical and judicial/ legislative power. Homosexuality would be absorbed into a eugenics-crime paradigm.

At the 1901 International Congress of Criminal Anthropology, Cesare Lombroso stated that any homosexual who acts out his "diseased" urges should be thrown in prison.[31] Lombroso, like many doctors of his day, not to mention many without medical degrees, traced a wide array of criminal acts to heredity (and, by extension, to race and nationality). He asserted that criminals were biological regressives, ill-equipped to negotiate the modern world. Homosexuals were among these atavistic throwbacks, at a lower evolutionary stage than heterosexuals. This pronouncement fit nicely with Lombroso's model of the *delinquente nato* (born criminal) who, unable to restrain the baser animal instincts, was prone to violence, promiscuity, pederasty, and tattooing.[32] "[H]omosexual offenders who are born such, and who manifest their evil propensities from childhood without being determined by special causes... should be confined from their youth."[33] Even Moreau, who had advocated the decriminalization of homosexuality, came to call for a similar policy. The punitive zeal of eugenicists ensured that the discourses around punishment and cure would continue to converge.

In 1893, Dr. F. E. Daniel, of Austin, Texas, presented a paper that epitomized the intersection of legal and medical procedures, designating corporeal punishment as actual treatment. Over the next nineteen years, "Should Insane Criminals or Sexual Perverts Be Allowed to Procreate?" was reprinted in three separate medical journals. "I would," stated Daniel, "substitute castration as a penalty for all sexual crimes or misdemeanors, including masturbation." Daniel opted for castration over execution as a kinder means to an end.[34]

The aim of jurisprudence should be, in addition to the repression of crimes, a removal of the causes that lead to it, and reform, rather than the extermination of the vicious. It should comprehend both the therapeusis and prophylaxis in the widest sense; thus, drunkenness should be cured, and intemperance prevented.... So with the sexual sins; the offender should be rendered incapable of a repetition of the offense, and the propagation of his kind should be inhibited in the interest of civilization and the well-being of future generations.[35]

While Daniel spoke of the societal benefits of the elimination of "much that is defective in human genetics," and the eventual "sanitary utopia" that would be born of a union between "preventative medicine" and the "enactment and enforcement of suitable laws," he also propagated a belief in the implicit humanitarianism of eugenics brought to bear on individuals—namely the substitution of the "cruel execution of criminals" with castration. With proper medical and legal intervention, he explained, even that procedure would become obsolete a few generations hence. "It does seem that the scheme of our government, and the tendency of our civilization is to foster the criminal and mentally defective classes.... Is it not a remarkable civilization that will break a criminal's neck, but will respect his testicles?"[36]

Daniel's criticism notwithstanding, there was no shortage of physician participation in legislative incursions into matters of sexuality. Beginning in the last decades of the nineteenth century, a flood of state sodomy laws were passed or amended to encompass a greater array of sexual practices. Doctors, foreshadowing the Amendment 2 litigation strategy, provided a legitimizing presence among lobbyists. There was a certain reciprocity involved as castration and like procedures were transformed from court-mandated penalty to medically endorsed treatment. Physicians saw their diagnoses legally sanctioned and thus their esteem and power consolidated. At the same time, the judicial system was able to mete out corporeal punishment while still appearing to have the best interests of the defendant/patient, the public, and the national gene pool at heart.

One of the most sweeping manifestations of this dynamic (and a partial fulfillment of the prescription by Lydston and his colleagues) was the rash of sterilization statutes enacted by thirty states between 1907 and 1932. In almost every state that legislated sterilization, eugenics boards were convened. Essentially, these were medical panels established to grant or deny doctors the right to sterilize anyone with a real or imagined physical or developmental disability. Usually these were prisoners or patients at hospitals or asylums and sometimes they were members of the public at large. Fees related to enforcement were absorbed by institutional maintenance funds in some states and by state treasuries in others.[37] In a fusion of medical and legal discourses, individuals singled out for sterilization were referred to as *defendants*. In short, the legislatures had established a physician-run de facto court system whose jurors were not bound by any state bar and who were free to authorize, for "eugenic or therapeutic purposes," any number of sterilization operations on those unfortunate enough to be ensnared by either the criminal justice or mental health systems. Beyond *merging* with the judiciary, medical science had *become* the judiciary.

In 1909, California, the third state to pass a compulsory sterilization law, became the first to include in its provisions the sterilization of "moral degenerates" and "sexual perverts showing hereditary degeneracy." A commission composed of the secretary of the state board of health and doctors and superintendents of state hospitals, state homes for the "feeble-minded," and the state prison were empowered to have sterilized

> any inmate, patient or convict for therapeutic, eugenic and punitive reasons. This commission's judgment is conditioned by the fact that in the case of an inmate or convict, the individual must have been convicted at least two times for some sexual offense, or at least three times for any other crime, and shall have given evidence, while confined, that he or she is a moral and sexual pervert.[38]

Two years later, Iowa's sterilization statute targeted "criminals, idiots, feeble-minded, imbeciles, drunkards, drug fiends, epileptics, syphilitics, and moral and sexual perverts." Like the California statute, the Iowa law was amended and replaced several times over the decade that followed in response to legal challenges over constitutionality, particularly with regard to equal protection and broadness of scope. Still, the target groups did not change. If anything, the new laws were more efficient. Iowa's 1929 legislation called on the superintendents of state institutions to submit quarterly reports to the eugenics board, listing viable candidates for sterilization. Unlike California, Iowa extended this roundup beyond the hospital and prison walls to the general public and provided free legal counsel to those about to go under the knife.[39]

Oregon passed a law claiming to be both eugenic ("for the prevention of the procreation of the feeble-minded, insane, epileptics, habitual criminals, moral degenerates, and sexual perverts with inferior hereditary potentialities") and curative ("[a]ny sterilization operation may be ordered but emasculation, except as a necessary procedure to improve the health of the individual"). In the years that followed, homosexuals and other "sexual perverts" were explicitly equated with violent offenders as the law was supplanted (1923) and amended (1925) "so as to render those people who are convicted of rape, sodomy, or any other crime against nature amenable to the sterilization law."[40]

Nebraska's 1929 law made sterilization a prerequisite for discharge or parole in some cases, taking aim at "sexual perverts" and "moral degenerates," among others. "Criminals convicted of rape, incest or any other crime against nature may be castrated on court order, if the performance of this operation on the male would make him eligible for commutation of sentence."[41] Medical procedure, as decreed by the legislature, could yield freedom. Sterilization, even castration, was characterized as more benevolent, and more to the point, than internment.

Utah passed its bill, allowing sterilization of "habitual sexual criminals," in 1925. Ironically, the framers' insistence that the provisions therein were therapeutic and not punitive was what spared Esau Walton from having to submit to the operation. Walton was born in Georgia but left the state after his mother's death in 1923. While a minor in the South, he was arrested on various charges of theft, including one for stealing silk shirts. When he was nineteen, he was sentenced to Utah State Prison. Four years later a guard accused him of having sex with another inmate, and the warden petitioned for his sterilization on this basis. Given the disproportionate number of African-Americans that came under the knife of many states' sterilization programs, it is not unreasonable to assume that warden R. E. Davis's eagerness to have Walton sterilized had as much to do with codified racism as with an institutional zeal to rout out homosexuality. Defying the popular wisdom of the day that decreed sodomy, or at least the degeneracy that begot it, to be hereditary, the Supreme Court of Utah in *Warden Davis v. Walton* issued the following ruling in 1929:

> The Utah act is in no sense a penal statute. The operation provided for is
> not punishment for a crime. Its purposes are eugenic and therapeutic. . . .

As a general rule, members of judicial tribunals are not well informed as to the law of heredity. Even though they may be so informed, they may not take judicial notice that, if Esau Walton should have offspring, the same shall be socially inadequate offspring likewise afflicted. Those who have made a thorough scientific study of the law of heredity are not in entire accord as to the operation of the so-called law, but doubtless people so trained may lend valuable aid to judicial tribunals in determining the probable nature of the offspring of a given person. *We doubt, however, that even the most ardent advocate of the immutability of the law of heredity would wish to determine the probable nature of the offspring of Esau Walton without more facts than appear in the record before us.* Be that as it may, the record before us does not support the finding that "by the law of heredity Esau Walton is the probable potential parent of inadequate offspring likewise afflicted."[42]

In essence, the court did not challenge Utah's law in any significant way. It ruled that the evidence did not support the order to sterilize, but upheld the law's constitutionality. The Supreme Court, in *Buck v. Bell*, had already decided the constitutionality of compulsory sterilization two years earlier, and the tribunal's decision was consistent with that judgment. The last line of the above excerpt was very nearly a direct quote from Justice Holmes's 1927 ruling in the case of Carrie Buck, a young white woman from Virginia who came to the attention of the authorities after being raped and impregnated by a member of her foster family. The Utah law called only for eugenic and therapeutic sterilization. It was not intended, Judge Elias Hansen wrote, as a penal statute. Furthermore, the court ruled that neither the heritability nor habituality of Walton's actions had been established and that sterilization would not promote his health or welfare.[43] Had the attorney for Davis provided a more exhaustive argument for the innateness of sodomy, not difficult to do given the eugenics treatises that flourished during that era, the case might have gone a very different way.

Though the court spared Walton, it more than condoned the participation of eugenicists, a position it shared with other judicial and legislative bodies of the day, including the U.S. Congress, which welcomed eugenicists as medical activists and advisors on issues of immigration. In fact, while most measures that addressed sodomy and other "sexual offenses" were devised and implemented at the state level, the federal government did weigh in on the subject. The 1917 Immigration and

Naturalization Act denied entry to would-be immigrants who had been convicted of, or who admitted to, crimes involving "moral turpitude."[44]

It is unclear whether medical, judicial, and legislative institutions pursued sterilization out of a real belief in a fluid (or coerced) human sexuality that would enable homosexuals to procreate, resulting in the propagation of "sexual miscreants," or whether this was mere justification for a eugenic and moral crusade against the socially and economically vulnerable. Certainly, the "treatment" literature did not preclude or seriously challenge out-and-out punishment. It just reconfigured it. The location of incarceration changed. Judges had greater flexibility in lengthening or shortening prison or hospital terms as "treatment needs" could now offset earlier, more rigid sentencing practices.

Foucault wrote that "criminal justice functions and justifies itself only by this perpetual reference to something other than itself, by this unceasing reinscription in non-juridical systems."[45] By absorbing extrajuridical elements such as what passed for medicine, the judiciary and the legislature were able to reinvent themselves with a veneer, if not of leniency, then of compassion and a turn-of-the-last-century rendition of tough love that continues to this day. This reinvention coincided with cataloging sexual outcasts, the result being continued subjection to punitive measures.

In the midst of all this there was a parallel move to use the same science that penalized to exonerate. Seemingly unaware of the reactionary uses of scientific theories of homosexuality going on around them, some gays and well-intentioned allies—most notably Magnus Hirschfeld and Havelock Ellis—argued the liberatory potential of biological causation theories.

Biological Apologists

Appeals and Miscalculations

In 1992, Randy Shilts (author of *And the Band Played On*) told *Newsweek* that the discovery of a genetic basis for homosexuality "would reduce being gay to something like being left-handed, which is in fact all that it is."[1] It was an interesting choice of metaphor, in light of the fact that scientists have indeed been hard at work documenting left-handedness among gays and lesbians. Researchers at McMaster University claim that lesbians are twice as likely to be left-handed than straight women due, perhaps, to an "atypical brain organization" created by irregular hormone levels.[2] Another study asserts that while the fingerprints of gay men do not contain more ridges than those of straight men, the former were more likely to have a leftward asymmetry.[3] What Shilts was suggesting was that a reduction of the importance accorded to sexual orientation to the level of left-hand preference would eliminate the stigma attached to being queer. Setting aside his characterization of sexuality as incidental, and there is much to challenge there, and without issuing a reminder about what was done in the recent as well as distant past to left-handed individuals, Shilts's response to the research was extremely disappointing. He had spent much of his career covering gay issues, yet his remarks betray an inattentiveness to the consequences of genetic assertions faced by lesbians and gays during the past one hundred years.

In fairness, Shilts, not unlike the CLIP legal team, stood in a long tradition of those who looked to science as a gateway to civil rights. The medicalization and mediation of nonprocreative sexuality has long been seen not as encroachment, but as potentially progressive, righting old wrongs. Among these were Magnus Hirschfeld and Havelock Ellis, whose faith in the liberatory potential of medical models of homosexuality should serve as a warning against similar, contemporary impulses.

Hirschfeld classified homosexuality as congenital and as midway between total maleness and total femaleness. Since it was biological, he said in 1899, "nobody can be morally blamed for having such a disposition."[4] Hirschfeld was among the many physicians and thinkers who joined in the fight to repeal Germany's infamous Paragraph 175, which criminalized homosexuality. Defenders of Paragraph 175 in the Reichstag, however, also claimed to have biology on their side, insisting that if it were natural, "Nature would have put homosexuality to the service of procreation and the maintenance of the species."[5]

Hirschfeld's investment in the redemptive power of scientific theory was well-intentioned, but ultimately posed a grave threat to other gay men. As Gunter Schmidt noted, in "trying to counter certain prejudices, his theory opened the door to others. It became increasingly dangerous as more advances were made in the field of medical research and technology."[6] While Hirschfeld did not have treatment on his agenda, he was not opposed to surgical intervention. In 1916, E. Steinach, an anatomist and early endocrinologist from Vienna, teamed up with a surgeon to effect a "cure" for homosexuality in a test group of approximately a dozen men. After their gay subjects were castrated, they received transplants of testicular tissue from heterosexual males who had had undescended testicles removed. Doctors opted for unilateral rather than bilateral castration to enable their newly heterosexual patients (as they anticipated them to be) to marry and procreate.[7] In a few instances, the medico-juridical complex was able to satisfy supply and demand simultaneously, as donor tissue was obtained from men who were castrated for excessive (hetero) sexual drive or delinquency. Doctors were so zealous in their surgical crusade that one man received tissue from a donor with tuberculosis. Hirschfeld reportedly endorsed the operation in his journal, going so far as to print Steinach's address and even refer one or two patients to doctors performing the surgery.[8] A few years later, when he heard of a physician who was proposing brain surgery on gay men, Hirschfeld declared, "Let us hope they discover the brain center for homosexuality; only then when we have come to see homosexuals in the right light [can we] be certain that such operations are superfluous."[9]

There is no denying the intent of Hirschfeld's work. He was a gay rights champion—and an out one at that—a socialist and an advocate for women's rights (Hitler called him "the world's most dangerous Jew"). But Hirschfeld's acceptance of biological models meant that he still subscribed to prevailing, and increasingly popular, notions about homosexuals, most centrally those relating to physical deviance. Similarly, Ellis tried to defend lesbians and gays while leaving unchallenged—in fact promoting—the view of homosexuality as dysgenic and pathological. He is credited with bringing much of the medical literature on homosexuality to the attention of English-speaking readers.[10]

Ellis's *Sexual Inversion* was the second volume in his *Studies in the Psychology of Sex*. He defined "inversion" as a "sexual instinct turned by inborn constitutional abnormality toward persons of the same sex," as distinguished from the more general "homosexuality," which could be merely situational. The latter referred to behavior, not necessarily instinct. Inversion was simply "one of those aberrations which we see throughout living nature." Ellis categorized Oscar Wilde as a case "in which a heterosexual person apparently becomes homosexual by the exercise of intellectual curiosity and esthetic interest." Such instances, he said, were rare.[11]

Hereditary theories figured significantly in Ellis's construction of homosexuality. He believed that a *predisposition* to inversion, if not inversion itself, was innate or at least congenital. Reporting on the families of one hundred "sexual inverts" in the United States and Britain, he wrote:

> In 28 cases there is more or less frequency of morbidity or abnormality—eccentricity, alcoholism, neurasthenia, insanity, or nervous disease—on one or both sides, in addition to inversion or apart from it. In some of these cases the inverted offspring is the outcome of the union of a very healthy with a thoroughly morbid stock; in some others there is a minor degree of abnormality on both sides.[12]

It was a position that situated him firmly in the ranks of the eugenicists of his era, as does his reportage of one "very radical case," designated "History XX." According to Ellis, this man's homosexuality had its origins in faulty maternal stock. In one fell swoop, he managed to offer a eugenic argument, promulgate stereotypes about gay children, and affix blame to the mother:

> His father and father's family were robust, healthy, and prolific. On his mother's side, phthisis, insanity, and eccentricity are traceable. He

belongs to a large family, some of whom died in early childhood and at birth, while others are normal. He himself was a weakly and highly nervous child, subject to night terrors and somnambulism, excessive shyness and religious disquietude.[13]

Ellis's focus on the family led him to oppose marriage for inverts. While he did not go as far as August Forel, who called for a legal ban on homosexuals marrying lest they produce physically and mentally undesirable offspring,[14] Ellis was clearly concerned with the eugenic risks of procreation. "For the sake of possible offspring," he warned, inverts should not marry. "Often, it is true, the children turn out fairly well, but, in many cases, they bear witness that they belong to a neurotic and failing stock." "Cures" were risky, for they might prove to be superficial. He pointed to the case of a gay man, urged by his doctor to marry. He did so, and though he married a woman who was perfectly strong and healthy,

and was himself healthy, except insofar as his perversion was concerned, the offspring turned out disastrously. The eldest child was an epileptic, almost an imbecile, and with strongly marked homosexual impulses; the second and third children were absolute idiots; the youngest died of convulsions in infancy.[15]

It may be that Ellis used eugenics rhetoric strategically, as a check against those who would pressure gays and lesbians into treatment and marriage. Nevertheless, his commitment to eugenics was strong and he was among its stalwart champions. Many that came after him, whatever their motives, employed similar tactics. In 1935, Dr. James P. Winsco issued a call for tolerance in the pages of *Sexology* magazine: "It is time for us to sit up and take notice of the fact that we are in need, not of more, but of better children. For the marriage of a homosexual, followed by unsuccessful sex relations, develops more of his type. I cannot help but feel that such a union is a failure."[16]

It is important to note that though his adherence to eugenic tropes went beyond issues of sexuality, Ellis differed from those who would use medical models as a rationale for surgical intervention. At a time when physicians were prescribing and some gay men actively seeking castration as a curative, Ellis decried the practice, citing more than one case of men who were castrated with intolerable consequences, both physical and emotional. "Sexual Inversion is not a localized genital condition. . . . Castration of the body in adult age cannot be expected to produce castration of the mind."[17]

This did not mean Ellis favored full sexual rights for homosexuals. His condemnation of the social ostracism leveled at gays—which has led many to regard him as one of the more enlightened minds of his circle—is complicated by his advocacy of "self-restraint," and his belief in the implicit abnormality of inverts:

> The invert is not only the victim of his own abnormal obsession, he is the victim of social hostility. We must seek to distinguish the part in his sufferings due to these two causes. When I review the cases I have brought forward and the mental history of the inverts I have known, I am inclined to say that if we can enable an invert to be healthy, selfrestrained and selfrespecting, we have often done better than to convert him into the mere feeble simulacrum of a normal man.[18]

Despite his professed, and no doubt genuine, desire to protect gays and lesbians, and his calls for an end to legal and social harassment, Ellis never adequately resolved the contradictions inherent in this eugenic paternalism.

Ellis and Hirschfeld's failure lay in their belief that homosexuality needed an identifiable, explicable origin. They underestimated the capacity of that yet unnamed hostility, homophobia, to absorb and exploit any causation model at hand. Further, they failed to consider the price being paid by other marginalized and pathologized groups to whom science had turned its attention during this same period—immigrants, African-Americans, poor women. That being the case, they could not see the cumulative hazards facing lesbians and racialized gays.

Gender, Race, and the Strategy of Metaphor

The journalist covering Hamer's and LeVay's findings for *Newsweek* reported that lesbians "are warier of the research." He attributed this hesitancy not to questions about *why* there is such a fixation on finding a gay gene, anger over funding priorities that finance causation studies and underfund AIDS and breast cancer research, or a panicky sense of déjà vu, but to lesbians' "conspicuous absence from most studies." The article cited Penny Perkins of the Lambda Legal Defense and Education Fund, who called the omission "part of society's intrinsic sexism,"[1] and Frances Stevens of the lesbian magazine *Deneuve* (now *Curve*), who remarked, "My response was: if the gay guy's [hypothalamus] is smaller, what's it like for dykes? Is it the same size as a straight male's?"[2]

Hamer and LeVay notwithstanding, lesbians' gray matter has not been overlooked by a medical establishment bent on discerning a fundamental cause of homosexuality. In 1913 Paul Nacke posited a masculine libido center in lesbians. An advocate for the examination of the brains of all homosexuals, male and female, he theorized the presence of an anatomical "homosexual center" in the brain.[3] During a 1921 debate on the floor of the British Parliament, lesbians were declared victims of an "abnormality of the brain."[4]

Just months after *Newsweek* ran its story, psychologist J. Michael Bailey and psychiatrist Richard C. Pillard presented the American Psychological Association (APA) with their "evidence" that genes have a substantial impact on the development of lesbianism.[5] One hundred and fifteen sets of twins and thirty-two sets of adoptive sisters were interviewed as the basis for the study (conducted in an identical manner to one on gay male twins, performed by the duo in 1991 and funded in part by a grant from the National Institutes of Mental Health).[6] Bailey, Pillard, and their associates claimed that in almost half of the identical twin probands, both sisters were either lesbian or bisexual.[7] According to Bailey, genes may influence the development of "sexual orientation centers" of the brain in the fetus.[8]

More significant than Bailey's location of lesbianism in the brain was his suggestion that genes that "masculinize" heterosexual males may be "masculinizing" homosexual females.[9] The transformation of the adjective "masculine" to the verb "masculinizing" reveals not only the extent to which totalizing stereotypes operate as scientific premise (in this instance lesbian equals malelike, and maleness equals attraction to women), but also the tenacity with which researchers have hung on to the notion of naturalized, biologically ordained gender. In 1998, the American Psychiatric Association (APA) followed the American Psychological Association's lead and declared its opposition to any portrayal of lesbians, gays, and bisexuals as mentally ill. However, Gender Identity Disorder is still listed in the APA's *Diagnostic and Statistical Manual of Mental Disorders* (DSM-IV), and diagnoses of Childhood Gender Nonconformity (CGN) continue to pathologize young children who fail to assume "traditional" gender roles. Psychologist Joseph Nicolosi, of the right-wing National Association for Research and Therapy of Homosexuality, claims to be treating twenty-five "pre-homosexual" boys, including two three-year-olds who are "denying their masculinity."[10] Nicolosi's approach is certainly extreme, but there is no shortage of more mainstream practitioners willing to validate and treat Gender Identity Disorder in children and adolescents.

In Bailey and Pillard's 1991 and 1992 studies, subjects were retrospectively assessed for CGN. They were asked about their interest as children in "masculine and feminine activities." Had each respondent been comfortable with her or his biological sex? Had the men been interested in sports as boys? As girls, had they desired to be boys? An affirmative answer could just as easily have indicated an early attraction to the *advantages* of being male—a possibility that, even if acknowledged by researchers, could not have been controlled for.[11] CGN, they wrote, was unrelated to genetic loading for homosexuality. However, unwilling to dispense with or problematize CGN as a concept, Bailey and Pillard stated that if both twins were gay, their "degree of CGN" would be similar.[12] The fact remains: gender nonconformity cannot be considered genetic when gender conformity, and gender itself, are themselves social constructs.

Bailey and Pillard's confidence in CGN may be the most ominous aspect of their work, because designations of gender nonconformity, phrased in whatever popular verbiage a given era dictated, have always imperiled women in general and lesbians in particular. This was especially true in the late nineteenth and early twentieth centuries as scientists sought to delineate the biological roots of sexual and gender "deviance." Increased lesbian vulnerability to the resulting treatises and treatments had a lot to do with the lines physicians, and later psychiatrists, drew between insanity, sexuality, and female anatomy. To a large extent, documentation of visible, anatomical differences substantiated claims of aberrance and, ergo, justified violent medical interventions. Further, medical pronouncements on gender roles were succored by race- and class-based configurations of homosexuality (and vice versa).

Somerville has observed that scientific inquiry and pronouncements on homosexuality were reliant on previous and contemporaneous studies on race, the latter fueling the former. "[D]iscouses of race and gender buttressed one another, often competing, often overlapping, in shaping emerging models of homosexuality."[13] The last hundred years of "treatment" literature on lesbianism is marked by the use of women, queers, and people of color as metaphoric foils for one another. This connection meant that physicians could depend on an arsenal of mutually reinforcing "scientific" tropes to condemn lesbians, uphold gender roles, and bolster racism.

In 1931, Berlin neurologist Albert Moll attributed sexual inversion in women "to an hereditary neuropathic condition . . . Moreover, it is often discovered that besides sexual inversion, there are other disorders such as periodic insanity and hysterical epilepsy."[14] The link between sexuality and "insanity" in women had, by this time, received considerable attention from U.S. medical authorities. The question was which part of the body to blame and, consequently, which to treat. As Jennifer Terry has written, the demarcation between the realm of the psyche and the biological realm of the body during this period was not exactly stark.

Rather, early sexologists conceptualized intellectual, biological, physical, and moral qualities as "deeply reflective of each other and embedded in an individual's body."[15]

Lydston was concerned with the cerebellum. In *The Diseases of Society*, he asserted his belief that disuse had resulted in "physiologic atrophy" of women's cerebral organization. The rare woman of strong will, he argued, "is, unfortunately, usually a degenerate with virile tendencies" and "criminal propensities." Men, according to Lydston, had more powerful sex impulses than women due to their greater development of the cerebellum and cerebral lobes. "Women with large cerebellar development approximate the male sexual desire." In an 1889 lecture, he cited the work of a colleague who "recommends the application of the actual cautery to the back of the neck."[16]

> Basing this treatment upon the theory that the disease takes its origin in over-excitation of the nerve fibers of the cerebellum or some of the ganglia in the neighborhood, he also suggests blisters to answer the same purpose. Dry cupping to the nucha [nape of the neck] is also serviceable.[17]

His focus on the cerebellum did not, however, mean that Lydston was opposed to "the removal of the irritation of the sexual apparatus," in treating a host of "female sexual excesses," including masturbation, "satyriasis" (both buzzwords for lesbianism), and "nymphomania." He authorized both "extirpation of the ovaries" and clitoridectomy.[18]

This was a not uncommon course of treatment. In an 1899 editorial in the *Journal of Orificial Surgery*, C. A. Weirick (who believed in the necessity of examining all the organs—reproductive and otherwise—of women declared insane), wrote that the ovaries "seem to act in a capacity very similar to the governor of a steam engine, and in some unknown way regulate nervous force and energy."[19]

F. E. Daniel, in an 1893 tractate on the alliance between judge and surgeon, cited research endorsing excision of the ovaries as a cure for "hysteroepilepsy" (hysteria) in women.[20] E. H. Pratt was of a different opinion: "Granted, that in insane women the ovaries and tubes are more or less affected, the fact must also be recognized that the lesions found in these organs are in themselves merely effects, and the original trouble lies in the terminal nerve fibers *lower down*.... Orificial surgery is unquestionably the panacea for insane women, and it will most certainly cure nine out of ten of even the most aggravated and discouraging cases."[21] Yet another author declared the clitoris "a source of nerve waste in women."[22]

Many subscribed to the belief that masturbation caused insanity and that the former was more common among girls than boys.[23] For his part, Weirick blamed "excessive frequency of masturbation" on heredity. "We think it is very largely due to the sensual use of the sexual organs by the parents. We think very few masturbators will be found among the offspring of those who indulged in coition solely for the purpose of propagation."[24] Weirick was thus able to kill two ideological birds with one stone, explaining the origin of masturbation/homosexuality and holding the sexual practices of adults hostage in exchange for the promise of nondeviant offspring.

A reading of this record can be confounding because the line between masturbation and homosexuality was not clearly delineated in discussions of the sex lives of women and girls. Because masturbation was used as a general term for nonprocreative sex, it can be difficult, often impossible, to ascertain precisely which sexual offenses were being condemned in various medical writings.[25] At times, the allusions to lesbianism are clear, as when physicians wrote of the "solitary vice" as one that could be "practiced between two girls," or of one girl learning to masturbate from another. But either way, the punishment could be horrific. Given the climate, it is no surprise that women were singled out for some of the most invasive surgical procedures and brutal mutilations unleashed by nineteenth-century medical crusaders.

In the March 1899 issue of the *Journal of Orificial Surgery,* Dr. Julia Holmes Smith reported on a nineteen-year-old patient who was "guilty of masturbation." Regardless whether her reference was to lesbianism (it is ambiguous), the actions taken by the doctor and her colleagues reveal a certain vehemence, a compulsion to control and correct women's sexuality at any cost.

[I]t was decided that the best thing was, if possible, to remove the cause, and as masturbation was evidently a factor, a surgical operation was needed.

. . . A thorough curetting and dilation was done. All ragged bits cut off from the vaginal walls, the clitoris unhooded and tissue stitched in place, so as to leave the gland free.

Holmes titled her report "Three Disappointing Cases" because none of the operations yielded the desired effect. Her first "failure"—which may have signaled deliberate resistance on the part of her patient—tragically did not keep her from trying again. The description of the second case, involving a still younger girl, is more suggestive of a lesbian relationship:

> This girl was sent to boarding school at an early age, and from her schoolmate learned to manipulate the genitals, and at seventeen was a confirmed masturbator....
> ... We [Holmes and another doctor] decided upon the operation, dilating orifices and unhooding the clitoris.... She was at a sanitarium from the second week after the operation.
> ... I am informed that the surgeon who now has her in charge has removed the clitoris and all erectile tissue and applies some causta to the wound in order to keep the parts sore and to prevent the patient from masturbating.[26]

Holmes expressed little regard for the consequences visited upon the girls and absolutely no remorse for either procedure. What she lamented was the inefficacy of the operations.

Concomitant with treatment—and central to its justification—was the invention of a visibly distinct lesbian body type. All sorts of bizarre physical characteristics were attributed to lesbians by doctors and lay-men alike. Ellis lent credence to many such claims in *Sexual Inversion*. In his account of "History XXXVII," he noted that at the age of four "Miss M." had an "abnormally large" mouth, nose, and ears. He further asserted that inverted women had a "masculine distribution of hair," as illustrated by one lesbian with "a marked dark down on the upper lip, ... hair on toes and feet and legs ... also a few hairs around the nipples," and thick pubic hair. Foreshadowing the hormonal experimentation on lesbians and gay men that would take place decades later, he posited that such anomalies suggested that an "abnormal balance in internal secretions" might be responsible for inversion.[27]

Ellis wrote that muscles in inverted women were firmer and soft connective tissue comparatively absent. Sexual organs remained at "the infantile stage of a girl of 10." Breaking with some of his cohorts, Ellis concurred with Hirschfeld and Krafft-Ebing that "women with a large clitoris ... seem rarely to be of the masculine type." However, the clitoris would tend to be "deeply hooded," the uterus small, and the hymen thick.

Ellis cited the work of researchers who claimed that lesbians had "a very decidedly masculine type of larynx." Lesbians had a different tone of voice than heterosexual women, a variation that rested on a "basis of anatomical modification." In a related observation, he declared inverted women to be proficient whistlers.[28] The issue of whistling capacity comes up several times in his book, lesbians being highly adept and a "considerable minority" of gay men failing miserably.[29]

Then, as now, discourses on difference drew heavily on one another. At the time Ellis was formulating his theories and reporting on the postulations of others, phrenology, criminal anthropology, and craniology were gaining adherents.[30] Skull size and shape, facial features, and so on, were considered legitimate indicators of criminal tendencies, including homosexual leanings. In an 1884 article in the *Detroit Lancet,* James Kiernan, a professor of "legal psychiatry," wrote of a lesbian patient whose face and cranium, he maintained, were asymmetrical. He treated her with cold sitz baths and "a course of intellectual training," but did not dismiss eugenic causes, stating that the woman had "a neurotic ancestry on the paternal side."[31] Ellis recounted the 1892 case of Alice Mitchell. Mitchell, who cross-dressed and took a male name, murdered her sister after she foiled her plan to marry another woman. Her face, Ellis was careful to note, was "obviously unsymmetrical," and "she had an appearance of youthfulness below her age."[32]

Later writers went directly to what they considered to be the source, measuring vulvas, classifying nipples, and cataloging "genital findings." Dr. Robert Latou Dickinson took copious measurements for his study, "The Gynecology of Homosexuality," published in 1941. His case studies are much more descriptive than most of those recorded by his predecessors, to the point of voyeurism. Nearly every case comments on whether the woman's nipples were flat or "normal," and whether one or both became erect when stimulated.[33]

Much of the medical literature differentiated between gender-conforming and gender-transgressing lesbians in ascribing physical characteristics to women. Maurice Chideckel's 1935 book, published by the Eugenics Publishing Company, fostered the claim made by some psychologists that menstruation ceases in lesbians "taking the part of a man." This "conversion symptom" was the physical manifestation of a presumed desire to be a man: "the womb simply cooperates with its possessor."[34] Such arguments served to label one partner queerer than the other. In this instance, a short-circuited menstrual cycle signaled the loss of yet another tenuous link to true womanhood. Some researchers did not even consider "feminine" women to be lesbians. Chauncey notes that both medical and newspaper accounts tended to find "the 'womanly' partner unremarkable, as if it did not matter that her 'husband' was another female so long as she played the conventionally wifely role."[35] They were not lesbians, but lesbians' quarry. The 1908 English translation of August Forel's *The Sexual Question* recounts the story of a woman who, passing as a man, became engaged to "a normal girl":

> Soon afterward the woman was unmasked, arrested and sent to an asylum, where she was made to put on women's clothes. But the young girl who was deceived continued to be amorous and visited her "lover," who embraced her before everyone. . . . I took the young girl aside and expressed my astonishment at seeing her continue to have any regard for the sham "young man" who had deceived her. Her reply was characteristic of a woman: "Ah! you see, doctor, I love him, and I cannot help it!". . . This is how it happens that a normal woman, systematically seduced by an invert, may become madly in love with her and commit sexual excesses with her for years, without being herself pathological.[36]

Years before "butch" and "femme" would enter common verbiage, medical literature set the sexual normalcy epicenter for women. The "deceived" (as Forel, perhaps naively, believed her to be) was not only not pathological, but "normal." Even her reply, to the doctor's bewilderment, was "characteristic of a woman." Accordance with gender ideals could offset the stigma of attraction to another woman, especially if that woman was more than a few concentric circles away from femininity's hub.[37]

Science has held fast to such designations, the parceling out of the pathological and the situational. In G. Dörner's 1968 reports on homosexuality in rats, only the mounting female and the mounted male were characterized as homosexual. Wendell Ricketts commented, "It scarcely needs to be said that this is a model based on reproduction, and that in the case of animal experiments the animal that deviates most strikingly from its 'proper' reproductive role is the one considered homosexual."[38]

Attempts at quantitative measurements for "dominant" and "passive" lesbians continued in the research of Muriel Perkins. Based on the notion of a standard heterosexual female body type, she interviewed, weighed, and measured 241 white lesbians between 1974 and 1976. She concluded that lesbians had "statistically narrower hips, increased arm and leg girths, less subcutaneous fat, and more muscle" than the control group. More to the point, she maintained that "psychosexually dominant groups" were significantly taller, more muscular, and had broader shoulders, greater arm and leg girths, and narrower hips than the "psychosexually intermediate and passive groups." Whereas the women identified as dominant (whether by themselves or by Perkins is unclear) were heavier, "the average fatness for the passive homosexual women most closely approaches the mean . . . of the control group."[39]

As assessments of lesbian anatomy have differentiated between butch and femme, they have also rested on and promoted a rigid stratification along racial lines. The two have frequently merged as prevailing constructions of biological differences borrowed heavily from one another. Homosexuals, like members of "primitive races," were cast as throwbacks, their evolution incomplete, as evidenced by claims that their bodies were not as sexually distinct as they should be.[40] The medical literature by this time had already endowed lesbians in general, but Black lesbians in particular, with a masculinized anatomy. In 1921, P. M. Lichtenstein wrote that "an abnormally prominent clitoris" is almost always seen in lesbians, especially in "colored women."[41] Physicians easily projected their ideas of masculinity and femininity onto interracial couples. Further, they were more than amenable to racist fantasies about who pounced and who was prey.

Maurice Chideckel, an associate in medicine at Johns Hopkins University in the 1930s, wrote that white lesbians "even consider themselves heroic in having a colored 'husband.'" In a sketch accompanying his chapter "Institutionalized Homosexuals," two women, one African-American, the other white, are shown on a prison cot. The white woman, long-haired, eyes closed, legs crossed and wearing heels, reclines limply. The Black woman's hair is cropped short, her eyes narrow as she leers at her lover's exposed breast. She is barefoot, legs and back sculpted, with muscles virtually rippling through her garment, the embodiment of "manly," "animal" lust, and, in Chideckel's mind, predatory victimizer.[42]

Chideckel did not discuss lesbian relationships between women of color, perhaps because he had no burning desire to "save" either party. After all, white homosexuals, once cured, could still be called upon to perform their eugenic, procreative duty to the state—if one believed their offspring would not be similarly tainted with perversion. As much as Chideckel loathed lesbianism (a "monstrous craving for the unnatural by phantasy-fed, mostly fanatically unhappy women"), he cultivated a special hatred for interracial relationships and for those that crossed class lines. He placed the blame for such unions on the connection "between the sense of smell, the nose, and the genital organs," stating that institutionalized white women were attracted to nonwhite women "due to their low moral stamina or to the pathology of smell."[43] He did not, however, limit himself to lesbian couples:

> Why does a white woman marry a negro? Why does a young cultured woman elope with a gypsy? Why does a woman of wealth and refinement give up a man to whom she is ideally mated, mentally and culturally, and run away to marry a truck driver, a taxi driver, or her father's chauffeur? . . . The olfactory nerve . . . is abnormal and hence smell itself is distorted and misdirected. The sense of smell is closely related to the sexual organs.[44]

These were no mere words for Chideckel. He convinced a wealthy, white woman in love with a Puerto Rican man to have her tonsils and adenoids removed. Chideckel's diagnosis and treatment protocol were products of larger cultural anxieties over cross-class and cross-race fraternization in non-work-related situations. In Jim Crow America, Somerville wrote, "interracial desire" was perceived "as a type of congenital abnormal object choice."[45]

The race- and class-based doctrine that augmented the documentation and "diagnosis" of homosexuality was tailored to comply with dominant society's working myths about itself. One such fiction was that rich and poor, white and racialized, did not share common impulses when it came to "vice." While Albert Moll believed sexual inversion to be more common among "the higher levels of society," Ellis invoked metaphor to claim that Europe's "lower classes," like the "lower races," demonstrated a certain lack of revulsion when it came to homosexuality. "In this matter, as folklore shows in so many other matters, the uncultured man of civilization is linked to the savage."[46] This statement not only damned both the poor and the nonwhite by linking each with a reviled group, namely each other, it also nourished hereditary explanations of class by equating low-earning, presumably white Europeans with non-Westerners, already deemed eugenically (and therefore socially, economically, and politically) backward by scientific racists.

Some writers sought to literally map out sexual acts. Ellis attributed tribadism to southern Slavic women of the Balkans.[47] In 1886, a Russian writer named Tarnowsky declared that Armenia was the birthplace of pederasty, which then spread to the rest of Asia and the Middle East.[48] While Moll did not necessarily subscribe to this, he did believe that tribadism and other related practices were rampant in the Arab world, whereas "[i]n all civilized countries this instinct has remained hidden."[49] Restraint, like vice, was racially determined. The idea that warm climates produced more than their share of homosexuals also resonated with Paul Moreau, who stated in 1887 that inhabitants of those zones were more libidinous than those in colder regions and were therefore more likely to succumb to homosexuality.[50] A few years later, James Foster Scott offered a supporting opinion:

> Climate, race, vigor of constitution, heredity, and social conditions have a marked influence on the period of life at which the earliest active manifestations of sex appear. Thus it occurs earlier in warm countries and in the class of society which lives luxuriously than in cold countries and among the poorer classes.[51]

The association of tropical climates with sexual openness, lasciviousness, and hypersexuality among children in the popular racist imagination lasted well into the next century.[52] In 1935, Dr. Glen Wadsworth chastised the readers of *Sexology*, a self-described progressive publication, writing that "among primitive races masturbation is simply a step in the sexual education of the individual," that the "little savage learns ... and then ... loses interest in," while "civilized" children are made nervous and ill by parental attempts to end their sexual exploration.[53] Another contributor, unnamed, penned the following lines:

> Just as vegetation attains excessive growth and ripens early where heat and moisture is extreme, man appears to exhibit this same characteristic. It is an established and accepted fact that sexual desire is markedly increased in the races that inhabit the tropical zones. Heat plays an important part in the development of these functions. This has been attributed to the indolent and inactive existence that prevails in these territories. Whether this is the cause, or whether heat stimulates the function of the sexual glands, causing an increase in their secretions, is yet to be determined.[54]

Writers in Europe and the United States were no less attentive to those cast as racial outsiders in their own midst. While demasculinized descriptions of Jewish men abounded, many accounts decreed that homosexuality was rare among Jews.[55] Even Chideckel, reiterating statistics on insanity that had gained currency in xenophobic treatises twenty years earlier, grudgingly concurred, "It is a fact worth noting that the Jewish race which supplies so many inmates of our insane asylums is practically immune to sex perversities. With the exception of prostitution, one hardly meets a Jewish homosexual woman, sadist, masochist or nymphomaniac."[56] Scott, too, wrote that Jews were almost free of venereal diseases and masturbation (a term still inclusive of homosexuality at the time he published the first edition of his book in 1898), calling the Jewish people "that circumcised nation who to this day remain as the 'standing astonishment of the world.'"[57] Given the orificial-surgery-as-cure-all mentality, it is not difficult to see how Scott and others came to this conclusion. Indeed, several years earlier a writer in the *Journal of Orificial Surgery* went so far as to re-envision the biblical covenant:

> If Moses had prescribed the amputation of the hood of the clitoris as well as the removal of the foreskins for Jews, they would undoubtedly by this time have had the qualities which, crystallized, produced these formations, so thoroughly eradicated from their natures that the forms themselves would have disappeared and the Jews today would be born without foreskins and without hoods to the clitoris. It is not uncommon, I am told, even as it is, to encounter Jewish male children who are without a foreskin.[58]

Of course, the reason Jewish male children without foreskins were encountered was because Jewish boys are circumcised eight days after birth. His creative recasting of evolution and Jewish anatomy merely joins a long list of morphological fantasies regarding Jews and other racialized minorities.

Discourses on homosexuality and Jewish difference were linked, despite the exoneration offered by Scott and others. As Sander Gilman has noted, the medicalization and sociological classification of both Jews and gays emerged at the same time in history. Karoly Benkert coined "homosexuality," a clinical term, a term for a "disease," during the same period that Wilhelm Marr came up with "anti-Semitism." According to Gilman, the purpose of the latter term was "to create a new scientific discourse for the hatred of Jews.... It has no validity except as a marker of the discourse of Jewish difference."[59]

But even those who maintained that there were no Jewish homosexuals, a claim that flew in the face of Hirschfeld's public prominence, saw Jews as at least culpable, along with gays, of weakening the state.

Eugenics and nationalism were, after all, inextricably entwined—the one easily supporting the other. George Mosse writes that in nineteenth-century Germany both groups were charged with eugenic travesties: homosexuals for refusing to sire workers and soldiers, and Jews for lusting after German women, who, thus corrupted, would forever bear unhealthy children. Additionally, Jews were blamed for inventing birth control, part of a plot to weaken and destroy the Aryan race. Mosse underscores the seriousness of these accusations, given the dread of falling birth rates at that historical moment. Perhaps, he posits, the only reason Jews escaped being branded inherently and genetically homosexual at a time when every crime, every pathological behavior was ascribed to them, was the visibility and general acknowledgment of Jewish family life, an image that did not jibe with the world of homosexuals concocted by imaginative physicians and others. Nevertheless, a certain consolidation of the scorned evolved, and a corresponding metaphor emerged. In 1930, when the Reichstag was considering a bill calling for the castration of gay men, homosexuals were branded "that Jewish pestilence."[60]

The use of metaphor in furthering eugenics enterprises is strikingly evident in writings and experimentation on queers. It was revealed to a still greater degree when those subjects were women and/or racial outsiders. In some cases, as evidenced by the Reichstag debate, the mere implication of depravity and sexualized degeneracy—considered givens in every gene pool that deviated from the white, Aryan, or Anglo-Saxon European and American stock—was enough to link the two, no matter who one chose as an erotic object. Metaphors, as Nancy Leys-Stepan has suggested, bring two different and usually separate entities into "cognitive and emotional relation with each other." In constructing similarities, analogies construct new knowledge. Metaphor, she writes, is no longer an agent of description, but metamorphosed into scientific fact. Its power rests in its ability to "bring together a *system* of implications, whereby other features previously associated with only one subject in the metaphor are brought to bear on the other."[61] It speeds the process along if both subjects are marginalized, loathed, or deemed genetically, biologically, and/or anatomically defective.

Metaphor enabled other prevailing discourses, most importantly scientific racism, to come to the aid of physicians eager to administer surgical curatives and/or establish the credibility of their observations. Pronouncements on the "masculinized libidos" of lesbians, the effeminacy of Jewish men, and the hypermasculinity of Black lesbians all served to reinforce associative links between despised populations, and, just as critically, validated the classification of certain orientations, expressions, and behaviors as the natural domain of particular gendered or racialized groups. In this context, rating the degree of Childhood Gender Nonconformity among LGBT people is but one more step on a historical continuum where metaphor and simile are factored into the scientific equation.[62]

The treatment of lesbians relied on a series of medical dicta that not only marked any departure from gendered sexual norms as aberrant, but constructed insanity in women as both manifested by, and dependent upon, sexual "deviance" and discernible physical anomalies. The lines drawn between heredity, anatomy, and insanity in late nineteenth- and early twentieth-century treatises on lesbians were forerunners of the consolidation of biology and psychiatry that would imperil queers throughout the entire century to come.

Homosexuality and the Bio/Psych Merge
An Additive Model of Causation Theories

Pathology is but physiology working under new conditions
—Havelock Ellis, *Sexual Inversion*

Popular wisdom holds that, post-Freud, eugenic and other biology-based theories of homosexuality gave way to psychoanalytical explanations. As perilous, and sometimes fatal, as psychiatry and psychology have been to gays and lesbians, it would have been a luxury had this been the only framework queers had to contend with.[1] It did not dislodge hereditarian or evolutionary theories any more than biological explanations eclipsed the rhetoric of sin and vice—biology might delineate a propensity, but it did not mean that the afflicted should not be expected to exert self-restraint. Sexual outcasts were stigmatized anew each time another causation theory emerged. At times, emotional development and hereditary models competed for legitimacy. Over the last seven decades, they have frequently worked in concert. Characterizations of homosexuality as inborn (biological, hereditary, congenital, biological, rooted in the body) did not fade with the ascension of acquired (environmental, psychological, psychiatric) models. If anything, they fortified one another.

The amalgamation of biological and psychotherapeutic discourses meant that even psychiatric diagnosis could mandate medical assaults: aversion therapy (chemical and shock), hormone inundation, psychosurgery (lobotomies), or any combination of these. The record of these medical interventions, like that of orificial surgery, is agonizing. As in other eras, treatment was sometimes sought by lesbians and gays—made complicit and desperate by a climate of ostracism and malice—and some-times foisted upon those least able to resist—the young, the imprisoned, the institutionalized. Resting on the same premise as preceding hypothe-ses (the aberrance of homosexuality), those formed by the melding of psychiatric and biological treatises on homosexuality did nothing to obviate the enmity and physical harm unleashed on queers. Even med-ical models emerging in the midst of a visible gay rights movement in the 1970s did not serve that movement. On the contrary, they exacted a horrible toll.

The bio/psych merge typifies the additive nature of all causation theo-ries—none of which was, or is, mutually exclusive. Dean Hamer, for ex-ample, has synthesized genetic and psychiatric explanations of homo-sexuality. In *The Science of Desire* he wrote,

> [The] gay subject's sexual direction was, in most cases, evident long
> before puberty, an observation well known to psychologists and psychi-
> atrists, most of whom believe that sexual orientation is established
> within the first six years of life. This early manifestation that sexual
> orientation is a deeply ingrained component of a person's psychological
> makeup . . . is consistent with a genetic predisposition.[2]

Hamer has incorporated psychiatry's atlas of homosexuality in much the same way that punitive legal responses greedily absorbed medical misinformation on gays and lesbians. His is a throwback to the 1930s when psychoanalysts and medical doctors were negotiating the distance between their respective postulations on human sexuality.

Freud himself spanned that chasm. Greenberg reports that he drew extensively on nineteenth-century writings on neurology and psychiatry. While Freud stated in his "Letter to an American Mother" that homo-sexuality was not a sickness, and refused to declare it a sign of heredi-tary degeneration, he did believe in a possible constitutional element to its formation. The psychoanalytical viewpoint was elastic enough to al-low for a congenital *predisposition* to homosexuality that, combined with an acquired tendency, would produce inversion. He signed his name to a statement issued in 1930 decrying the punishment of homosexuals as an "extreme violation of human rights."[3] Nevertheless, generations of gays and lesbians and their mothers were physically ensnared and emo-tionally dissected by a psychiatric establishment that bent to a Freudian interpretation of the development of sexuality.

Sexology magazine, having begun publication in 1933, was poised to absorb treatises all along the causal spectrum. In the March 1934 issue, Dr. George Lake wrote that gays were *psychically* "deformed," yet his analogies were all with *physical* disabilities: sexual inverts "are no more to blame for their condition than are those persons who are born deaf, blind or physically distorted."[4] Articles such as Kermit Reidner's "Cure for Homosexuals?" allowed for congenital and acquired homosexuality, the latter lending itself more easily to treatment. "Normality," he noted, could also be acquired with the proper gender-role training.[5] The journal's editor himself weighed in on the issue, managing to consolidate psychological, biological, and moral indictments. "Homosexuality," he wrote, "is a *psychopathy*, often with a constitutional basis; the homosexual has pronounced anti-social attitudes, and anti-social individuals lack moral stamina."[6] Despite the fact that nearly all contributors to *Sexology* who wrote on the subject deemed homosexuality deviant in some way, the journal professed to be extremely enlightened in its attitudes toward gays. A 1940 article called for a shift in society's attitude. Differentiating between acquired and congenital homosexuality, it argued against the incarceration of gays and lesbians: "Punishment will not drive out what is ingrained at birth."[7] Whether or not modern psychiatry offered a more enlightened alternative to overtly eugenic models of homosexuality, it did not quash the medical profession's abuse of gays and lesbians. There was coexistence or double jeopardy, depending on which side of the diagnosis and treatment one fell on.

Albert Moll's intellectual evolution is testimony to the cumulative nature of causation paradigms and the flexibility of axioms on the "sexual instinct." For a good part of his career, he concentrated on a corrupted heredity, though he did acknowledge acquired tendencies. By 1936, Moll had altered his perspective considerably, stating that homosexuality was acquired by improper sexual experiences and that only an insignificant number of cases were innate. Still, he never fully abandoned hereditary factors.[8]

Like all eugenicists, Moll held to the belief that man should control his own evolution through selective breeding. "The role of the individual in good health is to perpetuate the species for the rest of his life," he averred, concluding, "we ought to consider as diseased all men who are incapable of contributing to the propagation of the species." In contrast with the individual in good health, the Uranist, possessed of an anatomy gone awry (small feet, "well-developed" pubic hair, a "virile member [which] attains to dimensions no greater than that of a little boy" or else is "abnormally long"), was guilty of shirking his procreative obligations.[9]

Moll was highly adept at traversing the acquired-congenital terrain, reconciling these potentially contradictory explanations by subscribing to a belief not uncommon for his era, namely, that acquired traits could be passed on to the next generation. "I think," he wrote, "that the morbid desire for individuals of the same sex acquired by an ancestor is transmitted to the progeny." As possible eugenic origins, he posited too great an age difference between parents of homosexuals and "psychosexual hermaphrodites" (bisexuals in today's vernacular) or consanguineous marriages. "I know several cases of consanguinary marriages in which the father of a Uranist child was a liver in the full sense of the word, celebrated for his success with women." For Moll, hereditary degeneracy could manifest itself in any number of ways. "We do not know why epilepsy is found in the one, and sexual inversion in the other; no more than we know why of two persons catching cold one has only catarrh while the other has articular rheumatism." Moll even toyed with another theory circulating at the time, that homosexuality was somehow linked to epilepsy.[10]

Moll recognized that any characterization of homosexuality as acquired (and therefore something that could be restrained) could be used to justify legal and criminal penalties.[11] It may very well have influenced his ideological growth. However, he still urged, "when this perversion exists the impartial observer ought to regard the act not as criminal but as morbid."[12] While this might lend credence to the notion that biology-based models of homosexuality tend toward tolerance, Moll advocated medical intervention even when he believed firmly in the innateness of homosexuality. Furthermore, while the lingo and the treatment suggest psychiatry's imprint, the diagnosis was unequivocally rooted in eugenics. Moll claimed that most experts on the subject were in full accord: the roots of congenital and acquired inversion were the same.

> We may characterize them as neuroses or psychoses due to overtension of the central nervous system.... Consequently when there is a light form of degeneration of the central nervous system in the parents, hysteria for example, serious psychic trouble may be found in the children.

It is certain that in all cases of sexual inversion whether acquired or
congenital, where perversion is involved, we find a very burdened
heredity.[13]

This was the brilliance of what were essentially additive exegeses. Heredity itself could not be "treated," but an inborn bent or a burdened parentage could ready the ground for acquired perversion—acquired and
therefore able to be unacquired.

Moll's treatment of choice was "association-therapy," a sort of grad-
ual orientation replacement program (though Ellis pronounced him "a
great authority on hypnotism, and with much experience of its application to homosexuality"). Gay men who were attracted to young boys
were Moll's most encouraging cases. Such men, he reasoned, had the
homosexual urge closest to "normal sexuality" because "boys resemble
women, and therefore it requires a less profound organic twist to become
sexually attracted to them." In almost all Uranists there was a bridge to
heterosexuality. Properly navigated, it could lead a man to love a "boyish woman."[14]

A penetrable boundary between the physical, hereditarian domain
and the psychogenic realm enabled Moll to locate the origins and manifestations of homosexuality in biology, but the path to eradication lay
in the psyche. Dr. Louis Max of New York University took the inverse
approach to homosexuality and treatment. At the 6 September 1935 meeting of the American Psychological Association, four years after Moll's
Perversions of the Sex Instinct was published in the United States, Max
presented "Breaking Up a Homosexual Fixation by the Conditional Reaction Technique." The source of the "problem" might be psychological,
but the treatment was very, very physical. What Dr. Max related to his
colleagues was the first documented use of aversion therapy on a gay
man or lesbian: electric shock therapy.

Low shock intensities had little effect but intensities considerably higher
than those usually employed on human subjects in other studies
definitely diminished the emotional value of the stimulus for days after
each experimental period.[15]

Sixty-odd years later electric shock therapy is again gaining legitimacy
as a viable therapeutic method for a range of "disorders." As lesbian,
gay, bisexual, and transgender youth continue to be committed to insti-

tutions that employ shock therapy, it is painfully obvious that psychiatric and biological treatises do not operate independently of each other.

The history of science's approach to homosexuality, including the amalgamation of psychiatric and physiological treatment paradigms, reveals the disheartening truth that queers, beaten down by societal hostility, have often cooperated with our own pathologization. In the 1930s and 1940s, the union of medical and psychogenic discourses became so pervasive, so insidious, that overt coercion was not always necessary. Renowned poet, writer, and activist Pauli Murray expended significant time and energy and intellect trying to a find cause and cure for her lesbianism. Her personal papers reveal a journey that exemplified the prevailing additive approach to homosexuality.

Reading Murray's notes from meetings with doctors is excruciating, in part because they demonstrate so clearly how lesbians and gays, including the most well-informed and politically minded, were willing to defer to medical "expertise." Before her death in 1985, Murray worked most of her adult life in civil rights struggles. She attended Howard and Hunter and Yale. In the 1930s she taught in the WPA and was the executive secretary of the Negro People's Committee to Aid Spanish Refugees. She corresponded with Eleanor Roosevelt from 1939 until the latter's death in 1962. She ran for New York City Council on the Liberal Party ticket in 1949. She agitated against the poll tax and for the Equal Rights Amendment. Yet, even Murray, whose political work rejected the marginalization of the disfranchised, took medical pronouncements on homosexuality as axiomatic.

As some gay men had once actively sought out castration, there were those among Murray's generation who believed they had little choice but to willingly subscribe to a medical model that now included hormonal theories. The range of questions Murray put to doctors (at least one of whom was a psychiatrist), in a 1937 meeting at the Long Island Rest Home in Amityville, New York, and the corresponding answers illuminate the extent of psychobiological rhetoric and guesswork being propagated at the time: Did she have a mother fixation? Others seemed capable of suppressing the emotional discord they felt, why couldn't she? Nervous condition. Could it be that one of her genitalia was actually male in constitution? No possibility. Why, she asked, did she resist men's sexual overtures? Why, in the parlance of the day, was her sex instinct inverted, that is to say, why did she prefer pants, why did she want to do what men did, be one of the guys? Why could she bear being controlled by women, but only those she liked? Glandular on all counts.[16]

Murray wrote that the solution to genuine homosexuality lay in experimental science, meaning hormonal remedies. As her notes from the following day indicate, she was more than willing to subject herself to untested and unwarranted "research." She asked if Dr. Ruth Fox would be open to experimenting with male hormones. Did the doctor believe the conflict originated in glands, the brain, elsewhere in the body? If she didn't have the answer, where, Murray asked, could she go to get one? What branches of science, hospitals, medical institutions (here or abroad) were properly equipped and actively engaged in experimentation? Lastly, she returned to the possibility of unseen male genitals: was there an issue of pseudohermaphroditism?[17]

Two years later, Pauli Murray was actively inquiring into hormone "treatment." In November 1939, two New York papers, the World-Telegram and the Amsterdam News, ran front-page stories on four young men who had synthetic tablets of pure crystalline testosterone subdermally implanted between their ribs. The World-Telegram's banner proclaimed: "Pill Planted in Body Turns Weak Effeminate Youths into Strong Virile Men." The tablets had reportedly "transformed them from tragic youths with piping voices and few masculine traits into normal men.[18] The Amsterdam News promised that these "sex tablets" were the "magic formula that transforms effeminate males into normal men, strong and virile; a scientific achievement designed to have far reaching influence and effect in the human race."[19] Murray clipped both articles, sending a copy of the World-Telegraph report to a Dr. Richards with a note saying that if experimental treatment was already at this stage, perhaps something could be done for her. She urged Richards to put her in touch with Dr. Joseph Eidelsberg, who conducted the research.[20]

In a 1940 document in which she outlined symptoms and side effects of her disquietude, Murray wrote of irritability, fear, and anxiety. It's little wonder, given what she subjected herself to. In January 1938, tests at a neurological clinic declared both her basic metabolism and "female" hormone levels to be normal. Her "male" hormone levels were reported low. The following year there were more hormone measurements, this time at New York Hospital. A Dr. Schorr took nude pictures of Murray. Her physique was pronounced symmetrical and boyish, but not especially masculine (though her abdominal muscles did generate some attention). Dr. Schorr, or perhaps Dr. Charles Richardson, advised psychiatric treatment, but Murray felt that if the cause of her homosexuality was not physical, she should be able to handle the remedy herself. She was still speculating that perhaps tumors or concealed testicles were to blame for the way she was feeling and toying with the idea that her physical and emotional turmoil might be the result of having one ovary and one male organ. Later that year, in November 1939, Murray visited the endocrine clinic at Post Graduate Hospital. She had read of the work being done there on male hormones and wanted to know if the medical staff had done any experiments on masculine women or girls. Would they be willing to test male hormone therapy on her? The doctor at Post suggested using female hormones, but Murray did not return for treatment, in part because she began work as secretary for National Sharecroppers Week. But her reluctance ran deeper. Her longing to be male was so intense, she wrote, that she could not yet bring herself to submit to hormonal therapy. Ultimately, she was not willing to adapt to life as a stereotypical, normative female.[21]

The hormonal therapy Murray referenced was a product of the fusion of biological and psychiatric discourses, a two-pronged attack able to satisfy psychologists, medical doctors, and researchers. It began in earnest in the United States in the late 1930s and 1940s and remains one of the most frightening additive treatment approaches. Endocrinology (the study of hormones, glands, and their associated disorders) was carving a niche for itself, and the climate was right for an embrace of hormonal experimentation on humans. In 1939, an unnamed writer declared in *Sexology*'s pages that while "congenital" sexual inversion was incurable, the fundamental cause was no mystery. "The invert individual is so because of a deficiency in one of his sources of personality—his native glandular endowment."[22] Another of the journal's contributors described a "force, as yet undiscovered, which in some individuals destroys the female *hormone*, and makes those persons predominantly male; while in other persons, it destroys the [male] hormone and makes them predominantly female." He named this the "X-force," and explained that it sometimes failed. If it did fail, female hormones in a male body might not be destroyed, and the formation of male hormones might be retarded, resulting in homosexuality.[23] Not everyone was convinced. The

magazine's editor wrote that trying to alter a man's sexual tastes by injecting hormones into him was "like giving him drugs to turn him from
wanting to vote the Communist ticket."[24] His skepticism did not, however, keep him from running a story of Kathleen Latham—on trial for
murdering the husband of the woman she loved—under the heading
"Endocrimes."[25]

Hormone theories breathed new life into old biological models. In
April 1935, *Sexology* ran a letter from an eighteen-year-old gay man who
seemed on the threshold of self-acceptance. He wondered if "I am normal and the majority abnormal?" Wouldn't it be wiser, he asked, to
"lead a life according to the dictates of my conscience, living fully and
freely what is to be a 'normal' life?" But the magazine's self-proclaimed
openness applied only to unencumbered reassertions of old taboos, not
a radical departure from them. "D. L." received the following response
from the editor: "sexology is not inclined to favor the abnormal as a
form of human behavior.... the greatest amount of satisfaction can
only come to the individual through leading a sexual life that is biologically correct." The editor went on to explain that homosexuality was in
many instances the result of "changed gland secretions, rather than a
simple mental determination to live that kind of life."[26]

Practitioners' turn to hormonal curatives marked a new and uncertain
era for gays and lesbians. Wendell Ricketts related the chemical regime
inflicted upon a forty-six-year-old African-American man referred to as
"A. D." Over the course of a six-month period, from October 1939 to
April 1940, he was given a total of seven substances: "oral doses of Stilbesterol; a subcutaneous testosterone implant; injections of gonadotropins
from pregnant-mare serum, for which pituitary gonadotropic hormone
was later substituted; desiccated thyroid to increase his receptivity to
the other hormones; injections of testosterone propionate; emenin, an
estrogenic preparation; and estriol."[27] A. D. suffered from nausea, but
his doctors continued administering the drugs. At the end of six months,
they were forced to admit that A. D.'s behavior had not changed. He still
failed to pass masculine muster except in the measurement of his genitals, which were described as extremely large at the outset—perhaps in
keeping with the prevalence of racist stereotypes in such reports. The
experimentation loosed on A. D. should be seen not only within the
context of how gays suffered scientific research, but also as emblematic
of an era in which subjects were chosen by virtue of their race. At the

time this chemical deluge was unloosed, the infamous Tuskegee experiments had been underway for seven years.[28]

The doctors concluded that the failure of the treatment was due to the fact that A. D.'s "homosexuality is of long-standing—35 years in this case—and in which the personality structure has been altered at an early date [so that] autogenous hormonal factors are no longer of significant potency."[29] With this pronouncement, they were able to negate any reading of A. D.'s resistance or comfort with his homosexuality, absolve hormonal therapy of its inefficacy, and cement the bond between developmental psychology and biological hormonal processes.

Many researchers posited that hormone inundation was, if not superior, at least complementary to psychotherapy—the former targeting the sexual drive itself and the latter merely concerning itself with the neuroses of homosexuality. Despite the fact that many researchers in the 1940s saw little or no change in sexual orientation resulting from these methods (except in some cases there was an increase in sexual activity), physicians continued to manipulate the hormone levels of their patients for the next four decades.[30] In a tangle of needless choices proffered by the medical establishment, hormone treatment was considered a more humane response than surgical castration, as sterilization had been so deemed years earlier. In the mid-1950s, a writer in the *British Medical Journal* recommended hormones over surgical castration to suppress sexual urges in a fifty-two-year-old gay man, advocating a deadened sexuality to a nonprocreative one.[31] Similarly, doctors Charles Bery and Clifford Allen set down these guidelines for chemical castration:

> Chemical castration is permissible only in men over fifty years of age in whom there is no hope of psychological cure. In them it can be used to destroy their sexual urge and give them mental peace, but even so they should not be allowed to continue occupations which put them in the path of temptation.[32]

Bery and Allen warned that giving a gay man testosterone, as recommended by some of their colleagues, "may make him more homosexual" by "accentuating the patient's urges in the wrong direction!" Neither was chemical castration to be entered into lightly. "The use of 'chemical castration' by giving female hormones is an expression of despair. It means that one has given up hope of making the patient a normal person."[33]

Because none of the psychological, biological, or hormonal approaches challenged the heterosexual-as-norm/homosexual-as-disease paradigm,

medical assaults were often waged simultaneously. Even as hormone injections were being administered by some physicians, others were subjecting their gay patients to LSD-25, hypnosis, and variations on Max's
aversion therapy. Ralph Blair related a 1953 study in which the twenty-
five gay male subjects drank coffee or tea with emetine (an antibacter-
ial, rarely used anymore due to its toxicity). Ten minutes later they were
given an injection of emetine, ephedrine (a bronchodilator, used to treat
asthma, which increases heart rate), pilocarpine (used in the treatment
of glaucoma and causing pulmonary edema, a drop in blood pressure,
sweating, vomiting, twitching, and a slowing of heart action), and apomorphine (a respiration-depressing narcotic that also causes nausea
and vomiting). As nausea and then vomiting set in, the subjects were
shown slides of males in various stages of undress. Later, while feeling
the effects of a 50 mg dose of testosterone, they were shown "provoca-
tive" films of women. This cycle was repeated ten times, with little of
the impact researchers were so desperately seeking.[34] Other mutations
of aversion therapy were also tried. In one instance, the patient was given
brandy along with his apomorphine injection to increase its effect and
made to listen to a tape of his own case history.[35] This is not to suggest
that inducing nausea replaced electric shock treatment. The latter continued, sometimes, as reported in the *British Medical Journal*, by court
order.[36]

The increased visibility of the lesbian, gay, bisexual, transgender (LGBT)
community in the 1960s and 1970s altered neither the course of "scientific
inquiry" nor the dominant view that there was something inherently
"curable" in queers. Not even the uprising at Stonewall, credited with
ushering in a militancy that had actually been cultivated over many,
many years, led researchers to reconsider the sine qua non of their research, that queers were somehow tainted—biologically, hormonally,
genetically. Psychobiological hypotheses, even in an era marked by the
community's intensified struggle for civil rights, did nothing to further
the goals of gay liberation, a fact lost on too many civil rights strategists
today. In hindsight, the difference was that after 1969 and during the public, organized, accelerated campaign for gay rights in the 1970s, research
and "treatment" of homosexuality took on an even more reactionary
stance. In the 1940s and 1950s, physicians worked against a backdrop of
post–Kinsey Report panic and a cold war purge of gays and lesbians

from government.[37] In the 1970s, they worked in opposition, not only to homosexuality but also to an entire mass movement.

In 1971, M. Sydney Margolese released a set of two studies in collaboration with Oscar Janiger and Richard Green. The doctors contended that, in heterosexual men, the level of androsterone (A), a breakdown product of testosterone present in urine, was higher than that of another breakdown product, etiocholano (E). But if the amount of E exceeded the amount of A, the man in question would likely be gay. According to Margolese, the excess of etiocholano was not a causal factor, but an indicator of a biochemical difference between gay and straight men. "I am close to saying homosexuality is genetic.... We have not found anyone who is a strongly motivated homosexual and who is *otherwise healthy* whose E level does not exceed his A level" (emphasis added). This pronouncement reinscribed biology/hereditary-based theories and reinforced the notion that homosexuality was symptomatic of ill health. Again, the use of metaphor came into play, as equivalences were drawn between gay men and unhealthy heterosexual males. As for the implications of his research, particularly the prospect of a "cure," Margolese stated, "Well, that's a long way in the future. But if we can make this stick, it's certainly going to push that along, all right."[38]

The Margolese study and the biased coverage it was given in the *Los Angeles Times* prompted a small but significant protest—an important reminder that medical models have never gone wholly unchallenged by the LGBT community. Demonstrators carried signs reading "The LA Times Needs Treatment and Prevention" and "Prevent Normalcy." Don Kilhefner, one of the organizers, told the *Advocate:*

> What we're objecting to is the value judgement that goes along with this. They're starting with the basic assumption that people with the E hormone somehow have to be cured or prevented from existing, while those with the A hormone are allowed to exist....
>
> By isolating this E hormone in the fetus even, they might be able to... prevent homosexuals from being born.... what you have here is the basis for a form of sophisticated genocide, if you read between the lines.[39]

Kilhefner's fears were by no means unwarranted or exaggerated. Several months earlier, internationally acclaimed geneticist H. B. Glass told the American Association for the Advancement of Science that mandatory abortions by the state would be necessary to purge humankind of "uncontrollable defects such as mongolism and sex deviation." A "far more regulated society of man" would be needed.[40] As with previous eugenic assaults, this one was built on the unchallenged and violent dehumanization of the disabled. Once the unworthiness of the disabled was established, it became a simple matter to enlarge that category or use it as metaphor for other unwanteds, including lesbians and gays.

Perhaps it was the insistence on the genetic and/or hormonal failings of lesbians and gays that led to the 1973 removal of homosexuality from the American Psychiatric Association's list of emotional disorders, downgraded in DSM-III to "sexual orientation disturbance"—a move that still left queers in the realm of the deviant. The vote was hardly unanimous. Thirty-seven percent of the association's membership voted against the measure, and a survey of 2,500 psychiatrists revealed that 70 percent disapproved of the reclassification.[41]

This gesture by the APA did not, however, curtail the abuses heaped upon queer patients of psychiatrists and psychologists. In particular, those ensnared by the criminal justice system during this period experienced firsthand the resiliency of a century-old partnership between the state and medical science. In 1972, *Gay Sunshine* reported that over four thousand lobotomies, many as curatives for homosexuality, had been performed on prison inmates at the Atascadero State Hospital in California. According to the article, further plans were underway to transfer gay prisoners from Vacaville to the University of California Medical Center in San Francisco for brain surgery.[42] In 1975, Senator Sam Ervin's subcommittee released the results of a three-year investigation on constitutional rights in prisons and other federally funded projects. The findings disclosed that Atascadero State Hospital had been subjecting gay prisoners to aversion therapy. The Ervin subcommittee also discovered that the Department of Health, Education, and Welfare had been financing a UCLA study of "childhood gender problems" focused on the prevention of homosexuality, as well as a Mississippi study of gay teenagers and adults that used "various conditioning techniques, aversion therapy, and systematic desensitization."[43]

Still, there were those in the gay community who continued to welcome biological theories. It was, after all, the decade of Anita Bryant.[44] Socialization arguments were being leveraged to deny gays and lesbians custody, teaching jobs, and more. Hereditary hypotheses might be the antidote. If homosexuality was innate or congenital and not communicable, no one was at risk for "contamination" who was not already biologically inclined.

Norman C. Murphy, a gay psychologist, was especially receptive to science's not-so-new promise of a quick fix. In the introduction to a lengthy article he wrote for the *Advocate* in 1978, he and his coauthor Dean Gengle did warn of "massive psychobiological conformity." However, as the essay progressed, Murphy became less critical of measures that, if enacted, would ultimately lead to such conformity.[45]

Murphy was concerned with the "cross gendered," those whose "neurogender" differed from their "morphological/physical/genital gender." According to Murphy, gays and lesbians had "cross sexed brain differentiations."[46] Harkening back to Carpenter and Hirschfeld, who plotted homosexuals along a sexual and gender continuum, Murphy wrote,

> When one looks at the behavioral and genetic groups, it is difficult to dispute the existence of a continuum of sexual brain development that plays a part in the manifestation of predominantly male or female stereotypic sexual and/or gender behaviors.[47]

Murphy anticipated Bailey and Pillard when he emphasized a correlation between the selection of "female-associated games and toys" by boys, an adult sexual orientation toward men, and "androgen insensitivities." He cited surgical and hormonal treatments performed on girls with hypertrophied clitorises. "Those who have their penis [*sic*] left intact are reared as 'males' and exhibit 'heterosexuality' in the adult phase. Those who have a penis excised and who also undergo hormonal treatments—and have a vagina created surgically—are reared as 'females' and also exhibit 'heterosexual' behavior."[48] Here identity was completely constructed on the basis of anatomy, a surgically induced anatomy to which young children were forced to submit. Medicine would choose not only the biological sex, but sexual orientation as well. Nature and nurture arguments converged into a single regimen, as child rearing was called upon to reinforce surgical mandate.

Certainly, Murphy was not the first or the last to conflate gender identity, biological sex, and sexual object choice. What was different was his forum. He was reiterating the claims of rabidly homophobic physicians of the 1800s in the 1970s gay press, albeit with the benefit of a more sophisticated medical vocabulary. Lesbians not only had "excessive testosterone" compared to heterosexual women, but were "more masculine as measured on scales of androgyny" and, he related, with "greater stature and shoulder width and look older than they are." Not surprisingly, Murphy claimed a lowered testosterone level in gay men. What's more, androgen level differentials in gays and lesbians were present during the "critical hypothalamic sexual differentiation period *in utero.*" Most significantly, Murphy was able to salvage a genetic link, telling an interviewer that while homosexuality "appears to be genetic... genetics *per se* cannot be manipulated." Hormone levels, however, can be, and without changing the "underlying genetics that went into creating the difference."[49]

Murphy's faith in a science that would change straight America's image of queers seems to have been unwavering. He went so far as to claim that "from a neurological viewpoint, it is doubtful that many homosexuals engage in rape or murder." He was speaking here of gay men, who, researchers decreed, were sure to be less aggressive than "the androgenized heterosexual."[50] Murphy's espousal of a queer model minority was reminiscent of Dr. George Lake's assessment, over forty years earlier, that sexual inverts had a "strong and well-balanced character," rarely committed "sex offenses," and were possessed of self-control and social responsibility.[51] Murphy's contemporaries were also pondering the scientific merits of homosexuality. According to John Kirsch, James Rodman, and E. O. Wilson, writing in the mid-1970s, the persistence of homosexuality through the generations must mean that it serves some positive evolutionary function for the species, perhaps offsetting overpopulation. Homosexuality, Wilson submitted, might be a manifestation of the genetic component of altruism. As a "sterile caste," gays and lesbians aid and support relatives who do have children, thus ensuring the survival of the gay/altruism gene among scattered offspring of siblings, cousins, and so on.[52]

Murphy recommended "low emission tomography be used to map the sexuality of the brain by using irradiated hormones" in order to facilitate research in the field. He stated that he himself would never undergo surgery or aversion therapy "so as to be acceptable to a homophobic world," nor would he support outside engineering by making or denying that choice for/to another person. Yet he believed that sexual orientation could be preempted by altering sexual hormones during the fourth or fifth month of fetal development. He cited G. Dörner's rat study.[53]

In the 1960s, Dörner, working at the Institute of Experimental Endocrinology in Berlin, was experimenting with androgen levels in female rats.[54] In the mid-1970s, he suggested that "inborn errors of metabolism," as he classified homosexuality, could be prevented by injecting pregnant women with steroid hormones.[55] In 1980 he fixed the blame for male homosexuality on prenatal stress:

> Out of 865 homosexual males who were registered by venerologists in 6 districts of the GDR, highly significantly more homosexuals were born during the stressful war and early postwar period. This finding suggests that stressful prenatal (or perinatal) events may represent an aetiogenetic factor for homosexuality.[56]

The men in Dörner's study were born between 1932 and 1953, but he failed to note the dates of their examinations. Their age during their examination, as well as the current sociopolitical climate in the year they were treated, would factor heavily into a decision to come out to a physician and thereby be listed as gay in medical records. Dörner called the findings striking because "there was not found such an increase in heterosexual males born during this critical period who were registered by venerologists during recent years." Of course, these men may have gone to different clinics. Doctors, operating on their own biases, may have been quicker to diagnose gay men with venereal disease than straight men, or they may simply have failed to register heterosexual men as patients, thereby inflating, deliberately or unconsciously, the relative number of gays who sought medical attention. None of these possibilities were acknowledged in Dörner's report. Instead, he concluded,

> [T]he highly significant peak of relative frequency of homosexual males born during the war and early post-war period suggests a possible relationship between prenatal stress due to bomb attacks or other stressful war or early post-war events and sexual differentiation in the foetal brain.[57]

Dörner seemed to imply that with the right prenatal care, none of these women would have borne gay progeny. Like Hamer's location of the phantom gay gene on the X chromosome, Dörner's theory provided a biochemical corollary to the absent-father-overbearing-mother model of homosexuality: if only these women could have been stronger, calmer.[58]

According to Gunter Schmidt, Dörner saw his work as furthering teratophysiology, the study of malformation and physical damage. In other words, he believed he was working to isolate and correct a gross malfunction of nature. Dörner, whose laboratories developed methods for diagnosing fetal hormone levels, considered his research results to be "cornerstones of preventive medicine."[59] Like many of those who contributed to the body of "treatment" literature on homosexuality, Dörner omitted any mention of ethics or the substantial harm to women and fetuses caused by hormone injections during pregnancy.

What becomes evident in explicating the consolidation of various biological and psychiatric treatises is the way that unproblematized science feeds on itself. As Ellis once praised Moll, and Murphy reported uncritically on both the Margolese and Dörner studies, Dean Hamer would later cite the research of Richard Green and credit Pillard and Bailey's twin studies with assuring him that the search for a gay gene was a valid and plausible undertaking.[60]

At the close of his article in the *Advocate*, Murphy declared, "Current science points toward the liberation of all psychobiological minorities."[61] It was not, however, the promise of liberation *of* minorities that was so appealing, but liberation *from* minorities—as is the promise of all eugenic campaigns.

AIDS, Backlash, and the Myth of Liberatory Biologism

If scientific hypotheses during a militant gay rights movement could not offset homophobia in the 1970s, the designation of queers as a distinct psychobiological type could not be anything but disastrous to the LGBT community during the AIDS crisis. The early responses to AIDS in the United States and the way AIDS was explained to the American people were logical extensions of the construction of queers up to that point. The medical paradigm of homosexuality accommodated new information but was not seriously challenged by it.

Before AIDS became part of the medical and then popular lexicon, newspapers reported on a strange "gay cancer." Most cancers are named for the part of the body they attack—lung cancer, breast cancer, cervical cancer, etc.—not for the demographic group that suffers the highest initial losses. More importantly, while a susceptibility to cancer may be hereditary, cancer itself is not communicable. Members of a community or population who may be linked culturally, socially, politically, but not genetically, could only be diagnosed with "gay cancer" under the operative assumption that queers shared some specific bodily anomaly unknown to heterosexuals. Gay-related immune deficiency (GRID), which replaced "gay cancer" (until the Centers for Disease Control settled on acquired immunodeficiency syndrome in 1982), was not much of an improvement in this regard. Ideological precedents within both the research community and the larger culture that biologized homosexuality made the decidedly unscientific labels of gay cancer and GRID sound reasonable and accurate and set the tone for AIDS rhetoric for years to come.

At a time when AIDS research and services were drastically under-funded, there was financing for studies like Bailey and Pillard's, Hamer's, and LeVay's—studies that had nothing to do with saving lives. It be-came apparent that the twin surveys, the gene mappings, and the au-topsies were allocated money and legitimacy not despite the AIDS cri-sis, but because of it. After more than two decades of death and debilitation and loss, it has become evident that these scientists were re-sponding not to AIDS but to the mainstream discourse that enveloped it. For along with the claims that gays "asked for it" and that AIDS emerged from sexual overindulgence, there was talk of AIDS being divinely or-dained or, for the secular homophobes, AIDS as natural selection: a kind of passive eugenics designed to rid the earth of evolutionary misfits. There were even suggestions that AIDS might be nature's way of com-pensating for overpopulation.[1] In short, much of the initial political, scientific, and sociocultural (non)responses to AIDS outside of the gay community can be legitimately read as eugenic discourse.

In *Sex and Germs: The Politics of AIDS,* Cindy Patton wrote, "Disease reportage creates its market by stoking the residual fears of massive on-slaughts by disease and then calming the reader with incomprehensible 'discoveries' that will erect an even greater barrier against germs."[2] An emphasis on evolutionary or genetic differences between heterosexuals and homosexuals (the "discovery" of a variance in hypothalamus size or of a genetic component to homosexuality) can be extremely reassuring to a predominantly straight population in a country where HIV first surfaced among gay men. It certainly furthers a sense of normalcy and false protection in that AIDS is associated with a restricted class of "sex-ual miscreants," now made still more distant with claims of anatomical anomalies.

Much has been written on AIDS and the Other, whether that Other be a gay man, an African-American woman, a sex worker, or a heroin user. Richard Goldstein observed that fears conjured by difference and "deviance" have come to the surface and been made "hyperreal" by AIDS, which, he writes, has made society even more unwilling to confront this hysteria because the fears are seen as useful in distinguishing between who is "risky" and who is "safe."[3] This need for demarcation offers a par-tial explanation for the receptivity to genetic theories on homosexuality, for research that enshrines social categories as biological destinies. In turn, such research is marked and marred by a climate that insists on intrinsic difference and distance. Endeavors to chart the roots of homo-sexuality are but another way to single out queers, underscore an imag-ined, biologically immutable difference, and increase our political and social vulnerability.

To a substantial degree, AIDS activists, by the very nature of their work, were successful in combating the biological characterizations of gays that stamped the initial medical and media narratives of the epidemic. Most significantly, activists laid bare the institutionalized homophobia of the scientific establishment and the impact of such bias on research and social policy. These observations did not, however, prompt the rejection of flawed medical models by all LGBT rights advocates. To be sure, there was and continues to be strong dissent within the community, but by and large community mouthpieces have failed to relay, let alone forge, a queer critique of risks posed by political uses of current research.

On 12 December 1993, Judge Bayless ruled Colorado's Amendment 2 unconstitutional (a ruling upheld by the U.S. Supreme Court). While Bayless stated that the "nature versus nurture" issue was irrelevant to the case, he did write that the "preponderance of credible evidence sug-gests that there is a biologic or genetic 'component' of sexual orienta-tion."[4] This kind of pronouncement may not carry the weight of legal precedent, but the fact that it was issued from the bench at all is pro-foundly disturbing. More worrisome is the fact that Bayless was afforded the opportunity to make this statement by a legal strategy devised by gay rights advocates.

Basing a struggle for civil and human rights on the biological fixity of homosexuality ultimately does a disservice to the LGBT community. History reveals how dangerous such tactics have been to all marginal-ized groups, how often scientific hypothesis has acted as defender of the social, economic, and political status quo. Late nineteenth-century physicians responded to growing anxiety over economics, urbanization, immigration, and other perils of the modern world by linking sexual "perversion" with eugenics and crime, cementing the bonds between medical "treatment" and legislated penalty. Likewise, the 1990s ushered in the voices of LeVay and Witelson and Hamer, telling a homophobic nation that queers really are different from them at the most basic level—messages that coincided with both the increased visibility of queer ac-tivism and a groundswell of antigay legislation and referendums from coast to coast.

Ten years from now, researchers might reverse themselves entirely and pronounce homosexuality the result of purely environmental causes. It wouldn't be the first time. A political strategy tethered to the concept of biological immutability is precarious at best. Even if the research holds up, there is too much at stake for queers (but not only for queers) to privilege biology as arbiter and lone protector of civil rights. Lisa Duggan, coauthor of *Sex Wars: Sexual Dissent and Political Culture*, offers an alternative model. She suggests that the discourse of religious tolerance would serve as a better analogy for the political goals of gays and lesbians:

> We need strategies that do not require us to specify who is and is not a "member" of our group. If sexual desire is compared to religion, we can see it as not natural, fixed, or ahistorical, yet not trivial or shallow, as the term "lifestyle choice" implies. Religion is understood as not biological or fixed. . . . But it is also understood as a deep commitment. That commitment is seen as highly resistant to coerced conversion, and deserving of expression and political protection.[5]

Duggan may overstate the degree of religious tolerance in this country. "Tolerance" itself—bestowed and withdrawn at the whim of the majority—is a limited goal, and certainly the ways in which individuals come by their religion are very different than the ways they come to their sexuality. Nevertheless, Duggan effectively challenges the assumption that political equity can only be pursued by claiming genetic predetermination. It is the sort of challenge that is needed if we are to counter the assertion that genetic research is the gateway to liberation.

Magnus Hirschfeld, Havelock Ellis, and Karl Ulrichs all believed that scientific analysis went hand in hand with legal and social redress for sexual minorities. Hindsight is twenty-twenty, so perhaps they can be excused for their optimism. How, though, do contemporary researchers and strategists reconcile all that the last hundred years have witnessed with their own endeavors? After a century of evidence to the contrary, many still adhere to a science-as-savior paradigm, ignoring the additive nature of all causation theories, be they sin-based, hereditary, hormonal, or psychological.

The emergence of hereditary constructions of homosexuality in the late nineteenth and early twentieth centuries might have appeared to be a more progressive approach than church-based condemnations (or the explanation offered by an eighteenth-century English source blaming tea drinking and the "pernicious influence of Italian opera"),[6] but they were part of the same continuum, operating on the same premise and with the same mandate to intervene. The ideology of sickness did not eclipse the ideology of sin. Theories on the inborn nature of homosexuality and the consequent treatment enhanced, rather than ousted, morality-based constructions, often with much graver consequences.

In the end, it doesn't matter which pathology is bestowed upon queers: psychological, evolutionary, moral, genetic. The discourses of psychology, psychiatry, new and improved biology are not engaged in a linear replacement. It is unrealistic to think that one will unseat another when all rest on the same foundation and all have been brandished as weapons. New "discoveries" do not alleviate homophobia, do not right past wrongs, do not even stand up to the scrutiny of other scientists. They are simply subsumed into a network of causation theories. They displace nothing. Sodomy laws remain on the books in fifteen states. Shock therapy is still being performed on queers, particularly on our youth. Gays are still being lumped together with rapists and child molesters, as Montana's proposal in the mid-1990s to make all three register with the authorities demonstrated. The best that can come from an embrace of biologism is a modicum of pity from those softened by a belief that genetic programming has damned an unfortunate 10 percent of the population. But pity is not liberation. Ultimately, we can only win by aggressively opting out of nature versus nurture arguments, by rejecting any and all revivals of eugenic discourse inside and outside our communities.

The political consequences wrought by a reliance on genetic arguments are profound. Emphasis on biologically determined sexuality (or intelligence or life chances or anything else) is more than a distraction. It has a depoliticizing effect on marginalized groups. Even flawed medical models immobilize: why agitate against the status quo if lab work is the salient determinant of our political and social future? At stake here, to paraphrase Tony Kushner, is whether the queer community will choose to be agents or accidents of history.[7]

Timing is everything. Scientifically sustained political backlash flourishes when marginalized groups grow more visible in their demands for equity. The current round of biological determinism dovetails nicely with attempts to roll back political gains. In this case, we heard about "the homosexual agenda" and "special rights for gays." But there is a broader backdrop. Eugenics is hydralike in strategy and ideology: one tentacle entwined with nationalism, another extending toward reform-oriented liberalism, others to blatant homophobia, racism, misogyny, and white supremacy. Multiple identities and a shared demonization has meant that the consequences of eugenics for lesbians, gays, bisexuals, and transgendered people are now, and have always been, bound up with those for immigrants, people of color, and the poor. The liberal determinist rapture that has greeted this latest attempt to "rebiologize sexual orientation"[8] is not limited to technology's promise to prevent homosexuality or expel homophobia. It is part of a larger and longer hope that science alone can make fast work of inequity and poverty.

III

Sterilization and Beyond:
The Liberal Appeal of the Technofix

You, friends from foreign countries who have come here to our greatest city, must have noticed the intricate system of signals which regulates the crowded traffic in our streets and thoroughfares. By this system, the pedestrian is assured some degree of safety. But while the congestion of American cities has forced upon us a system to regulate traffic in city streets and country roads, America as a nation refuses to open her eyes to the problem of biological traffic and racial roads. Biologically, this country is joyriding with reckless carelessness to an inevitable smash-up. Is it too late to prevent national destruction?

—Margaret Sanger, Address of Welcome to the
Sixth International Neo-Malthusian and
Birth Control Conference

Liberal Loopholes

In the fall of 1996, four students at the University of Minnesota demanded the removal of a Margaret Sanger poster from the campus library. The four, all members of Campus Republicans, maintained that their objection to the display (and to the student center's birth control literature that exhorted Minnesota students to "Be Like Margaret Sanger") was unrelated to the Planned Parenthood founder's birth control activities. Junior Tom Gromacki stated, "We felt the university should not be glorifying her racist views."[1] The reference was to Sanger's promotion, and often aggressive embrace, of eugenic ideology and practice.

Sanger's contributions to "voluntary motherhood" and the separation of sex from procreation has often led liberals, including mainstream feminists, to project an association of birth control with reproductive rights and freedoms backward in history onto Sanger. But her own associations were more problematic. She courted eugenicist support, championed economically coerced sterilization of those deemed "unfit," endorsed medical experimentation on poor women, collaborated with Clarence Gamble (who zealously established sterilization programs in Puerto Rico, the southern United States, and elsewhere), published the work of Harry Laughlin and the like, and penned what can only be described as eugenic tracts herself.

Yet Sanger's name and her legacy are rigorously defended. Scholarship calling her to account has been routinely parried by attempts to rescue her from her own paper trail. She continues to be looked upon as hero and beacon, due in large part to a steadfast reluctance to call attention to liberal collusion with what has been seen as an essentially right-wing enterprise. As tempting as it may be to limit our scrutiny to right-wing eugenics enterprises, liberal propulsion of these assaults cannot be ignored. Glossing over liberal abetment is a viable option only for those who do not feel imperiled by the resulting historical revisionism.

While liberal philosophy can encompass a critique of institutional biases and injustice, it remains grounded in a belief that these can be offset by individual exercises of choice and responsibility, for example, birth control use. An emphasis on choice, in this context, is extremely constricted, obviating analyses of larger issues of racism, economic disparity, and other structural inequities. What's more, the liberal tradition provides a loophole: choice is only for the "rational." The "irrational," those deemed incapable of acting in their own best interests (including, at different historical moments, women, youth, the poor, the colonized, the enslaved), may be acted *on* for their own good. Janice Raymond has named this "the alibi of altruism."[2] Sanger's appeal to liberals has likewise rested on her advocacy of birth control as a palliative to poverty.

The concept of birth control (a set of reproductive technologies) and population control (the external application of those technologies to a designated demographic group) as societal balm has persisted through discourses on public relief in the 1930s and 1940s, "culture of poverty" arguments in the 1960s and 1970s, and into current Norplant/Depo-Provera welfare configurations. Faith that world hunger can be offset by population control or that physician-dependent contraceptives, administered to women on welfare, will redress economic disparity or that institutionalized barriers to the disabled will be rendered immaterial by prenatal testing, all illustrate the long-lasting liberal appeal of the technofix. "Fix" is operative here, referring both to the liberal *fix* ation on technology (in this instance reproductive technology) as a quick and allegedly painless alternative to deeper, more efficacious structural changes, and to the belief that such measures will *fix* the lives of the poor.[3]

Endemic to liberalism, as David Theo Goldberg has written, is the notion that there are pockets of injustice and that reason, embodied in rational reform, will set right all skewed social arrangements.[4] But the ideological bedrock of reform efforts is left untouched and uninterrogated by liberalism. Hence, many liberals genuinely believed, and continue to believe, in sterilization, Norplant, and Depo-Provera as mechanisms to "break the cycle of poverty." Liberal ideology emphasizes individual solutions (e.g., fewer children for welfare recipients) to systemic problems (a lopsided distribution of resources, institutionalized racism, ableism, and so on). The result has been an ends-justifies-the-means approach that often echoes right-wing endorsements of similar methods—methods that have included sterilization abuse, human rights violations, and high pressure population control programs. Robert Blank notes that while most proponents of technological fixes shy away from overt eugenic utterances, they will couch their agenda in more palatable, seemingly politically-, racially-, and class neutral language, citing instead: "(1) reducing social or family burdens, (2) benefits to future generations, (3) social responsibility, (4) cost containment, (5) individual health, (6) individual rights to a sound mind and body, (7) the best interests of the subject, and (8) cost-effectiveness."[5]

The hard truth is that science, and even mere scientific conjecture, has often functioned as consensus builder. The divide between reactionaries and reformers, nationalists and leftists, conservatives and feminists has not been the yawning chasm we suppose. As Loren Graham noted, the "natural alliance" between eugenics and conservative and fascist sentiments "was not logically preordained . . . and was not perceived in the early twenties by large numbers of radical social critics."[6] In some circles, it still isn't. Adolph Reed's diagnosis of liberal assent to eugenic discourse is useful here. In response to the legitimacy accorded to *The Bell Curve*, Reed denounced liberals for failing to condemn Charles Murray and for allowing him to set the terms of the debate. At heart, Reed declared, liberals and reactionaries like Murray share the same racism, the same fear and loathing of the poor, the same virulent scapegoating, and the same investment in and loyalty to capitalism. The only difference between the two, he stated, was that Murray blames a backward heredity and liberals a deficient culture.[7] The quartet at the University of Minnesota, despite the transparency of their motives (student association president Helen Phin reported that they were opposed to birth control altogether), merely rushed in where liberals feared to tread.

It is important to situate Sanger not as emblematic of the most virulent of eugenicists, but as bridge builder between these elements and reform-oriented birth control advocates. Her brand of eugenics must be examined against the medical and public policy backdrop of the era in which it emerged. This is vital, not only to the positioning of Sanger and her followers, but because there is a danger that intense focus on Sanger's culpability will obscure the history and preeminence of the more overtly racist and misogynist promoters of the eugenics cause. Early eugenic sterilization drives are of primary importance, as they were first forerunners, then contemporaries of Sanger's own campaign for involuntary population control. To that end an abbreviated overview of early twentieth-century eugenic assaults, particularly in the area of compulsory sterilization, is needed before proceeding to a refutation of historical readings that have exonerated Sanger, casting her not as a eugenicist but merely as a product of her age; a maker of unfortunate, but understandable, alliances. Most of these assertions are based on a denial of Sanger's racism and an inattention to her views on class and disability.

Over the years, Sanger's work and the work of her ideological cohort refashioned eugenics rhetoric into the more palatable language of population control.[8] Early eugenics attestations that society had a vested interest in which children were born of which women solidified in post–World War II decades: the continuing investment in the technofix as remedial to poverty in the United States and abroad, the singling out of entire regions for sterilization campaigns, and the resulting wave of reactionary legislation and welfare policies. Class bias, so central to eugenic policy (and a principal motivator for Sanger) came to the fore. This is not to say that class, in particular reliance on welfare, was a greater determinant than race, but rather that the invocation of economic rationales and the unchallenged vilification of the poor enabled eugenics to go unchecked and unnamed. Class is underscored here in an attempt to counter claims that Sanger and others were not eugenicists because they never publicly uttered racial slurs, and to highlight the vulnerability of low-income women who found themselves snagged in various institutional nets (i.e., relief, Medicaid, welfare). An attack on the poor may have seemed more genteel and more viable than an openly racist attack on people of color, but ultimately the same women were targeted.

After World War II, eugenicists merely bowed to post-Holocaust sensibilities, publicly—though not always—de-emphasizing race, while maintaining the same race-based practices. At least one historian has credited Clarence Gamble's efforts in the South during these years with moving sterilization into the purview of welfare reform. The resiliency and legacy of Gamble and Sanger's platform over the decades that followed is evident in the systematic sterilization abuse of African-American women and girls elsewhere in the South, Native American women and girls at Indian Health Service facilities, and Chicanas at Los Angeles County Medical Center. While these campaigns can hardly be said to have emerged from a liberal agenda, many mainstream feminists joined with conservatives in resisting attempts to curb the practice. More recently, Norplant and Depo-Provera have been cast as curatives to an unsatisfactory criminal justice system, welfare expenditures, and teenage pregnancy. Genetic testing and now the Human Genome Project promise to make good on eugenicists' unchallenged assumption that the disabled should be weeded out. Liberal endorsements of such "remedies" echo early calls for sterilization of the poor and "defective."

What resonates throughout this history is the convergence of eugenics and the liberal enchantment with the technofix, and the primacy of class (including welfare status), race, and gender as salient determinants of who would fall prey to the resulting cure-alls. While race, class, and gender were no less dominant in the eugenic constellation of Sanger's day than they are today, and while the paternalism toward and the vilification of poor mothers (both actual and potential) have been constants, in the intervening decades these biases have become increasingly encoded in legislation, welfare policy, and judicial renderings. While this has been resisted by feminist health care advocates, antiracism activists, and entire communities, liberals, including mainstream feminists, have too often united with conservatives in beating back that challenge.

The aim here is not to discredit birth control by noting its historical allure for eugenicists and population controllists, but to underscore the long-lasting repercussions of this mutual attraction. This early affiliation, with its increasing advocacy for economic-based contraceptive "incentives," allowed public policy to intrude more and more into the bodies of poor women of color. As Stephen Steinberg wrote, while liberals are not the enemy,

> the enemy depends on the so-called liberal to put a kinder and gentler face on racism; to subdue the rage of the oppressed; to raise false hopes that change is imminent; to moderate the demands for complete liberation; to divert protest; and to shift the onus of responsibility... from powerful institutions that could make a difference onto individuals who have been rendered powerless by those very institutions.[9]

It is certainly true that eugenics campaigns have been propelled chiefly by immigration foes, adamant white supremacists, and practitioners who held little more than paternalism and contempt for women, queers, and the disabled. Their efforts have been documented and cannot be excused or eclipsed by the mere fact of liberal acquiescence to particular components of their agenda. That said, an understanding of that ideological complicity is imperative, for it endowed eugenic prescriptions with a veneer of legitimacy, even benevolence, extending its viability well beyond what might have been a shorter public policy life span.

Buck v. Bell and Before

Several dynamics emerge in discussing the campaign of involuntary and eugenic sterilization of the "unfit" in the United States. First is the ideology of "racial hygiene," predicated on the "betterment and protection" of the white race by various means, including restrictive immigration, antimiscegenation laws, and surgical intervention. Dr. B. A. Owens Adair, writing in 1922, called sterilization "the only method by which the river of life may be purified." Next there is the use of sterilization as a punitive measure directed not only at the incarcerated and the institutionalized, but also at those pronounced guilty of burdening the state with unwanted children. The flip side of punishment is paternalism and the belief that the masses of poor people, particularly poor people of color, cannot be trusted to regulate their own fertility. Finally, as evidenced by the language of both eugenic tracts and court findings, there exists an ominous and overpowering preoccupation with the morality and sexuality of low-income women.

The call for the sterilization of the "manifestly unfit" flourished, but did not originate, among twentieth-century sociologists, physicians, scientists, and lawmakers. The first proposal of the kind published in the United States was written in 1887 by the superintendent of the Cincinnati Sanitarium, targeting prisoners.[1] Six years later, F. E. Daniel, the Texas doctor who recommended castration for gay men, proposed sterilization for "the purpose of race improvement."[2] The Michigan state legislature considered, but failed to pass, the country's first eugenic sterilization bill in 1897. That distinction went to the state of Indiana in

1907.[3] In the years and decades that followed, proponents of eugenic sterilization forged their rhetoric into public policy. In state after state, eugenicists saw their proposals voted into law and the constitutionality of their program ultimately upheld by the Supreme Court in the 1927 *Buck v. Bell* decision. At least 70,000 people are known to have been sterilized in the United States between 1907 and the close of World War II.[4]

Many proponents of compulsory sterilization denied that their work was motivated by race or class biases. Harry Laughlin, named "eugenics expert" of the U.S. Congress, maintained that his concern was not with "inferior nationalities, but inferior family stocks," a dubious claim, given his active lobbying efforts on behalf of the 1924 Johnson-Reed Act and his virulent attacks on immigrants. "If our standard of physical, mental and moral qualities for parenthood strike more heavily against one race than another," he explained, "then we should be willing to enforce laws which take on the appearance of racial discrimination but which indeed would not be such, because in every race, even the very lowest, there are some individuals who through natural merit could conform to our standards of admission."[5] Laughlin considered it to be "the business of the state" to encourage procreation of the fit and to deny marriage and reproduction to the "potential parents of hereditary degenerates." Compulsory sterilization laws should, Laughlin announced, begin slowly, targeting only the "most patent cases of hereditary degeneracy." In time, however, "as more experience is had by administrators and the courts, and the states become more confident of the use of sterilization in preventing degeneracy, the standard may be raised, and still greater portions of the degenerate sections of the population [will] be prevented from reproducing by sterilization."[6]

Laughlin knew whereof he spoke. By 1932, thirty states had sterilization statutes on the books. Though some were contested and rewritten, twenty-seven states kept their respective enactments on the books until the 1970s.[7] Illegal sterilization was a felony only in Utah. The majority of these statutes convened eugenics boards to review petitions brought by doctors at various correctional and mental institutions. In general, these boards only heard cases where sterilization was to be performed on prison inmates, patients at state and county hospitals, ex-convicts, and "at large" persons declared mentally ill by affiliated physicians. The panels did not review or record surgeries carried out by doctors in private without institutional ties. In most instances, the petitioning physi-

cian was himself a member of the tribunal. The boards were supposed
to hear from the potential recipients of the surgery, but only nine states
provided for free legal counsel in sterilization hearings. None of these
were in the South.[8]

Virginia was the third southern state, after North Carolina and Alabama, to pass compulsory sterilization legislation (1924). Its statute
was the first to be adjudicated by the Supreme Court. *Buck v. Bell* involved Carrie Buck, a young white woman, raised by foster parents in
Albemarle County. She was raped by her guardians' nephew and became
pregnant, at which point the family had her committed to a mental institution on the grounds of immorality, an accepted indicator of mental
impairment, just as her rape was an accepted indicator of immorality.
Carrie's biological mother, described in circuit court as a woman who
had led "a life of immorality, prostitution and untruthfulness" had also
been institutionalized.[9] Carrie became the test case for the Virginia Law
of 1924, which provided for the sterilization of inmates "inflicted with
hereditary forms of insanity that are recurrent, idiocy, imbecility, feeble-
mindedness or epilepsy."[10] In the now infamous 8-to-1 decision, the U.S.
Supreme Court upheld the state's right to sterilize, ruling that Buck "was
the probable potential parent of socially inadequate offspring." Justice
Oliver Wendell Holmes went so far as to suggest that since sterilization
operations "enable those who otherwise must be kept confined to be returned to the world, and thus open the asylum to others, the equality
aimed at all will be more nearly reached."[11] The ruling has never been
overturned by the Court.[12]

Buck v. Bell was part of a deliberate and determined effort to situate
women as the primary candidates for sterilization. According to data
culled by the Human Betterment Foundation, established in 1929 by sociologist and *Journal of Heredity* editor Paul Popenoe,[13] there was a radical shift in surgical sterilizations between 1928 and 1932. At the end of
1927, 53 percent of all people who had been legally sterilized in the United
States were male. During the next five years, however, this percentage
dropped to 33 percent. In institutions, where the procedure was even less
likely to be voluntary, the percentage of women and girls who were sterilization recipients climbed from 47 percent to 67 percent.[14] These numbers
represent the corporeal fulfillment of eugenic constructions of women
up to that point, both as hostages to their reproductive organs and as
de facto regulators of the national gene pool. This first categorization,

particularly in the late nineteenth century, left women and girls vulnerable to "orificial surgery," including clitoridectomies and hysterectomies for the prevention and treatment of epilepsy, melancholia, pulmonary tuberculosis, eczema, rheumatism, hip injuries, stammering, headaches, fainting, "violent temper," kidney disease, insanity, "idiocy," and, of course, masturbation and "hysteria."[15] As for the second categorization, Laughlin and his colleagues designated women as either eugenic or dysgenic breeders:

> It is contended without being confuted that the degenerate blood of the country is controlled largely by the number of degenerate women; that in the lower strains of humanity the degenerate women reproduce to full natural capacity; that if reproduction were made impossible for degenerate members of this sex, eugenical requirements of the situation would be met.[16]

Laughlin was advocating eugenic sterilization, but many in the emerging birth control movement saw a common cause. Given the timing of their emergence, it was perhaps inevitable that birth control and eugenics would find each other and become entangled in a complicated relationship—their respective adherents sometimes allies, sometimes at odds. Though eugenicists recoiled at the notion of "fit" couples using birth control, and Sanger deplored eugenicists' constant appeals to the "fit" to reproduce out of national or civic duty, both constituencies left unchallenged the notion of science as societal caretaker. This conviction served, in fact, as consensus builder among individuals across the political spectrum.[17]

In 1924, a contributor to Sanger's *Birth Control Review* proclaimed "birth control eugenics" to be a "divine command" from God.[18] This assertion was significant, not only for its familiar "God on our side" refrain, but also for its fusion of "birth control" and "eugenics" into a single, chilling term. Furthermore, as was true with other eugenics enterprises, the "reason" of science succored, rather than subverted, religious-based injunction.

Sanger, by this time, had already written on the "eugenic value of birth control." The rhetoric she employed, the political road she traveled, the homage she continues to be paid—go a long way toward explaining the lasting accord between eugenicists pursuing a "clean" national gene pool via scientific racism, and liberals, with their own brand of racism in tow, in search of a quick fix to deep social and economic inequities.

Margaret Sanger and the Eugenic Compact

Margaret Sanger's involvement in eugenics is no secret. Linda Gordon, Angela Davis, and others have challenged the idealized image of Sanger and problematized her motivations.[1] While many feminists and women's health care advocates were instrumental in exposing the fallout from Sanger's early alignment with eugenics, there were many who did not rise to the challenge—some out of fear of exposing birth control's political history to hostile lawmakers and anti-choice lobbyists, others out of deference to Sanger's perceived labors on behalf of gender equity, self-determination, and redress of economic and personal privation. This admiration is a very liberal impulse, for, as Goldberg has noted, liberals measure progress "by the extent to which institutional improvement serves to extend people's liberty, to open up or extend spaces for free expression."[2] On the surface, Sanger's contribution to "choice" and to the primacy of the individual does indeed imply such an opening up. However, in keeping with liberalism's Enlightenment roots, "choice" was never meant to be universal. "Subpersons," Charles Mills has written, were accorded "a different and inferior schedule of rights and liberties."[3]

Carole R. McCann and Ellen Chesler, whose books came out in 1994 and 1992 respectively, have provided similar explanations for Sanger's alliance with the eugenicists of her day. While recognizing Sanger's "problematic association with eugenics," McCann has argued in *Birth Control Politics in the United States, 1916–1945* that the relationship was advantageous for the early birth control movement, particularly in its dual struggle against church-based opposition and professional encroachment from the medical establishment. Eugenicists, wrote McCann, spoke of "genetic capacity, reproductivity, pregnancy wastage, expressed fertility, and birth rates." Such terminology armed Sanger and her followers with "a language with which to discuss contraception without vulgarity, and Sangerists used this sexually neutral language to legitimate birth control as just one more public health and welfare issue."[4] Chesler, author of *Woman of Valor: Margaret Sanger and the Birth Control Movement in America,* likewise concluded, "Eugenicists were largely responsible for having introduced explicitly sexual topics into the boundaries of acceptable scientific discourse."[5] This is certainly true, but the cost of eugenicists' assumed authority in the realm of sexuality was tremendous. McCann, citing the usefulness of "evolutionary social thought," has suggested that Sanger further sought out allegiances with eugenicists as a bulwark against the Catholic Church's opposition to birth control.[6] Chesler wrote:

> She deliberately courted the power of eugenically inclined academics and scientists to blunt the attacks of religious conservatives against her.... there is no denying she allowed herself to be caught up in the eugenic zeal of the day and occasionally used language open to far less laudable interpretations.[7]

Here, Sanger is at once a shrewd recruiter of eugenics movers and shakers and an unwitting reiterater of their propaganda. The assertion that she was swept up in the eugenic fervor of the day (elsewhere Chesler refers to eugenics as "a popular craze")[8] implies that Sanger lacked the intellectual sophistication and fortitude to resist eugenics doctrine, that she was overly impressionable and incapable of orchestrating any other response to reactionary attacks on birth control dissemination.

McCann, too, has insisted that Sanger used eugenicists to offset physicians' skepticism, stating that she exploited eugenicists' legitimacy to counteract "charges of amateurism" leveled at birth controllists by doctors. "Eugenicists were powerful allies against the medical hegemony," she writes. "The statistical expertise and scientific legitimacy that eugenicists brought to the movement's claims proved to be valuable counterweights to the medical profession's derision."[9] But the "statistical expertise" Sanger borrowed from eugenicists included the flawed, and ultimately discredited, findings of IQ tests conducted by Robert Yerkes and Lewis Terman, both of whom she cited in her 1922 book, *The Pivot of Civilization.* "The Binet-Simon intelligence tests ... present positive statistical data concerning the mental equipment of the type of children brought into the world under influence of indiscriminate fecundity."[10]

According to McCann, Sanger was also motivated by fear of usurpation of birth control by the medical profession and the need to fend off professional encroachment. But contrary to these assertions, Sanger supported legislation that would limit the prescription and dissemination of contraceptives to doctors in exchange for physicians' public support of birth control. Others opposed the move, citing the barriers such an arrangement would pose to women without access to clinics or medical care.[11] For Sanger, the payoff came in 1937, when the American Medical Association finally pledged its support of birth control and affirmed "the right of a woman to contraceptive assistance within the privacy of a doctor-patient relationship."[12] To get there, she generated some of the most menacing class- and disability-based eugenics doctrine of the last one hundred years, and endorsed the race-based surgical experimentation unleashed by her colleagues.

In April 1925, less than a year after passage of the restrictive 1924 National Origins Act, Sanger welcomed attendees of the Sixth International Neo-Malthusian and Birth Control Conference with these words:

> While the United States shuts her gates to foreigners, and is less hospitable than other countries in welcoming visitors to this land, no attempt whatever is made to discourage the rapid multiplication of undesirable aliens—and *natives*—within our own borders. On the contrary: the Government of the United States deliberately encourages and even makes necessary by its laws the breeding—with a breakneck rapidity— of idiots, defectives, diseased, feeble-minded and criminal classes.[13]

Sanger's own words controvert McCann's assertion that the Planned Parenthood founder accepted eugenicists' claims on the magnitude (and therefore the "fact") of "racial deterioration," but did not concur with their pronouncements on its root causes. "Sanger's articulations were not biologically determinist. She located the causes of racial decay in economic environmentalism and conventional sexuality."[14] Even if it were true that Sanger never employed biologically determinist rhetoric, there is little doubt as to who bore the brunt of the "economic environmentalism" she pursued. As it was, Sanger's belief in birth control as a cure-all for economic inequity and other ills led her to declare in 1922, "In recognizing the great need of education, we have failed to recognize the greater need of *inborn health and character.*"[15]

Sanger attributed everything, from child labor to world war, to unchecked breeding.[16] Eugenic considerations and "solutions" overtook all other remedials. She quoted Havelock Ellis, "If it were necessary to choose between the task of getting children educated and getting them well born and healthy... it would be better to abandon education."[17] A belief in the primacy of "dysgenesis," in the debilitation of society and race through improper breeding, and a desire for government intervention were defining characteristics of eugenics. Sanger's list of the hereditary offenders (as well as the language she used to describe them) was identical with that of Harry Laughlin and other eugenic stars of her day:

> Modern studies indicate that insanity, epilepsy, criminality, prostitution, pauperism, and mental defect are all organically bound up together and that the least intelligent and thoroughly degenerate classes in every community are the most prolific. Feeble-mindedness in one generation becomes pauperism or insanity in the next. There is truly... a feeble-minded peril to future generations—unless the feeble-minded are prevented from reproducing their kind. To meet this emergency is the immediate and peremptory duty of every state and of all communities.[18]

All this is not to say that Sanger identified as a eugenicist. She alternately marketed birth control as the ultimate in eugenic fulfillment and distanced her goals from those of both her eugenicist backers and detractors. Her chief objection to eugenics was not its racism, its ill-will toward the disabled or its pathologization of the poor. It was "positive eugenics," which encouraged the "fit" to have large families in service to race and nation, that she decried.[19] However, she considered "negative eugenics," the prevention of procreation among the "unfit," tremendously valuable.[20] "Those least fit to carry on the race are increasing most rapidly," Sanger wrote.[21]

> [T]he example of the inferior classes, the fertility of the feeble-minded, the mentally defective, the poverty-stricken classes, should not be held up for emulation to the mentally and physically fit though less fertile parents of the educated and well-to-do classes. On the contrary, the most urgent problem today is how to limit and discourage the overfertility of the mentally and physically defective.[22]

Though Sanger objected to "cradle competition," she was an inconstant opponent. Her 1921 article entitled "The Eugenic Value of Birth Control Propaganda" declared the "differential birthrate" between the "fit" and the "unfit" to be "the greatest present menace to civilization."[23] There is evidence that on at least one occasion she advocated financial incentives not only for the "unfit" to refrain from having children, but also for the "fit" to breed more. In 1938, she told a reporter that those concerned that a decrease in the U.S. birthrate would lead to Americans being out-bred, "overrun and wiped out should and could increase that native birth rate by fellowships allowing young Americans of good stock to borrow on such a foundation sufficient to give them a start in marriage and to have a family."[24]

Despite such proposals, Sanger and the birth control movement came under fire from some eugenicists in the 1910s and 1920s for not adequately addressing the threat of "race suicide."[25] Paul Popenoe believed that birth control, in concert with the "differential birth rate," would bring "racial deterioration." Only when birth controllists supported more children from the "fit" would eugenicists support the contraceptive cause.[26] Mc-Cann notes that Charles Davenport (zoologist, geneticist, and vice chairman of the Eugenics Committee of the United States) declined an invitation by the American Birth Control League (ABCL) to preside at their 1925 conference and that Madison Grant (world renowned for his book *The Passing of the Great Race* and instrumental in framing U.S. immigration legislation) opposed Sanger's efforts altogether.[27] In 1930, Grant received a letter from Hans F. K. Gunther, anthropologist and author of *The Racial Elements of European History*, inquiring about the stance on immigration and the preservation of the Nordic race taken by North American women's organizations. Did they, he queried, grasp the grave danger of family limitation? "Birth control seems, indeed, not yet to have differentiated selectively in the matter of births and contributes thus further to the impoverishment of the Nordic race."[28] Other eugeni-cists voiced similar fears in the pages of the ABCL's own journal, the *Birth Control Review (BCR)*. Albert E. Wiggam echoed the concerns of many *BCR* contributors when he wrote:

> The question of our whole racial future is bound up in this: Do the abler, more energetic sections of the population love children the more and are they more willing to undertake the burdens of rearing them? It is plain that, so far, Birth Control has been utilized almost entirely by these more competent sections of the community. It is their very pru-dence, their foresight, their far-sighted love for their children for whom they wish to make a better place in the world even than they have had themselves, and who wish to make the world a better place for these

children to live in—it is these *noble moral traits that lead them to utilize Birth Control* and to limit their children to such numbers as they can properly care for, educate and give a start in the world. But Birth Control has been withheld from those who are less gifted with the abundant natural endowments....

... When Birth Control is not universal it acts to decrease intelligence and character and increase incompetence and poverty.[29]

Wiggam's words carried weight. He was, according to Mehler, one of the best-known popular science writers of his time. A member of the executive committee of the Eugenics Research Association, Wiggam also sat on both the board and the advisory council of the American Eugenics Society (as well as its Committee on Popular Education, responsible for organizing the "fitter family" contests at county fairs). The *New York Post* and various other newspapers began publishing his syndicated column in 1935.[30] Unlike Madison Grant, Wiggam's reservations did not compel him to oppose birth control. On the contrary, he considered it "our greatest moral privilege, our most religious duty and our loftiest patriotic obligation to place this great evolutionary force at the service of all humanity." Wiggam stressed birth control's complement to negative eugenics, underscoring the need to universalize birth control, but maintained his commitment to positive eugenics:

If we extend it [birth control] to the able, far-sighted and competent and withhold it from the witless, shiftless, and incompetent, these lower classes will soon outbreed the upper and our civilization will tumble into ruins. If we make it universal, then it becomes the most dramatic race between the intelligent and the stupid that the world has ever staged. Instead of a war of deaths, it becomes a war of births, a benign war... but nevertheless one in which the whole destiny of the race is at stake.

This is what eugenics really means, the birth by natural processes and by the determination of man's highest emotions, of better, stronger, happier creatures than those who now people the world. Birth Control... will insure, as we look into the "long realities" of the future, that the good, the virtuous and the intelligent will outbreed the bad, the foolish and the incompetent, and that they and they alone shall eventually inherit the earth.[31]

Sanger herself responded to critics within the eugenicist ranks by constantly reassuring them that eugenics and birth control were not only compatible, but inseverable. She informed *BCR* readers that "the campaign for Birth Control is not merely of eugenic value, but it is practically identical in ideals with the final aims of Eugenics."[32] She went on to draw parallels between the two movements, urging her eugenicist brethren to recall their own struggle for legitimacy:

> Eugenicists may remember that not many years ago this program for race regeneration was subjected to the cruel ridicule of stupidity and ignorance. Today Eugenics is suggested by the most diverse minds as the most adequate and thorough avenue to the solution of racial, political and social problems. The most intransigent and daring teachers and scientists have lent their support to this great biological interpretation of the human race. The war has emphasized its necessity.
>
> The doctrine of Birth Control is now passing through the stage of ridicule, prejudice and misunderstanding. A few years ago this new weapon of civilization and freedom was condemned as immoral, destructive, obscene. Gradually the criticisms are lessening—understanding is taking the place of misunderstanding. The eugenic and civilization value of Birth Control is becoming apparent to the enlightened and the intelligent.[33]

Sanger appealed to eugenicists' desire to disseminate their gospel. The wide-scale demand for contraceptive information was a sign that "the masses themselves today possess the divine spark of regeneration." She therefore called upon "the courageous and the enlightened to . . . kindle that spark, to direct a thorough education in Eugenics based upon this intense interest. Birth Control propaganda is thus the entering wedge for the Eugenic educator."[34] Sanger reiterated this assertion in *The Pivot of Civilization,* stating that birth control was "really the greatest and most truly eugenic method, and its adoption as part of the program of Eugenics would immediately give a concrete and realistic power to that science." Moreover, she wrote, the most "clear thinking and far seeing of the Eugenicists" (and here she named Havelock Ellis, William Bateson, and Leonard Darwin) embraced birth control as integral to "racial health."[35] In short, if eugenics was the theory, birth control would be the practice; the *BCR*, during and after Sanger's tenure as editor, would be the forum.

The *Birth Control Review* became the "Official Organ of the American Birth Control League" (though it predated the founding of the ABCL). Betsy Hartmann has noted that by 1917, following a split with Sanger over control of the journal, radicals left the birth control cause and "eugenicists came to fill their shoes."[36] The rift, according to Gordon, can be traced to Sanger's single-issue approach and her social and sexual conservatism. This is not to imply that all radicals successfully avoided the eugenics trap. Some contested eugenic platforms, others did not. Additionally, many leftists, both then and now, have objected to the casting of Margaret Sanger as *the* pioneer, pointing instead to the leadership and activism of free love advocates, Wobblies, and others who articulated a class analysis that Sanger lacked. As Gordon so succinctly put it, "If any leader could have drawn Socialist party and feminist support together behind birth control, Sanger was not the one."[37] Consequently, the *BCR* roster of contributors included such eugenic luminaries as C. C. Little and Harry H. Laughlin.[38] Little was president of the American Eugenics Society, director of the ABCL, president of the International Neo-Malthusian League, research associate in genetics and cancer research and assistant dean at Harvard, and later president of the University of Michigan. In addition to his responsibilities as "eugenics expert" for the U.S. Congress, Laughlin was director of the Eugenics Record Office and coeditor (with Charles Davenport) of the *Eugenical News.*[39] The *BCR* published favorable reviews of Leonard Darwin's *Eugenic Reform* and *What Is Eugenics,* as well as Havelock Ellis's positive report on ABCL board member Lothrop Stoddard's *The Rising Tide of Color against White World-Supremacy.*[40] Reprints from eugenics gatherings, such as the annual neo-Malthusian conference, also appeared. Sanger and the *Birth Control Review* were not one and the same, of course, and she no doubt published some articles with which she disagreed. However, her decision to print not only eugenicist, but also explicitly racist material implies that she did not consider such espousals beyond the pale of acceptable social policy discourse. Given her own eugenically informed treatises, it is not surprising that Sanger accepted these works as well within the realm of the birth control debate.

Linda Gordon has contested readings of Sanger as a merely misdirected, slightly insensitive, largely reluctant recruiter of eugenicist support. She reports that in 1925 Sanger began to record the nationality, religion, heredity, occupation, and trade union affiliation of her clinic patients at the behest of eugenicist supporters. In 1929, Sanger received a request from Edward East of Harvard:

> I suppose it would be a delicate matter, but it would be a very interesting thing, from the standpoint of science, if your clinical records . . . show the amount of racial intermixture in the patient. Perhaps, without

embarrassing questions, it would be possible to make a judgment as to whether the person was more or less black, mulatto, quadroon, etc.[41]

Such thinking lay at the heart of the eugenic belief system: a preoccupation with "blood quantum" and white panic over interracial unions, and the need to correlate human behavior and reproduction with race. Sanger responded that, as "already colored patients coming to our clinic have been willing to talk," there would be little difficulty in obtaining the information East sought. McCann has noted Sanger's race and national origin record keeping and its repercussions:

> Through this practice, the clinic could become a source of data on the differential in fertility rates between the races. With this revamping of the clinic's record-keeping system, eugenically framed racial thinking was brought into the center of America's contraceptive delivery system.[42]

This kind of activity, conducted under Sanger's command, has posed a significant challenge to those seeking to exonerate her from charges of racism. Sanger "was never herself a racist," Chesler has asserted, "but she lived in a profoundly bigoted society, and her failure to repudiate prejudice unequivocally—especially when it was manifest among proponents of her cause—has haunted her ever since."[43] This type of disclaimer suggests an ideological monolith and erases the history of dissent, including the competing birth control agendas offered by free love advocates, the Black press, and various African-American organizations. There is also an inconsistency in such attempts to rescue Sanger from her own racism. She cannot be simultaneously cast as both the cagey, realpolitik courter of eugenicist muscle and the unsophisticated, uncritical consumer of scientific racism.

Sanger cannot elude the designation of eugenicist simply because there is no recorded evidence of her uttering blunt statements linking specific ethnic groups to economic, social, or intellectual shortcomings. Several of her campaigns, both in the United States and abroad, were intentionally race-specific. Furthermore, even a cursory glance at who constituted the urban and rural poor who fell prey to the convergence of birth control and scientific racism reveals that her work was clearly informed by ethnic/racial preoccupations having more to do with eugenics practices than reproductive rights.

Liberating Sanger from the category of eugenicist would require the total elimination of class as an axis of inquiry. But class position was always uppermost in the minds of eugenicists, who forged a threefold link between economics and fertility: First, they argued that pauperism was hereditary. Second, they preached a neo-Malthusianism that blamed individual and global strife on the "excessive" breeding of the poor—a paradigm that still attracts adherents from various points on the political continuum. Third, eugenicists constantly cited the fiscal drain posed by the fecund poor in society's midst, emphasizing the personal and state monies "wasted" on caring for the disabled. In 1925, Sanger ran a paper on the "differential birth rate" in the *BCR*. First delivered by Raymond Pearl at the Sixth International Neo-Malthusian and Birth Control Conference, its appeal was its explicit, statistically bolstered, cognitive link between reproduction and vocation/socioeconomic standing:

> [R]ecent figures on the relative fertility of different occupations in England and Wales show that the number of children per 100 married couples, when the age of the occupied husbands is below 55 years, are as follows: for teachers 95, non-conformist ministers 96, Church of England clergy 101, physicians and surgeons 103, authors and editors 104, policemen 153, postmen 159, carmen 207, dock laborers 231, barmen 234, miners 258, and general laborers 438.[44]

Most efforts to conjoin class and heredity relied on an equation of the poor (be they employed or out of work) and the "unfit" that was absolute. The alleged disabilities of the impoverished were constantly invoked in economic rationales for eugenic policies.[45] Over the years, the *BCR* published a steady stream of articles that linked the poor and the disabled if not in body, then in public expenditure. Pearl continued:

> [I]t is not only desirable in the eugenic interest of the races to cut down, indeed completely extinguish the high birth rate of the unfit and defective portions of mankind, but it is also equally desirable because of the menacing pressuring of world population, to reduce the birth rate of the poor, even though they be in every way biologically sound and fit.[46]

"The Purpose of Eugenics," penned by John C. Duvall, appeared in the *BCR* in December 1924. Duvall began by declaring that the "human flood" could be "arrested at its source" and sickness, crime, poverty, and "hereditary pauperism" wiped out through the application of eugenic principles (namely a commitment to eliminate the "degenerate classes by making their propagation impossible"). The author's emphasis then shifted from alleviation of human misery to the conservation of untold dollars in custodial care and other public spending before returning to the issue of "racial betterment":

> The enormous amount of expense involved at present in caring for and controlling our hereditary misfits could thus be saved and utilized to encourage the most desirable elements to carry out a plan of positive eugenics which primarily involves encouraging the mentally, morally, and physically fit to marry and reproduce, to the ultimate end of racial improvement if not perfection.[47]

In 1925, an unattributed article appeared in the July issue of the *BCR*. "The Story of a Subsidized Family, or How to Populate the Earth with the Unfit" brought together several common eugenic themes. The family, a composite of stock eugenic villains, was described as "subnormal" (disabled), dependent on "many agencies" (poor), and headed by an immigrant couple (alien). The author argued against eugenically informed marriage bans (they would only lead to out-of-wedlock births among the "unfit") and custodial care (too expensive and "impracticable"). The solution lay in "[c]utting off the stream of life at its source by the practice of Birth Control—or, where the particular case calls for it, by eugenic sterilization."[48]

The disabled, to Sanger's mind and the minds of the *BCR* staff, were becoming increasingly costly. Letters from poor women—some disabled, others sick—were grouped together in the *BCR* under the distinctly Christian banner, "Unprofitable Children: Are These Bodies Fit Temples for Immortal Souls?"

> Every year millions of dollars are collected in taxes and spent on the maintenance of the defective, the feeble-minded, the insane and the criminals. This means that every man who is able to earn his own living has not merely to support himself and his family, but also to help carry these expensive members of society.[49]

Sanger told neo-Malthusian conference goers in 1925 that "[b]illions of dollars are expended by our state and federal governments and by private charities and philanthropies for the care, maintenance, and *perpetuation* of these classes."[50]

The American public is taxed—and heavily taxed—to maintain an increasing race of morons which threatens the very foundations of our civilization. More than *one-quarter* of the total incomes of our States is spent upon the maintenance of asylums, prisons, and other institutions for the care of the defective, the diseased and the delinquent.[51]

She called for economically coerced sterilization to rectify the situation:

If the millions upon millions of dollars which are now being expended in the care and maintenance of those who in all kindness should never have been brought into this world were converted to a system of bonuses to unfit parents, *paying* them to refrain from further parenthood, and continuing to pay them while they controlled their procreative faculties, this would not only be a profitable investment, but the salvation of American civilization. If we could, by such a system of awards or bribes or whatever you choose to call it, discourage the reproduction of the obviously unfit, we should be lightening the economic and social burden now hindering the progress of the fit, and taking the first sensible step toward the solution of one of the most menacing problems of the American democracy. It is not too late to begin.[52]

The onset of the Depression only cemented anxiety over government expenditures on the poor and disabled with the call for economically mandated birth control. Gordon wrote that by the mid-1930s "relief babies had become a public scandal." In an accusation that has not faded with the years, relief recipients were charged with having more children for the sole purpose of receiving increased benefits (an increase of as little as $1.15 a week). One insurance company claimed that the Federal Emergency Relief Administration was causing an increase in the birthrate. "Taxpayers' money," Gordon reported, "was not only being used to support the poor, but to produce more of them—this was an implicit charge being made in a variety of publications, from *Time* to the *Birth Control Review*."[53] Certainly, eugenics in this era was in part a response to New Dealism. Sanger faulted the Roosevelt administration for refusing to recognize that "as long as the procreative instinct is allowed to run reckless riot through our social structures . . . grandiose schemes for security may eventually turn into subsidies for the perpetuation of the irresponsible classes of society."[54] In this, she found an ideological kinsman in Clarence Gamble, graduate of Harvard Medical School, president of the Pennsylvania Birth Control Federation (PBCF), and heir to the

Gamble (as in Proctor & Gamble and Ivory Soap) fortune.[55] Theirs would be a long association with disastrous consequences for poor women worldwide.

Like Sanger, Clarence Gamble exploited the desperation of the Depression to promote medical intervention as the solution to economic privation. Fund-raising in 1935, Gamble promised prospective donors that "each dollar will multiply itself by hundreds in lessening the load of debt that the next generation must pay. Our children will have to care for those who are now being born and raised in the crippling atmosphere of public relief." The previous year, he had calculated that sixty-four thousand babies were born to Pennsylvania families on relief at a cost, before their first birthday, of $10 million to the state's taxpayers. These were babies "who could hardly be said to have been wanted by anyone" (unlike Gamble's five children). According to a pamphlet put out by the PBCF, the parents of these sixty-four thousand babies could have spared the taxpayers expense had they been given "$10 worth of birth control advice."[56]

Gamble was resolute in his belief that wide-scale birth control, including sterilization, would offset poverty.[57] Over the years, he was active in India, Japan, South Africa, Pakistan, Egypt, Israel, Ceylon (now Sri Lanka), Puerto Rico, Hawaii (pre-statehood), and the United States. In 1936, as part of a three-year study supported by the American Birth Control League and gynecologist Robert Latou Dickinson, Gamble dispatched a social worker to Logan County in West Virginia. Chemical jelly was dispensed to 1,345 Appalachian women, courtesy of Ortho Pharmaceutical.[58] What is significant here is the willingness of Gamble and the ABCL to use poor women as guinea pigs; the availability of these women, who were easily identifiable through their enrollment in a public health project (sponsored by the American Friends Service Committee); and an underlying belief that contraceptive technology would deliver one of the most underserved areas in the country from economic devastation.

The following year, Gamble wrote to Sanger from Pasadena. He told her of Dr. Nadina Kavinoky, who had arranged for her students to instruct Mexican-American relief recipients in contraception at a time when restricted county funding placed limitations on birth control clinics. This was not, however, so much an issue of poor women's access to contraception as it was of Gamble's access to poor women. Like their Logan County predecessors, women on relief in Los Angeles had been targeted by Gamble for experimentation. He informed Sanger that Kavinoky was interested in testing the "foam-powder-sponge outfit" in Mexican/Mexican-American neighborhoods of the city. If a nurse could be given a two- or three-week supply (a sample of which Gamble had shown the doctor), Kavinoky would supervise her work and check for irritation during follow-ups at the county health clinic. From his perspective, this was an ideal opportunity to study the device closely and publicize its success throughout the southern part of the state.[59] Gamble contributed $150, plus supplies, to see the plan realized. Research subjects were available because of their reliance on public health care. Class position and a racial identity that cast them as less than full citizens rendered them approachable and expendable. By this time, and on the same basis, Gamble had singled out another demographic group: the women of Puerto Rico.

As early as 1928, eugenicists inside the birth control movement were lobbying for population control in Puerto Rico. The *BCR* 's "Reason IX—The Preservation of Civilization" of "Ten Good Reasons for Birth Control" read:

> The "family" and the "home" do not exist among the poorer classes of Porto [sic] Ricans in the sense in which these terms are used ordinarily. The degree of poverty which prevents a family from having more than one small room, and that virtually without furniture—with perhaps a hammock or a poor bed for the man, no chairs and no other conveniences—makes of the "home" only a room where the family sleeps in mass on the floor at night. Privacy does not exist. Life is lived on the street, and only a people of unusual kindness and clean instincts could make of the situation one in which sordidness was not the rule.[60]

BCR editors were not talking about birth control as a freely chosen option to increase women's sovereignty, but rather as a population control device to be imposed externally, in this instance by a colonial authority. As Betsy Hartmann has noted, U.S. officials casting an eye on island poverty overlooked their own culpability in undermining the Puerto Rican economy. The devaluation of the peso after the 1898 U.S. invasion and the eviction of ranchers and farmers by American sugar interests left 70 percent of the population landless by 1925. Two percent of the population owned 80 percent of the land,[61] but Puerto Rico's problems were laid at the door of "population pressure." Herbert Hoover appointees at the American Child Health Association reckoned that there were 174,650 children too many in Puerto Rico.[62] After the Puerto Rican Emergency Relief Administration bowed to pressure from the Catholic Church and cut its own contraceptive program in 1936, Franklin Roosevelt's Puerto Rican Reconstruction Administration established what were supposed to be birth control clinics.[63] These sites were staffed by Gamble's own fieldworkers, and he kept Sanger abreast of their progress, writing in early 1938, for example, of the fifteen clinics that opened in the past year. She responded a few days later, "Certainly Puerto Rico is a feather in your cap for what you have done there."[64]

Gamble used the clinics as recruitment centers for sterilization and testing grounds for pharmaceutical companies. In 1976, the Department of Health, Education, and Welfare (HEW) reported that 37.4 percent of women of childbearing age in Puerto Rico had been sterilized. Clarence Gamble had helped to secure that outcome. In 1939, he began financing trips to New York for Puerto Rican doctors to learn the latest in sterilization techniques.[65] In 1946, he received the following report from one of his fieldworkers:

> The policy of the hospital is to carry out sterilizations if the woman has three living children. In his [the acting director's] private practice two are enough. . . . It is the unofficial policy of the hospital not to admit (uncomplicated) multiparae [women who have given birth at least two times] if they do not submit to sterilization.[66]

Compulsory sterilization was but one component of Gamble's Puerto Rico plan. As with Ortho's involvement in Logan County, Gamble was able to secure donations from pharmaceutical companies who bid against one another for access to Gamble's clinics.[67] What followed is perhaps one of the most notorious abuses of medical power in birth control technology's history: the trial run, among poor women in Puerto Rico, of what would become known as "the pill," and was, at the time, a highly experimental drug administered without controlled dosage.

Gamble became involved in testing oral progesterone in the 1950s. In 1954, Sanger wrote to Gamble telling him of Dr. Gregory Pincus's trip to Puerto Rico. Two years later, Pincus, along with Dr. John Rock, would publish preliminary findings on the pill. In her 1954 letter, however, Sanger notified Gamble that Pincus would first administer injected progesterone and was preparing to "make a test of one hundred to two hundred women."[68] Gamble did not work under the auspices of Planned Parenthood, having, he wrote, been forced out of the organization; however, he remained on good terms with its president, Sanger, throughout the Puerto Rico venture.[69] He kept Sanger informed of his fund-raising successes and his expansion of oral progesterone trials, writing in 1957 of his plans to launch a new center in another part of the island with another physician on hand to broaden and verify the results.[70]

Sanger wrote that she was glad Gamble was on his way to Puerto Rico and solicited his opinion of Pincus's work. Gamble reported that tests of Pincus's birth control pill were going well. Thus far, no method failures had been reported among the approximately one hundred women who had been taking it for the last seven months.[71] There were complaints of headaches, dizziness, and nausea among participants, but Gamble dismissed these as "coincidences."[72] Undaunted by the drug's negative impact on women's systems, Gamble began distribution of the pill in a new location, under the supervision of a doctor at the Congregational Mission Hospital of Humacao, which he described to Sanger as a dense slum.[73]

Over the next few years, Gamble continued to send Sanger updates on the "rural work" in Puerto Rico. Along with his dispatches to Sanger, Gamble also kept major donors apprised on the efforts of visiting nurses and social workers in rural areas and the housing projects of San Juan, referring to these fieldworkers as "your army of occupation."[74] There was more truth than Gamble would acknowledge in this quip. It was, after all, U.S. colonialism that made Puerto Rico a viable locale for pharmaceutical experimentation.

While there can be little doubt that Gamble's endeavors in Puerto Rico constituted population control—coerced, externally imposed, eugenically informed, with only the most superficial trappings of real choice or autonomy—there has been a general reluctance in literature, both descriptive and critical of twentieth-century contraceptive history, to name similar campaigns within the U.S. borders as population control. In 1939, at the outset of his involvement in Puerto Rico, Gamble joined Sanger and the American Birth Control Federation (ABCF, successor to the ABCL) in launching the "Negro Project." Gamble wrote the project proposal, which argued that southern poverty could be remedied by lowering the birthrate, particularly among Blacks, who were pronounced "careless" and "disastrous" breeders.[75] Here, eugenic efforts (used to justify first slavery, then disfranchisement and violence) rested on constructions of unbridled (and therefore dysgenic) sexuality and intellectual incompetence. However, as Jessie Rodrique has written, there was a long and active resistance to these endeavors, evident in the Black press. By the time Gamble set his sights on southern Blacks, the *San Francisco Spokesman* had urged African-American clubwomen to become actively involved in the antisterilization movement (1934).[76] In 1935, the *Pittsburgh Courier* responded to Georgia's passage of a bill allowing sterilization of "feeble-minded and criminal incurables" warning, "If this law is found effective in Georgia as another club over the heads of black serfs, the other Southern States will be quick to follow."[77]

> Almost half of Georgia's population is Negro. A large percent is on relief or unemployed through no fault of its own. People who are hungry and homeless are very likely to be criminally inclined.... All such will be candidates for the sterilization operation, penalized for being victims of as vicious a system of economic exploitation as can be found.[78]

W. E. B. Du Bois and the rest of the editorial staff consistently alerted readers of the hazards of "so-called reforms," voted into law in state after state. "Just now, the thing we want to watch is the so-called eugenic 'sterilization.'... The pretense is that this procedure is based upon science; but it is not.... The burden of this crime will, of course, fall upon colored people, and it behooves us to watch the law and the courts and stop the spread of the habit."[79]

This vigilance was precisely what concerned Gamble. In a letter to Florence Rose, Sanger's secretary, Gamble wrote that the Black communities of North and South Carolina might view the "Negro Project" as a design for their extermination. To diffuse such suspicion, Gamble suggested the projects should *appear* to be run by Black people. In the margin, someone, presumably Sanger or Rose, scrawled, "Incorporated in present plan." Gamble called for a preacher to lead a "revival," and Sanger was of the same mind. In a letter dated the same day as Gamble's, Sanger notified him that she wanted "an up and doing modern minister, colored, and an up and doing modern colored medical man" to come to New York and train at her clinic and at the ABCF "until they are oozing with birth control *as well as population*" and then go south and "preach and preach and preach."[80]

Gamble would leave his mark on the South, but not for mere birth control advocacy. During the 1940s, the doctor established over twenty sterilization clinics throughout the South and the Midwest. Like Puerto Rico, North Carolina endured Gamble's eugenic calling. In 1945, he financed intelligence testing for Orange County children. According to Philip Reilly, Gamble used the findings, which suggested that 3 percent of the population was "mentally defective," to lobby for an extension of the state's sterilization program.[81] Later that year, the North Carolina eugenics board approved a proposal to hire (with Gamble's money) a sociologist charged with determining who in Orange County should be sterilized. In 1947 he wrote to Sanger of their efforts to boost sterilization among the state's "feeble-minded" and "insane." North Carolina, Gamble explained, was chosen for the experiment in part for its low percentage of Catholic residents. He reported that progress was slow, but good.[82] In truth, there was nothing slow about Gamble's progress. North Carolina, with 249 sterilizations performed in 1949, was second only to California.[83]

Meanwhile, Sanger became increasingly vocal in her embrace of sterilization—a position rooted in a fusion of neo-Malthusian certainty of overpopulation and a belief in the fiscal toll exacted by the poor. According to Sanger, "the inflationary number of human beings" leads to "the widespread devaluation of human lives."[84] In 1950, upon receiving the "Award in Planned Parenthood of the Albert and Mary Lasker Foundation" (Mary Lasker also funded the "Negro Project"), Sanger told those gathered at the Thirtieth Annual Meeting Luncheon of the PPFA:

> I do not intend to torture your ears with statistics, but it is well to repeat one point, one fact:... the human race now numbers 2,344,727,000 living creatures who must be fed. Let us not forget that these billions, millions, thousands of people are increasing, expanding, exploding at a terrific rate every year. Africa, Asia, South America are made up of more than a billion human beings, miserable, poor, illiterate labour slaves, whether they are called that or not; a billion hungry men and women always in the famine zone yet reproducing themselves in blind struggle for survival and perpetuation.[85]

Sanger's totalizing image of the southern and eastern hemispheres was certainly not unique. Nor was her finger pointing at the government's "maudlin, benevolent extravagances," echoing exactly early twentieth-century eugenicists' chastisement of the "maudlin sentimentalism" that allowed immigration of "inferior stocks." Sanger faulted the United States for "throwing billions of dollars away" on these "teeming continents."[86]

Domestically, she renewed her calls for the sterilization of the disabled (once again, employing the standard eugenic trope that the most imperiled and abused pose the greatest danger to the rest of the population):

> It should be the duty of the government not only to supply the facilities which will enable these unfortunates to be sterilized, but to protect them afterwards. There should be a provision for a pension of subsidy for those couples whose procreation would be dangerous to the community. If they are denied the normal family, not only for their own benefit but for that of society and above all their children, society could well afford to see that they are well-protected. If anyone thinks that this is a bribe to encourage sterilization of the merely poor, he mistakes human nature.[87]

Shortly after the Lasker award luncheon, Gamble wrote to Sanger commending her on her speech and asking if he might include some quotations in a pamphlet he was preparing on sterilization. Sanger, having already sent him a copy, responded in the affirmative, and underscored her central premise:

> The point I wanted to make . . . was that it would be a good investment for the US Government to pay a pension for life to couples with transmissible diseases who would submit to sterilization. In other words, our Government to pay dysgenic populations not to have children while Russia, Hitler's Germany and Mussolini's Italy, war mongers all were paying people regardless of quality to have children. Our plan would not only save the suffering of millions within a few years of parents as well as children, it would be a far cheaper investment than keeping these unfortunates in institutions.[88]

Sanger did not, however, want to limit sterilization to the "physically or mentally inadequate." The "couple who had as many children as they believe they can rear" and "the woman who as a girl was taught so sternly that any touching of the genital organs is wicked that she never overcomes her aversion" were also candidates. "For all these people, sterilization is the common sense procedure. They are entitled to it as surely as any person who is mentally or physically inadequate."[89]

Sanger's endorsement of disability- and class-determined sterilization continued with her involvement with the Human Betterment Federation (HBF).[90] In December 1950, Gamble asked Sanger to record an address promoting sterilization that might be played at the upcoming HBF meeting the following February. He praised her "pension plan" and asked her to place special emphasis on protecting those unlucky children who might be born to "feeble-minded" parents. She might, he suggested, also mention the value of "therapeutic" sterilization for sufferers of heart or kidney disease.[91] Sanger agreed and her address was played at the gathering in Des Moines:

> It is a cruel and wasteful system in our National Economy to keep on generation after generation retaining in institutions insane, feeble-minded, mental defectives. . . . Institutions are overcrowded and [there are] waiting lists of some who seriously need the care but are excluded for want of room. There are thousands of inmates in these institutions throughout the U.S.A. whose energies and abilities could be used outside the institutions were they rendered infertile by sterilization.[92]

Sterilization, Sanger told HBF conventioneers, should be prescribed not only for the "mentally inadequate," but also

> should be used when inheritable defects of heart or damage to heart, lungs, or kidneys make motherhood dangerous. There is also an inheritable blindness and other diseases when parents-to-be should be informed of the dangers which may be passed on to innocent children. Sterilization should be advised and recommended by medical advisors. Society has the right to expect our Public Health officials to protect it from transmissible diseases just as it protects us today from contagious diseases like diphtheria, measles, small pox, etc.[93]

The HBF continued to pursue Sanger in the years that followed. Her advice was solicited by both Irene Headley Armes and Ruth Proskauer Smith during their respective terms as executive director. Smith wrote to Sanger in 1960 asking her opinion on sterilization by way of "an overdose of x-ray treatment," having been told by a doctor that "this would be very practicable in India." She continued, "I do hope you will take a moment to look at the enclosed material. We have long wanted to enlist your interest in Human Betterment, which is working more and more closely with Planned Parenthood."[94] In 1961, Sanger responded to another request from Smith, writing that she would be "Delighted and honored" to be listed as a sponsor for Human Betterment.[95] Thus, her name appeared on the letterhead when Dr. H. Curtis Wood Jr., president of the board of directors, sent out his fund-raising letter the following February. He told would-be donors that "the world gained 500 million *new* human beings this year" and that "since World War II our country is experiencing a population increase at a higher rate than India," the latter being *the* signifier of fertility run rampant for population controllists.[96]

There is little doubt that Sanger and her supporters were highly receptive to Wood's rhetoric. But more than being amenable, Sanger was a key architect of such discourse and vital in bridging the span between reactionaries and reformers (and later, liberals) in mustering support for eugenics. This may be her most enduring contribution to that field. In life and in legacy, her work repeatedly placed birth control not in the realm of rights and freedoms, but rather in that of neo-Malthusianism. Despite the discriminatory pronouncements and the practices that resulted, Sanger packaged her stance as charitable, pragmatic, "for-their-own-good" benevolence.

Like Chesler, McCann has resisted the label of racist as applied to Sanger and her cohorts, choosing instead to describe their relationship to the African-American community as one characterized by "racial maternalism."[97] Sangerists certainly exhibited such maternalism, but where McCann errs is in not recognizing this as a racist practice. Chesler and McCann are steadfast in their claim that Sanger latched onto eugenics primarily as an avenue to professional legitimacy and a broader base of support. While this may account for some of her efforts to woo eugenicists in general, and for her relationship with Gamble (with his deep pockets) in particular, Sanger's espousal and promotion of eugenic ideology in the pages of the *Birth Control Review* and elsewhere point to greater culpability.[98] Her warnings about the "feeble-minded peril," her belief in "inborn character," her declaration of the inseverability of eugenics and birth control (even as she came under fire from those eugenicists who feared birth control would lead to "racial deterioration"), the pulpit she gave to scientific racists in the pages of the *Birth Control Review,* her at best inconstant opposition to "positive eugenics," her derision of the disabled, her calls for population control (class-based by definition in the United States and abroad), her involvement with Gamble's crusades in Puerto Rico and the Carolinas, and her ongoing commitment to economically coerced sterilizations, all secure her place among the ranks of U.S. eugenicists.

Though Sanger's political identification is elusive and somewhat periodized (many have argued she grew increasingly conservative over time), her commitment to sterilization remained unbroken. Moreover, Sanger's attraction to, and participation in, eugenic discourse is only part of the equation, and perhaps the lesser part. Equally telling is eugenicists' attraction to Sanger (Gamble was just the most ardent, lobbying hard in 1952, 1955, and 1960 to get Sanger nominated for the Nobel Prize)[99] and the resiliency of her philosophy unto this day. Sanger's dire and hyperbolic enjoinders to "save civilization," curtail expenditures, even safeguard democracy by means of eugenically informed birth control resonate with later "culture of poverty" prescriptions and current warnings on welfare and the federal deficit. Sanger continued to advocate eugenic sterilization "pensions" during the last decade of her life (she died in 1966), undaunted by revelations of eugenics campaigns in fascist Europe. The links between birth control, eugenic sterilization, and population control had been forged, and thanks in part to the foundation laid by Sanger and her colleagues, the postwar target groups looked remarkably like the prewar "unfit."

Physical Fallout

Racism, Eugenics, and Liberal Accomplices after
World War II

In the years and decades after World War II, the ideological association
between eugenically informed birth control and class standing was solidified. The physical fallout of sterilization's placement within the arena of
welfare policy was borne increasingly by women and girls of color. Nonconsensual tubal ligations and hysterectomies—proposed by legislators,
enacted by individual doctors, and often upheld by the courts—were
frequently rationalized by economic considerations. Fewer poor children,
the reasoning went, would translate into more money for individual
families and considerable savings to public coffers. This in no way pre-
cluded the selection of sterilization candidates based overtly on gender
or race. The assaults on African-Americans, Chicanas, and Native Americans, documented below, are illustrative of the longevity of scientific
racism and the consequence of its intrusion into the realm of health
care and welfare policies. Resistance to these campaigns came in the
form of court challenges (brought by and on behalf of women and girls
across the country) and the fight waged by activists (including feminists)
for federal sterilization guidelines. The response of liberals (including
more mainstream feminists) stands in stark relief. Still enamored with a
vision of a reproductive technofix and still eliding the differences in
vulnerability among women, they opposed the regulations and continued to cling to liberalism's belief in individual curatives to institutionally created and sustained injustices.

There is a persistent misconception that, in the wake of the Nuremberg Trials and the disclosure of the extent of Nazi eugenics, eugenicists

in this country folded camp or regrouped entirely, relying exclusively on economic rather than ethnic or racial arguments. Anything even remotely reminiscent of race-based eugenics was so distasteful, the thinking goes, that all but the most conservative, the most reactionary, the most supremacist of thinkers kept their distance. To some extent, this is true. Nazism probably did give some eugenics enthusiasts pause, and no doubt some abandoned the cause. For others, however, Nazi eugenics was such an extreme yardstick that U.S. proposals for race- or class- or disability-based sterilization seemed comparatively tame and therefore tolerable, especially given a rhetoric of "helping" the "unfit" rather than wiping them out. Still, overtly fascist scenarios continued to unfold, as scientific racism and its corresponding practices found new champions. In March 1945, just five months before V-J Day, the *New Leader*, a labor weekly, reported on a proposal by Representative Jed Johnson to sterilize Japanese-American internees:

> Our House of Representatives is supposedly composed of men and women who represent the sober judgment of the people, whose duty it is to exert a steadying, constructive influence consistent with the will of the people, on the running of this country. Representative Jed Johnson (D-Okla.) is a member of this house. Representative Johnson, supposedly speaking for his constituents, has proposed that Congress undertake specific legislation for the long-term disposition of the Japanese-Americans in relocation camps. He does not suggest that they be permitted to return to their homes in this country, nor that they be sent back [sic!] to Japan. He is more farsighted and thorough. He suggests that "we should make an appropriation to sterilize the whole outfit."
>
> The tragedy of such vicious imbecility is that it was spoken, not be [sic] a discredited fanatic . . . but by a member of the United States Congress. The responsibility goes directly to the people who put him there. When the people of this country fail to take seriously the obligation to insure for themselves and their fellow citizens a just democratic government, they . . . leave open the way for Fascists and fools.[1]

Jed Johnson was, of course, only one man. It was the aggregate of individual doctors, judges, and public health administrators that would prove the greater peril. Nevertheless, Johnson's remarks begin to dispel the myth that there was restraint on the part of eugenicists and sterilization advocates as revelations of Nazi atrocities began to surface. Alice Tanabe Nehira's 1981 testimony during hearings of the Commission on Wartime Relocation and Internment of Civilians further undermines such claims.

Nehira's mother, rounded up along with over 110,000 other Japanese-Americans under Executive Order 9066, was sterilized without her knowledge while interned at Tule Lake in 1943.[2]

In 1944, more than ten years after Germany's Hereditary Health Law had been enacted, Nils P. Larsen, medical director of Honolulu's Queen's Hospital, reported on the postpartum sterilization of Hawaiian women. Larsen found the Graefenberg Ring (a type of IUD) and sponge and foam powder to be ineffective because they required "a certain degree of intelligence for consistent and proper uses, which seemed beyond that of the plantation women." Sterilization, including postpartum sterilization, "was used for those mothers where other methods failed, or *where the doctor had every reason to expect failure*."[3] The women and girls at Queen's Hospital (though Larsen also sterilized men) were subjected to surgery because of a paternalistic belief that poor people, particularly poor people of color, are too inept to regulate their own fertility. In what had been and would continue to be standard procedure for uninformed sterilization, Larsen substituted his judgment for that of his patients and acted on an assumption of patient-controlled contraceptive failure. This modus operandi was excused, even mandated, by a liberal paradigm that reserved individual freedoms exclusively for those judged "mature in their faculties."[4] The poor and the racialized were not so deemed.

Poverty, Paternalism, and Scientific Racism: Eugenic Sterilization in the South

Perhaps the greatest display of indifference to recent history during the postwar years unfolded in North Carolina. It was an exemplar of the emerging nexus between sterilization and public relief or welfare.

Four years after the war, Gamble wrote to Sanger of a German sociology teacher who had agreed to translate his article on sterilization for submission to a German medical journal. According to the teacher, Gamble said, the "arbitrary" practice of sterilization under the Nazis had made the subject taboo, and German doctors were too scared to write such pieces now. In that same 1949 letter, he informed Sanger of a forthcoming manuscript on male and female sterilization in North Carolina.[5] The author was Moya Woodside, a psychiatric social worker brought from England by Gamble to the United States.

In her report on the state's sterilization program, which came out the following year, she asserted that the U.S. laws bore no resemblance to the "German experiment," stating that democratic countries "hold as axiomatic that individual liberty cannot be curtailed except by general consent." Woodside failed to acknowledge the danger "general consent" had posed to disfranchised groups, including the victims of Nazi Germany and the Jim Crow South. She dismissed apprehension among Jewish doctors who "have strong feelings on the subject, and . . . who would never sign a sterilization petition," and Blacks who "might conceivably fear that sterilization would be used in the South as an adjunct to doctrines of 'white supremacy.'"[6] Woodside managed to overlook the medical establishment's already long history of experimentation on African-Americans, made available as research subjects first by slavery, then by poverty.[7] She also succeeded in furthering three hundred years of white physicians' pronouncements on race and intelligence—thus situating herself firmly within their ranks.

Dr. Samuel A. Cartwright is perhaps the best-known antebellum subscriber to scientific racism. He measured bone length, hair texture, color of internal organs, and resistance to disease in his effort to biologically sanction slavery.[8] Nineteenth-century social Darwinists, believing African-Americans to be tubercular, syphilitic, and generally beyond medical help, assumed they would die out.[9] Early twentieth-century eugenicists favored a more proactive approach. Charles McCord, who penned *The American Negro as a Dependent, Defective, and Delinquent,* called for the "unsexing" of "unfit" Blacks. There were, he declared in 1914, thirty thousand "feeble-minded and mentally defective Negroes" in the United States. The "feeble-minded," he told his readers, multiply twice as fast as the "normal population." His panic jumped off the page: "[W]hen we find that . . . the breeding of the feeble-minded among Negroes is restricted only by their own instincts and the limits of their physical ability, the matter appears to be really of some significance." McCord held Black women in special contempt, declaring that as many as 90 percent were "unchaste."[10] G. Frank Lydston, a professor at the Chicago College of Physicians and Surgeons, who had earlier lectured on the dangers of "sexual perversion," wrote that "physical and moral degeneracy . . . with distinct tendency to reversion to type, is evident in the Southern negro . . . especially manifest in the direction of sexual proclivities."[11]

It was inevitable then, with the circulation and validation of scientific racism and the accompanying legacy of surgical intervention and experimentation, that African-Americans would be targeted for compulsory sterilization. Woodside herself reported that, in proportion to their respective percentages of the population, the rate of sterilization among Black men was two and a half times that of white men. This differential she ascribed to "certain selective factors, rather than discriminatory intent on the part of white administrators." As elsewhere, the institutionalized were at the greatest risk for compulsory sterilization, and in 1946 North Carolina, approximately one-third of institutionalized patients were Black. The disproportionate rates of institutionalization and selection for sterilization of African-American men were read not as indicators of bias by Woodside, but rather as the result of "good operative facilities" and hospital staff who were interested in the operation (particularly at the State Hospital for Negroes at Goldsboro).[12]

Woodside's report was characterized by an extreme emphasis on African-American women's sexuality, which the social worker viewed as a unitary construct, distinct from white women's sexuality. She was concerned with the "indifference among some lower-income Negro groups, to orthodox white sexual mores." For reasons never fully elucidated, she made a "racial breakdown of the figures for orgasm capacity," concluding, "It may be, that white stereotypes of Negro sexuality have some rational basis."[13] Woodside was concerned not only with alleged immorality and "feeble-mindedness," but with the public expenditures. Evidence suggests, she wrote,

> that the proportion of mental deficiency... is higher among Negroes. It is certainly true that the feeble-minded Negro woman, often with illegitimate children, is a familiar and recurrent problem to health and welfare agencies.[14]

The social worker attached tremendous significance to the number of Black children born to unmarried mothers or to parents in common-law marriages. One woman, who lost five of her fourteen children at birth or in infancy, was quoted as saying she wished all fourteen were living, "I enjoy all of them." Of this woman, Woodside wrote, "numbers of children are not a matter for concern since this type of Negro, lacking incentive and opportunity for the achievement of a higher standard of living, rarely envisages long-term goals and is content to live from day to day, taking whatever comes."[15]

Woodside's reading of this woman's longing for her dead children illustrates one of the most salient features of the North Carolina sterilization campaign and other postwar sterilization drives. Socioeconomic
considerations had long been cited by eugenicists, neo-Malthusians, and
other population control advocates. In North Carolina, "referrals" between
welfare bureaus and city and county health departments for outpatient
sterilizations were common. "They may also be active in arranging...
sterilization on medical grounds for individuals of normal mentality,
and in such cases it often happens that socio-economic factors are
taken into consideration."[16] Whether sterilization was rationalized as a
cost-saving benefit to the state treasury or to individual families, it took
on a punitive nature when foisted on poor people dependent on subsidized health care. "Referrals" is not quite an accurate term. It implied
that individuals targeted for sterilization had the option to refuse to
follow through with the procedure—an option that receded in the
wake of state laws mandating compulsory surgery and/or coercion on
the part of health officials. Sterilization was generally pitched during infant and maternal care, contraception, and immunization visits—all
services utilized primarily by women. By Woodside's own account, at
the time of her writing, three out of five sterilizations performed in the
United States were done on women. In North Carolina, the ratio was
four out of five.[17]

Woodside recorded with disdain a tenant farmer's desire to have more
children. In 1947, a nurse brought sterilization papers to the "X" home.
Initially, "Mrs. X" agreed, but a few days later explained that she was not
yet ready to be sterilized and that she would like to have another daughter.
"Living from day to day," Woodside lamented in her notes on this family,
"country people are all alike.... Sterilization seems a hopeless task."[18]

But what Woodside identified as a deficit in procreative planning could
just as easily have been attributed to any number of things, including
women's resistance to a seemingly omniscient network of state and county
agencies bent on exerting authority over their bodies and their families.
Also lost on Woodside was the long tradition of woman-controlled fertility regulation among African-American women. Frequently, the only
written testimony to this history is found in complaints against women's
actions, actions taken in response to the poverty of available options.
For example, in 1887 Eugene Harris of Fisk University railed against

"negro immorality." In a discussion of premature births and stillbirths, he anguished over self-induced abortions. "An official of the Nashville Board of Health, who is also proprietor of a drug store, tells me that he is astonished at the number of colored women who apply at his store for drugs with a criminal purpose in view."[19]

Moya Woodside's proclaimed "hopelessness" notwithstanding, her joint effort with Clarence Gamble in North Carolina did not go unrewarded. Philip Reilly credits Gamble with moving sterilization into the realm of welfare reform. The Pennsylvania doctor had been able to "refocus sterilization from the state's institutions to the welfare rolls of each county," thereby doubling the number of eugenic sterilizations in just three years.[20]

In the 1950s, North Carolina's legislature was paying close attention to the 20 percent of African-American babies in the state born to single mothers. Two bills were introduced as a panacea to the state's rising welfare costs. The first would have amended North Carolina's existing sterilization law to include giving birth outside marriage as proof of a woman's feeblemindedness. The bill's text described unmarried mothers as "sexually delinquent." Its eugenic intent was clear. State Senator Luther Hamilton praised the bill, saying it would be a check against "breeding a race of bastards." The second piece of proposed legislation would have placed the burden of proof on those individual women: "On the birth of the third illegitimate child, the Eugenics Board . . . shall order the mother to show cause why sterilization should *not* be ordered."[21]

Despite the defeat of these bills, the connections established by their sponsors between sterilization and welfare cannot be overstated. Not all sterilization procedures went through eugenics boards, and doctors were only too willing to comply with statutes that did not pass. Increasingly, their victims were African-American women. Between 1929 and 1940, 78 percent of candidates approved for sterilization by the Eugenics Board of North Carolina were women. Of these, 21 percent were African-American. By 1964, African-Americans composed 65 percent of all women sterilized in the state.[22]

The official and unofficial policies of North Carolina health care providers that secured this outcome had far-reaching effects. In the 1970s, an intern at Howard University Medical School would tell a reporter,

When I was at Howard, a woman came in through emergency and when I noticed she didn't have any children, I asked her why. She said she came from a town in North Carolina and that in this town they sterilize the daughters of every woman who was on welfare when the girl turned fifteen. That's what happened to her.

His friend had a similar experience:

I saw a few cases like that too. They were mostly from that area—North Carolina, South Carolina. They were routinely sterilizing the girls whose mothers were on welfare.[23]

Needless to say, neither these practices nor the popular and professional sentiments that fueled them were confined to the Carolinas. A Gallop poll conducted in the mid-sixties revealed a predominant view of Aid to Families with Dependent Children (AFDC) recipients as "dishonest, lazy, and lacking in initiative." Twenty-seven percent of respondents favored their sterilization.[24] The medical experts concurred. Ninety-four percent of obstetrician-gynecologists surveyed in 1972 favored withholding welfare from single mothers with more than three children if they refused to submit to sterilization.[25] Doctors, idealized not only as medical providers but as patient advocates, were actually more willing than the public at large to have the state intervene to compel sterilization. Unlike the general public, however, they had the means and the opportunity to enact their will upon the women they treated. By their own logic, doctors should have been targeting white women, who have always constituted the majority of AFDC recipients. While there were, indisputably, low-income and poor white women pressured into accepting sterilization, the most systematic campaigns for punitive sterilization were waged against women and girls of color—both in and out of the South. African-American women were not alone. Ninety-five percent of women sterilized in New York City in 1965 were Puerto Rican.[26] By the early 1980s, half the Puerto Rican women of childbearing age in Hartford, Connecticut, had been sterilized.[27] A few high-profile cases shed light on the confluence of racism and welfare status that endangered so many women during this time.

In 1965, Nial Ruth Cox was sterilized at Washington County Hospital in New Bern, North Carolina. Her "consent" was extorted by her caseworker, who threatened to withhold payments to her family. The operating physician assured her that the measure was temporary. It was later revealed that her doctor had classified her as an "eighteen year old mentally deficient Negro girl," thereby making her eligible for surgery under the North Carolina sterilization statute. However, like Carrie Buck, Nial Ruth Cox had never been evaluated. In a brief to the court, Cox's attorneys stated that she had never been examined or tested by a psychologist, psychiatrist, or anyone else qualified to make an appropriate determination. Racist decrees on the intellectual abilities of African-Americans and a mandate to sterilize the developmentally disabled, together with an atmosphere of racial and class-based paternalism, marked Cox as the perfect candidate for tubal ligation.[28] She recounted her story in the American Civil Liberties Union newsletter:

> I was living with my mother and eight sisters and brothers. My father, who was married to my mother until I was six is dead. My family was on welfare, but payments had stopped for me because I was eighteen. We had no hot or cold running water, only pump water. No stove. No refrigerator, no electric lights. It got cold down there in winter. I got pregnant when I was seventeen. I didn't know anything about birth control or abortion. When the welfare caseworker found out I was pregnant, she told my mother that if we wanted to keep getting welfare, I'd have to get my tubes tied—temporarily. Nobody explained anything to me before the operation. Later on, after the operation, I saw the doctor and asked him if I could have another baby. He said I had nothing to worry about, that, of course, I could have more kids. I know now that I was sterilized because I was from a welfare family.[29]

In addition to being faced with the loss of already inadequate welfare payments, women receiving Medicaid found that medical care, too, could be cut off should they fail to sign on the dotted line. Instead of coming under increased state scrutiny because of their ties to federal and state agencies, welfare-referred physicians were often shielded by them, empowered to serve these agencies by protecting "deserving" members of the tax-paying public. Clovis Pierce of South Carolina was among these ranks.

In 1973, Pierce was the lone obstetrician in Aiken, South Carolina. He sterilized Mrs. Virgil Walker after informing her that withholding consent would result in the termination of his medical services. She complied. Pierce issued an identical ultimatum to Shirley Brown, cornering her just one day after she gave birth. Brown refused consent and was subsequently discharged from the county hospital.[30] Angela Davis wrote that Pierce was unyielding in his drive to sterilize Medicaid mothers with two or more children:

> According to a nurse in his office, Dr. Pierce insisted that pregnant welfare women "will have to submit [sic] to voluntary sterilization" if they wanted him to deliver their babies. While he was "... tired of people running around and having babies and paying for them with my taxes," Dr. Pierce received some $60,000 in taxpayers' money for the sterilizations he performed.[31]

Walker and Brown sued Pierce, but his fellow doctors rallied to his defense. The South Carolina Medical Association passed a resolution declaring that "it is entirely ethical for a physician to inform a woman that he will require her to agree to sterilization as a condition to accepting her as a patient."[32] Dr. Curtis Wood stated, "I admire his courage."[33]

In federal court, Pierce testified that it was indeed his policy to prevent women on welfare from having more than three or four children. Mrs. Walker was denied an award, and while the court did conclude that Shirley Brown's civil rights had been violated, damages were assessed at the demeaning amount of five dollars. Antonia Hernandez, attorney for the plaintiffs in a 1978 sterilization abuse case against Los Angeles County Medical Center, commented on the Pierce decision, "Even though one half of the jury was composed of Blacks, the moral prejudice against women on welfare who bear children out of wedlock was ... a crucial factor."[34]

Links between AFDC/Medicaid and tubal ligation were not always articulated so blatantly. While economic threats were often unspoken, a perceived mandate to lessen public expenditures through population control programs impelled state and county workers to engage in lies of both omission and commission. In Alabama, deceit replaced outright threats in the sterilization of Mary Alice and Minnie Lee Relf, ages twelve and fourteen. One of the most striking features of sterilization demographics during this era, along with race and socioeconomic status, is the extreme youth of those selected for surgery. Many had not even reached puberty. In previous years, the American College of Obstetricians and Gynecologists (ACOG) had age-based *suggested* standards for contraceptive sterilization. These recommended that women considered for tubal ligations be at least twenty-five with five living children or thirty with four living children or thirty-five with three living children. These unenforceable guidelines were dropped by the ACOG entirely in 1969.[35]

The story of the Relf sisters has been recorded and recalled by scholars, feminists, reproductive rights and freedom advocates, and antiracism activists. It began in June 1973, when a nurse from the Family Planning Clinic of the Montgomery Community Action Committee picked up Mrs. Relf and her two youngest daughters and drove them first to the doctor's office and then to the hospital. Mrs. Relf could not read, and the hospital took full advantage. The staff presented her with a consent form, telling her it would grant them permission to inject her daughters with Depo-Provera, previously administered to all three of her daughters, despite its experimental status.[36] "I put an X on a piece of paper," Mrs. Relf later testified. In actuality, the form was a sterilization release. Mrs. Relf was taken home, and Minnie Lee (declared "mentally retarded") and Mary Alice were left alone. When their father came to see them that night, he was told that visiting hours were over and refused admittance.[37] The next day, Minnie Lee called her mother and asked to be brought home, but Mrs. Relf had no means of transportation. They were trapped. At this point, no one in the family knew that the girls were about to be surgically sterilized. Indeed, their mother didn't learn of the operation until her daughters were released three days later and told her themselves.

The Relf sisters, according to one welfare official, were targeted for tubal ligation because "Boys were hanging around the girls," and they believed the girls lacked the "mental talent" to take the pill.[38] Earlier in the year the Relfs' oldest daughter, Katie, had an IUD implanted over her objections and without her parents' knowledge.[39] The family never solicited any of the clinic's services. In each instance—the Depo-Provera injections, the IUD, and eventually the sterilizations—the clinic aggressively approached the Relfs. Their welfare status and the scrutiny that ensued left them unprotected in the face of these advances. Joseph Levin of the Southern Poverty Law Center testified before a Senate subcommittee, "They were already under this community action program, and Community Action, having moved them into the project, was aware of their existence and sought them out."[40] Pronouncements on the Relf girls' intellectual mettle, assumptions about their sexual activity, and the family's entanglement with social services made the children targets; their lack of resources assured that extrication was impossible.

Avenues of Redress? Regulation, Investigation, Adjudication

The hospital where Minnie Lee and Mary Alice Relf were taken operated first under the Office of Economic Opportunity (OEO) and then under the Department of Health, Education, and Welfare (HEW—the agency designated to fight the War on Poverty).[41] HEW began to include sterilization as part of its health program in 1971. Neither parental nor personal consent was mandated by its guidelines.[42] What put the Relfs and so many others at increased jeopardy was the fact that HEW's family planning projects, Medicaid, and AFDC were all funded through its Social and Rehabilitation Services.[43] It was the Relfs' very involvement with government-sponsored programs that put them at risk for sterilization abuse. Levin explained:

> They receive $156 per month from the Alabama Department of Pensions and Security; they receive food stamps; they receive subsidized medical assistance; and, I suppose, there are other forms of aid unknown to me at this time. In other words, each member of this family lives his or her existence under a microscope.
>
> They are visited on an almost daily basis by some social service person who either functions under the direction of the state or Federal Government.[44]

The Relfs, along with Mrs. Walker, took their case against HEW to federal district court. In *Relf v. Weinberger,* Judge Gerhard Gesell ruled that there was "uncontroverted evidence" that "an indefinite number of poor people have been improperly coerced into accepting a sterilization operation under the threat that variously supported welfare benefits would be withdrawn unless they submitted." Gesell further commented that "the dividing line between family planning and eugenics is murky." He barred federally funded sterilization of minors and ordered regulations safeguarding patients against intimidation or coercion.[45]

HEW began drafting regulations while *Relf* was being litigated. Various versions were in effect in the years that followed, some challenged by the National Welfare Rights Organization and rejected by Gesell for not going far enough. The finalized rules appeared in the fall of 1978.[46] The guidelines issued included a ban on hysterectomies for sterilization purposes in federally funded programs, prohibitions against implicit or overt threats concerning Medicaid or welfare, and an end to the sterilization of those institutionalized against their will. Consent could not be accepted during labor, directly prior to or after an abortion, or when the patient was under the influence of drugs or alcohol. Voluntary informed consent could be obtained only via a written form in the patient's preferred language, along with an explanation of the irreversibility and a presentation of alternate methods. A mandated thirty-day lapse between signed consent and the procedure was also stipulated.[47] Significantly, HEW abolished the distinction between "therapeutic" and "contraceptive" sterilization, fearing that the former designation might provide a loophole for doctors trying to circumvent the regulations. This was a credible concern, given the long history of eugenic sterilization and other invasive procedures performed under the guise of therapeutics.

The directives were limited in scope, applying only to OEO and HEW family planning programs. Medicaid programs operated independently under laws that varied from state to state.[48] The political will to adequately enforce the federal regulations simply was not there. A 1975 Public Citizen's report estimated that 3,600 out of the 4,800 hospitals with obstetrics-gynecology departments in the country were out of compliance nine months after the guidelines went into effect.[49] Public Citizen contacted department heads at the fifty largest teaching hospitals in the United States. Of the forty-two that performed nontherapeutic sterilizations, 76 percent in thirteen states were in violation. Furthermore, 33 percent in seven states were not even aware of the regulations. A full 31 percent of hospitals surveyed failed to advise patients that refusal of sterilization would not result in loss of benefits.[50]

Some of the most flagrant and blatantly racist infractions took place at Indian Health Services (IHS) facilities. In June 1977, the General Accounting Office (GAO) released a report disclosing multiple violations of the HEW guidelines at federally funded IHS hospitals. Thirty-four hundred Native American women were sterilized between 1973 and 1976 in and around Phoenix, Albuquerque, Aberdeen, and Oklahoma City.[51] A significant number for any demographic group, this figure was all the more alarming given the relatively small numbers of Native Americans— 0.4 percent of the population at that time.[52] Consent forms, when they were used, were not in compliance with the federal regulations. In addition, the GAO documented thirty-six violations of the moratorium on the sterilization of minors.[53] In a 1977 interview, Dr. Connie Uri of

Indian Women United for Social Justice estimated that 25 percent of Native American Women of childbearing age had been rendered sterile by the IHS.[54] At the IHS hospital in Claremore, Oklahoma, this translated into one woman sterilized for every four babies born at the facility.[55] Other sources gave even higher numbers. In 1979, Lee Brightman, president of United Native Americans, estimated that up to 42 percent of the women of childbearing age and 10 percent of the men had been sterilized.[56] Alen Rowland, Northern Cheyenne tribal chairman during this period, told a reporter that the mass sterilizations were, in his view, "just an extension of the extermination policies of the last century."[57] As the GAO was billed for each of the thousands of sterilizations, at a cost of several hundred dollars per procedure, it is unlikely the practice went undetected, suggesting, at the very least, government complicity. As Katherine Harris Tijerina, legislative aide to then Senator James Abourezk of South Dakota, said:

> You don't have to have a conspiracy to have the effect of one. The attitudes of racism and sexism have been translated into an excess of sterilizations. I would consider it genocide.[58]

The IHS scenario fit easily into a larger history of genocide, threatening a "gentler" form of eventual eradication. The language of physicians lent credence to these fears. Uri told of a doctor whose words harkened back to earlier eugenicists. "I asked how he could sterilize a 16-year-old girl, and he told me she was a 'polluter' because she already had several children."[59]

Certainly, not every IHS doctor was propelled by identical motives. What they did share was a paternalism and a belief that they were administering a cure-all to the "culture of poverty" that justified any action and any lie. The results can be gauged anecdotally: two fifteen-year-old Cheyenne girls given tubal ligations without their knowledge when they entered an IHS hospital for appendectomies; a sixteen-year-old given a postpartum sterilization and told she was "fixed" so that she would not get pregnant for another two years;[60] a woman told that sterilization would cure her headaches. Tribal judge Marie Sanchez related this last account to an interviewer:

> The doctor suggested she have a tubal ligation. And she did. She took his word. . . . and she had the operation, and her headaches came back. They found out later she had a tumor in her head.[61]

Stories such as these prompted Senator Abourezk to request a GAO investigation. While the GAO's report did state that the women and girls may have believed they had no choice but to consent to surgery, it stopped well short of conceding intimidation tactics on the part of IHS medical staff.[62] The report's validity, as well as the balance of the GAO's findings, is irreparably compromised by the knowledge that not one of the women or girls was interviewed by investigators. One official explained:

> We did not interview patients to determine if they were adequately informed before consenting to sterilization procedures. We believe such an effort would not be productive because recently published research noted a high level of inaccuracy in the recollection of patients four to six months after giving *informed* consent.[63]

This interview protocol suggests predetermined results and raises the possibility that the initial figure of 3,406 involuntary sterilizations may have represented a severe undercount. Likewise, the number of other violations enumerated in the report may have been deflated. The decision not to question the women involved was guided by the assumption that they were inherently irrational. In contrast, the operating physicians, who certainly had cause to mislead investigators, given their breach of federal regulations, were viewed as reliable informants. Dr. Jim Felsen, then deputy director of IHS program operations, contended that the women's reports were suspect:

> I'm sure there are cases of abuse. Some doctors play God, but on the other hand many of the stories you'll hear just reflect remorse after the fact. Women can't deal with their own feelings of guilt and they repress what the doctor told them.[64]

"Apparently," said Sanchez, "there is an immediate assumption that the Indian woman has no memory, that she is automatically unreliable."[65]

A year after the GAO released its findings, a California court began to hear arguments in *Madrigal v. Quilligan*. It was a bench trial, brought by ten Chicanas against the Los Angeles County Medical Hospital.[66] At issue was the compulsory sterilization of women admitted to the hospital over the course of the previous ten years. Between 1968 and 1971, the number of hysterectomies performed at the facility had gone up 742 percent. Tubal ligations had risen 470 percent, with a 151 percent increase in postpartum tubal ligations.[67] The women who came forward reported a variety of abuses: abortion requests made contingent on acceptance of sterilization; consent forms in English presented to Spanish-speaking women as cesarean section releases; operations performed immediately after delivery or abortion; sterilizations done even after explicit denials of consent.[68] One physician, who had left his position at the hospital, estimated that 20 to 30 percent of the doctors at L.A. County Hospital aggressively pushed sterilization on women "who either did not understand what was happening to them or had not been given the facts regarding their options."[69] Physical intimidation was also used. One staffer reported:

> I saw various forms of actual physical abuse used to force a woman in labor to consent to sterilization. There were incidents of slapping by doctors and nurses. A syringe would be shown to a woman in labor and she would be told, "we will give this to you and stop the pain if you sign."[70]

Guadalupe Acosta was punched in the stomach by her doctor while in labor (which he had induced by pressing down on her abdomen). The baby was stillborn, and before Acosta left the delivery room she was sterilized without her knowledge.[71] Mary Diaz entered the hospital, in labor, two years later on April 6, 1975. Doctors first tried to obtain her consent for a cesarean section, which she rejected, while she was still lucid. The paperwork was presented to her again after she was under anesthesia and this time Diaz signed—for the cesarean, not for the tubal ligation.

> I told them I could not accept that. I kept telling them no and the doctors kept telling me it was for my own good. . . .
> . . . they were still insisting that I would accept the tubal operation and I was still saying, "No, no, no." I was in great pain. I thought I was going to die. . . .
> . . . I *know* that I didn't [consent to sterilization] because there was a nurse who showed me the chart and there was no indication that I approved. The chart said I rejected all their efforts to sterilize.[72]

Like her coplaintiff, Mary Diaz did not learn that she had been sterilized until weeks after the fact. Her doctor told her not to cry, "It's best for you that you not have any more children. In Mexico the people are very poor and it's best that you have no children."[73]

Unlike the Relfs or Shirley Brown or the family of Nial Ruth Cox, the women of *Madrigal v. Quilligan* were not on welfare, though they were eligible for public medical assistance and marked racially and by their presumed immigrant status. One woman reported being verbally browbeaten by nurses who accused her and her children of "burdening the taxpayers."[74] Two years after the case, Adelaida Del Castillo wrote:

> Currently we are witnessing an accelerated campaign of aggression against Mexican people. Government authorities and mass media hold them responsible for overflowing the job market, increasing the crime rate, and endangering the public health. The medical profession, meanwhile, self-righteously feels no compunction in executing and rationalizing the sterilization of the country's poor and its ethnic minorities.[75]

The court found in favor of the hospital, perhaps due in part to a complainant strategy that foregrounded the cultural specificity of child-bearing among Chicanas and Mexicanas, rather than advancing an argument based on human rights violations and medical ethics. The ruling, issued by Nixon appointee Jesse Curtis, implied that the women were agitated to a degree far exceeding what their ordeal demanded. He sympathized, Curtis said, with the *plaintiffs'* inability to communicate clearly. His words underscore the vulnerability of actual or imagined immigrants and an intrinsic belief in the deficiency of non-Anglo culture.

> It is not surprising . . . that the staff of a busy metropolitan hospital which has neither the time nor the staff to make . . . esoteric studies [into Chicano culture] would be unaware of these *atypical cultural traits.*[76]

Embedded in the Curtis ruling, too, was the implicit acceptance of a two-tiered medical standard. County hospitals could not be expected to extend the same level of care or respect or even disclosure as facilities serving the middle classes and the affluent. The poor should neither expect nor demand comparable treatment, even in matters of such gravity as surgery.

Despite the *Madrigal* defeat, activists working against sterilization abuse made a tremendous impact. It was their documentation and publishing efforts that brought involuntary sterilization to public consciousness. It was their articulation of institutional race, gender, and class biases that situated these violations in a century-old eugenics tradition. The Committee to End Sterilization Abuse (CESA—whose core membership drew from groups such as the Puerto Rican Socialist Party, the Center for Constitutional Rights, and the Medical Committee for Human Rights), Indian Women United for Social Justice, the Committee for Abortion Rights and against Sterilization Abuse (CARASA—founded by radical and socialist feminists in New York City), the National Black Feminist Organization, and the National Welfare Rights Organization (NWRO), among others, succeeded in more than just securing federal and local government guidelines (which by itself would have been an important, if purely reformist, endeavor). Advocates and organizers placed involuntary sterilization within a larger analytical framework and problematized the liberal axiom of "free choice." CESA defined abuse as "the biasing of services along class, racial, and ethnic lines—limiting the poor's freedom of choice."[77]

> Forced infertility is in no way a substitution for a good job, enough to eat, decent education, daycare, medical services, maternal infant care, housing, clothing, or cultural integrity. We support the right of the individual to choose the method of birth control he or she prefers. But when the society does not provide the basic necessities of life for everyone, there can be no such freedom of choice.[78]

Activists' efforts garnered attacks from some quarters. The Association for Voluntary Sterilization, Planned Parenthood, and other population control organizations had opposed the HEW directive from the outset, as did the American Medical Association and the American College of Obstetricians and Gynecologists. A group of obstetrician-gynecologists brought suit, challenging the constitutionality of the federal guidelines (in particular the ban on sterilizing minors) as well as New York City and state regulations. The doctors claimed that it was their responsibility to gauge whether or not an individual minor had the maturity to give informed consent. Opponents of the regulations further asserted that reports of sterilization abuse were exaggerated, that denial of sterilization on basis of age constituted abuse of the poor, and even that doctors and hospitals should self-monitor to prevent coercion of women. Eventually, the suit was dropped.[79] Others warned of government infringement on the rights of patients and consumers, this last appeal being particularly resonant in a capitalist framework.

There was a certain irony in activists' recourse to the government for redress of sterilization abuse, particularly given HEW's culpability. Yet, the call for regulations was an invitation not for further government encroachment into women's bodies, but rather for the government to restrain doctors on its payroll from intruding in and exerting control over the lives of the poor. It must be seen in this light, contrary to opponents' invocation of the threat of federal paternalism.

While some regulation foes were clearly diehard reactionaries—such as the obstetrician who called the fight for regulations a "Puerto Rican communist plot"—many were not.[80] Objections to the new rules often relied on rhetorical devices borrowed from liberal ideology, namely, "free choice" and "right to access." An emphasis on "choice" and an unwillingness to interrogate the parameters of that choice made antiregulation arguments sound reasonable and masked the more substantive motivations of antiregulation medical professionals. As Rosalind Petchesky noted in 1979, it was not women's "right to access" that was being protected, "but rather the medical profession's privileged autonomy and the 'right' of providers to process people through sterilization procedures as expediently as possible."[81]

But there were other challengers, outside the medical profession. Mainstream feminist organizations, ideological progeny of Margaret Sanger, also stood against the regulations. Whereas the National Welfare Rights Organization opposed the early HEW standards because they didn't go far enough, the National Organization for Women (NOW) complained that they went too far.[82] NOW was particularly resistant to the mandated thirty-day waiting period, declaring it an infringement on the rights of women who wanted to be sterilized immediately.[83] This response exemplified liberalism's elision of differences between women. To borrow from Goldberg's critique of liberalism, NOW was concerned with what it perceived to be a broad, unifying identity—in this instance, gender.[84] Race and class, in contrast, were treated as epiphenonmenal rather than core determinants of women's life experiences. "The liberal tradition," as Alison Jaggar explained, "conceives of freedom as the absence of external obstacles to individual action and assumes that individuals are autonomous agents."[85] Mainstream feminists failed to problematize the autonomy they espoused. The imposition of "external obstacles" (embodied in this instance by the federal guidelines) may have been perceived as an attack on the self-determination of NOW's constituency (predominantly white, middle-class, and affluent women), but it offered at least nominal protection to those who ran the greatest risk of medical abuse—poor and low-income women and girls of color.[86] The autonomy of these women had been already constricted by their class status, by racism, and by their entanglement with government agencies that exercised varying degrees of control over their lives.

As Goldberg has noted, liberalism holds that the "particular differences between individuals have no bearing on their moral value, and by extension should make no difference concerning the political or legal status of individuals."[87] In the liberal paradigm, difference is to be transcended, collective identity and shared experience (such as being victimized by racism) de-emphasized in favor of individualism. Liberalism thus finds itself in "opposition to discourses and political movements which categorize people and privilege social determinations such as gender, race, and class over individuality."[88] While NOW was clearly attentive to the relevancy of gender, it was unwilling to recognize the prominence of racism and classism in the lives of women beyond its constituency. Liberal feminists assessed their own socioeconomic position as *the* referent for all women. NOW's objection to the HEW regulations rested on a belief that women *seeking* sterilization and women *targeted* for sterilization were, and had historically been, equally imperiled and equally in need of protection vis-à-vis the operation, that the inconvenience of the former was analogous to the peril faced by the latter. This was only arguable within a liberal discourse that treated systemic racism and economic intimidation as mere "regrettable deviation[s] from the ideal"[89] when they were, in fact, the salient causes of unequal treatment. This unwillingness to confront race and class issues and a long-standing belief in birth control as a palliative to poverty (grounded in a larger commitment to the technofix), meant continued liberal support for coercive population control tactics and silence in the face of newer threats.

New Technologies, Old Politics
Norplant and Beyond

In the last decade, new, temporary sterilization methods have come to the fore, assuming a role strikingly similar to their predecessors. Norplant and the freshly rehabilitated Depo-Provera were designated as tubal ligation and hysterectomy's heirs apparent, reaching out to liberals and conservatives by promising socioeconomic, if not medical, miracles.[1] The incorporation of Depo-Provera and Norplant into a framework that posits population control as remedial to poverty overseas is consistent with a liberal belief system that supports similarly premised policies in the United States.[2] These technologies have joined their forebears, making their appearance in courtrooms, welfare "reform" debates, and the bodies of low-income girls.

In 1979, a contributor to the *Hastings Center Report* wrote, "Depo Provera is now more attractive to these people [family planners, medical professionals, program administrators], since sterilization is more closely regulated under DHEW's new guidelines."[3] While the writer overestimated the efficacy of HEW regulations, she did not exaggerate the interest administrators, population control advocates, and others had in Depo-Provera.

As a result of pressure brought to bear by women's health care advocates and consumer protection organizations, the Food and Drug Administration repeatedly denied Upjohn's applications for approval of Depo-Provera over a twenty-year period. Early on, Depo-Provera was linked to cervical cancer, breast cancer (especially among younger women), and liver cancer, as well as long-term sterility (one study found that it took an average of thirteen months from the date of a woman's last injection for fertility to return).[4] This list grew to include osteoporosis, endometrial cancer, prolonged menstrual bleeding (lasting weeks at a time and putting women at risk of pelvic inflammatory disease), weight gain, severe—even suicidal—depression, loss of libido, abdominal pains, dizziness, headaches, hair loss, fatigue, nervousness, nausea, and potential hazards to breast-fed infants.[5]

Despite this litany, and despite its own twenty-year history of rejecting Depo-Provera—not to mention its 1978 ban on the drug—the FDA, under the Bush administration, approved Upjohn's product as a contraceptive in 1992. Planned Parenthood Federation of America applauded the move,[6] but the decision was actively opposed by the National Latina Health Organization, the National Black Women's Health Project, the Native American Women's Health Education Resource Center, the National Women's Health Network, the National Asian Women's Health Organization, the Women's Economic Agenda Project, and the San Francisco chapter of NOW.[7]

In the 1970s and 1980s, FDA hearings determined that there wasn't enough research to approve Depo-Provera. But it was the FDA that initiated reconsideration, soliciting another application from Upjohn. Approval came even though there was no change in the research findings. Luz Alvarez, of the National Latina Health Organization (NLHO), which called for a moratorium on Depo-Provera, explained the process:

> When these hearings are taking place the time for input from the overall community is very limited. We have about three minutes to present our case, while Upjohn, the company that produces Depo, basically get[s] the whole day to present their issues.[8]

The history of Depo-Provera's use on marginalized women in the United States and abroad—both pre- and post-FDA approval—parallels that of surgical sterilization. Because it had been okayed for women with advanced endometrial cancer,[9] it was available to doctors, many of whom dispensed it as birth control despite the existing ban. Developmentally delayed women, women with drug addictions, and women in prison were all targeted.[10] IHS continued to inject Native American women until at least 1987.[11] Hartmann reports on an Emory University Depo-Provera study involving 4,700 African-American patients. In 1978, FDA

auditors castigated the researchers for ignoring the administration's protocol, keeping poor patient records, losing track of subjects who dropped
out of the study, and failing to do long-range follow-up to determine
possible cancer risks.[12] In the early 1990s, HIV-positive Haitian women
held at the U.S. naval base at Guantánamo Bay were forced by doctors
to choose between Depo-Provera (introduced to women there before it
was approved for use in the United States) and more permanent sterilization.[13] Their status as would-be immigrants and consequently the acceptability of forcing Depo-Provera on them was determined by their
point of origin. Years earlier, in 1979, then FDA commissioner Donald
Kennedy stated, "Quite obviously, a drug that may not yet be suitable
for approval here could well have a favorable benefit/risk ratio in a less
developed nation."[14] He was not alone in this assessment.

Upjohn had been able to circumvent FDA regulations by exporting
Depo-Provera from a Belgian subsidiary. From the 1970s onward, International Planned Parenthood and the United Nations Fund for Population Activities (UNFPA)—both partially financed with U.S. money—
along with some American-supported private agencies were supplying
Depo-Provera to "less developed countries." By 1979, Depo-Provera had
received the World Health Organization's stamp of approval and was
commercially available in sixty countries. One and a quarter million
women worldwide were receiving shots.[15]

Norplant, approved two years earlier than Depo-Provera, followed a
similar trajectory. Like Depo-Provera, Norplant has been linked to osteoporosis, breast cancer, and cervical cancer,[16] and has been around
for some time. It was invented at the Population Council's laboratories
at Rockefeller University and has been in development since 1966.[17] Norplant was tested on women overseas long before it was approved by the
FDA. In the early 1980s, Brazil, under military dictatorship, became one
of the first countries to host Norplant trials. Two thousand women there
received the implants.[18] In Gazaria, Bangladesh, women were not told
that they were receiving an experimental drug. Women in Thailand
were informed that the implants would not be removed for "minor side
effects." In the Dominican Republic, Indonesia, and Egypt, some women
were not even told that Norplant *must* be removed after five years.[19] Failure to remove the implants can result in ectopic pregnancy, a leading
cause of death among pregnant women. Practitioners' reluctance and
often outright refusal to remove Norplant from women in targeted

demographics made it compulsory, as much an enforced population control device as eugenically informed tubal ligation.[20]

The World Bank, the U.S. Agency for International Development, and the UNFPA have been aggressive champions and distributors of Norplant.[21] Wayne Bardin of the Population Council said that Norplant was ideal for women in "some countries where there are traditional religions, [and] women are not allowed to be examined by any kind of health worker unless their husband is present" because they do not have to undress for the procedure.[22] The result is that some women are not screened properly before insertion. A Bangladesh source reported that while doctors there did check weight and blood pressure—two important assessors of Norplant's risks to an individual woman—prior to the procedure, they never heard of any woman *not* receiving Norplant on the basis of these measurements.[23] The absence of adequate prescreening for contraindicated conditions increases the likelihood of complications. It is especially hazardous for women with diabetes, epilepsy, depression, high blood pressure, high cholesterol, migraines, or gall bladder or kidney disease.[24] The gynecological specialist at one Bangladesh clinic dismissed these risks: "In order to have a good thing there is always a price to pay. If two or three women die—what's the problem? The population will be reduced."[25]

Shortly after Norplant went on the market in the United States, a *New York Times* editorial declared it to be "relatively free of side effects."[26] Dr. Katherine Sheehan of Planned Parenthood in San Diego claimed, erroneously, that fertility returns almost immediately following Norplant's removal from a woman's arm. The *Los Angeles Times* opened their story with the promise that women now had "a birth control method they can forget about for the next five years."[27] But Norplant's long list of side effects makes it extremely unlikely that recipients could forget its presence. The National Latina Health Organization has warned women about severe stomach pain, headaches, mood swings, changes in blood lipids, acne, and infection at the insertion site.[28] The *Journal of Long-Term Effects of Medical Implants* reported blisters, scabs, soreness, and itching at the implant location, and *Contraception* listed hair loss and arm pain among the side effects. Some women have experienced pseudo-tumor cerebri, an increase in intracranial pressure.[29] Less than four years after Norplant's approval, the *New York Times* reported that roughly

20 percent of all U.S. users had the capsules removed within the first twelve months.[30]

A 1994 survey of medical staff at thirteen Texas family planning clinics found that 37 percent of those surveyed believed Norplant was riskier than tubal ligation, 16 percent said it was more dangerous than an IUD, and 24 percent thought a full-term pregnancy posed less danger than the implants.[31] Yet 81 percent of surveyed staff members said it was a good birth control method for HIV-positive women, despite Norplant's potential to compromise the autoimmune system and the fact that it offers no protection from transmission.

Like sterilization drives, Depo-Provera and Norplant campaigns pose a special threat to women of color, who have long faced barriers to patient-controlled contraception. Because the underlying racism and class bias of medical and public policy that propelled involuntary sterilization has not been dismantled, women who previously would have been targeted for tubal ligation are now being singled out for Norplant and Depo-Provera. Commenting on Virginia's 1993 approval of state funds for Norplant, Linda Byrd Harden, then chief executive officer of the NAACP, said that given the disproportionate number of Black women undergoing the procedure, such sponsorship could translate into government sterilization of undesirables.[32] But the pressure is already on.

Marketers have honed in on African-American and young women, advertising in *Essence, Heart and Soul,* and *Young Women's Health* magazines. Andrea Smith, cofounder of the Chicago chapter of Women of All Red Nations (WARN) and cocoordinator of the Committee on Women, Population, and the Environment, told participants at a CWPE conference that Indian Health Services is currently using high pressure tactics to push Depo-Provera and Norplant as it once pushed sterilization. "It seems like all the Native women I know who have either gone to Indian Health or received public benefits were told to go on Norplant and that it has no side effects or contraindications."[33] According to Charon Asetoyer of Native American Women's Health Education Resource Center, not only does IHS withhold information on Norplant's drawbacks, it fails to administer required preprocedure pregnancy tests. A woman's mere presence at an IHS clinic, she stated, is read as consent to implants or injection. At the Cheyenne River, Pine Ridge, and Rosebud Reservations, women seeking removal of Norplant are frequently strong-armed by IHS workers into continuing with the implants. Some are refused outright by health care workers and doctors who claim to have their best interests at heart. But, as Asetoyer states, "You don't give a healthy woman a drug that will make her unhealthy and say you have her best interest at heart."[34]

Native American women who must rely on IHS-provided care have found that medical attention is tethered to an immovable, if unspoken, public policy agenda. Norplant intensifies their reliance on IHS. Women cannot remove the rods themselves, and so they are at the mercy of physicians. Latinas, who currently have the lowest rate of medical coverage and are the least likely to be covered by their employers, are often similarly trapped. Luz Alvarez Martinez tells of an Oakland, California, woman who was refused removal at the county hospital despite the fact that she was suffering from side effects of the implant. Because she had no coverage, she had no options.[35]

In response to all this, a coalition of reproductive rights groups and women of color advocacy organizations called on the FDA to require manufacturers of Norplant and other long-acting contraceptives to include consent forms with their products. The National Black Women's Health Project opposed FDA approval of Norplant altogether.[36] But liberal pro-choice groups have, for the most part, remained silent.

As Margaret Sanger framed birth control as panacea, without attending to the institutionally generated inequities borne by poor women, the pro-choice mainstream has concentrated almost exclusively on access to abortion (and to a lesser extent, RU486) and on existing birth control paradigms. "Abortion rights are important," says Andrea Smith, "but if 'pro-choice' means the right to choose among dangerous contraceptives, then it is not a movement for Indian women." Acquiescence to population control ideology and an adherence to the cult of the individual, together with what Smith has identified as a lack of interest in community self-determination for women of color, has once again put liberal feminists on the wrong side of the issue. Smith relates:

> I once served the state affiliate of a national pro-choice organization. We got involved in "informed consent" education around Norplant and Depo-Provera. We did not stand against these contraceptives; we merely tried to inform women about their side-effects and contraindications. We were told by the national office to "cease and desist." I was once asked to appear on TV with a representative from another pro-choice

organization that was distributing Norplant in only one of its offices in a primarily African-American neighborhood. The other organization, however, refused to appear on the show because they were "unprepared to discuss the political ramifications of Norplant." They were prepared to distribute it, but not to defend it.[37]

Norplant appeals to liberals and conservatives in much the same way as sterilization once did: as a curative to poverty and crime and as a means of curtailing welfare. It has all the superficial trappings of proactive reform and leaves the status quo undisturbed. Furthermore, as sterilization was once embraced as more humane than castration, Norplant is made to seem more benevolent than sterilization because of its reversibility. But Norplant's encroachment into the judiciary, into legislative debates on AFDC, and into urban high schools makes it anything but benign.

The Judiciary: Procreation and Punishment

In the last weeks of 1990, shortly after the FDA approved Norplant, an op-ed piece in the *Washington Post* warned readers, "The ethical debate is sure to get started the minute some benighted judge orders a woman to use it or go to jail."[38] The newsprint was barely dry when Judge Howard Broadman's gavel came down three weeks later in Tulare County, California. On 2 January 1991, Broadman sentenced Darlene Johnson, who had pled guilty to child abuse, to three years on Norplant. His ruling underscored the saliency of gender, race, and class in selecting recipients of the new sterilization technologies.

Johnson consented to Broadman's condition for probation, but the nature of that "consent" is highly suspect. First and foremost, she was given an ultimatum: accept Norplant or go to state prison. Broadman, appointed to the bench in 1988 by then Governor George Deukmejian (Republican), had made a similar offer to another woman the previous year. When she refused, he revoked her probation.[39] Second, Johnson was not fully informed as to what the procedure would entail. Like most of the population in early 1991, she had never heard of Norplant—it was not yet on the market. The word "surgery" was never uttered at the original sentencing.[40] Lastly, Johnson had no knowledge of the grave complications Norplant could pose to her diabetes, her heart ailment, or her blood pressure problems.[41]

By the California judiciary's own logic, the Johnson sentence never should have happened. In a 1984 case involving a woman convicted of child endangerment, the California Court of Appeals ruled that the court could not impose birth control as a condition of probation. The justices emphasized the need to protect an individual's right to privacy from punitive government intrusion (a right explicitly guaranteed in California's state constitution).[42] It is true that privacy rights can be problematic grounds on which to argue a case (all manner of brutality has been cloaked and protected under the banner of privacy, including rape, domestic violence, and child abuse). The danger Johnson posed to her children, however, was in no way mitigated by Broadman's sentence. Johnson's crime was not having her children, but beating them. Broadman's ruling failed then, even by his own specious rationale that corporeal punishment, in particular, government trespass into Johnson's body (declared an unconstitutional exercise of punitive power by the state Court of Appeals just seven years earlier), was permissible in order to protect the defendant's existing children.[43] In a *60 Minutes* interview, Broadman conceded that he had infringed on Johnson's constitutional rights:

> But what you have to do as a judge is, you must balance conflicting constitutional rights. And here, what I did was, I found that there were constitutional rights of the children, the born children and her *unconceived* children, and I balanced their rights against her rights, and they won.[44]

Protecting the "unborn" by blocking their conception has long been an imperative of population controllists and has surfaced repeatedly in pronouncements on disability and childbearing. In a more general sense, judicial deliberations that take not only "fetal rights" but now the "rights" of the unconceived into sentencing considerations—as if either constitute legal or material entities—set a chilling precedent.[45] Nightmare scenarios of court-ordered impositions or denials of childbearing, lifestyle monitoring, or other forms of punitive intervention loom large. There is already precedent for this in California. In 1966, Judge Frank Kearny, echoing *Buck v. Bell*, sentenced Nancy Hernandez to jail time because she refused a sterilization for probation deal. According to Kearny, the proviso was a logical one because Hernandez had a "propensity to live an immoral life." One of her two children was born while she was unmarried.[46]

Broadman's own sentencing practices are marked by the same sort of gender bias. Contrast Johnson's mandatory acceptance of a surgical procedure and denial of bodily integrity with the punishment meted out to a man convicted of child molesting. Broadman permitted this defendant to serve out his sentence in his own home on the condition that he post a sign on his door reading, "Do not enter. I am under house arrest."[47] Furthermore, as some critics have noted, Norplant's visibility in a woman's upper arm announces her reproductive status. Tamar Lewin, who covered the Norplant beat for the *New York Times,* noted that "unlike other contraceptive methods, the presence of Norplant could easily be monitored by a parole officer or a welfare official, anxious to prevent further pregnancies in a convicted child-abuser, a woman who carried the AIDS virus or a woman already receiving public assistance."[48]

In Darlene Johnson's case, hard-and-fast evidence of racial bias is more elusive. Was Johnson, African-American, consigned to the realm of constitutional unworthiness by a racist judicial shorthand? It seemed an obvious question, yet it was not posed to Broadman (or anyone else connected with the case) during the *60 Minutes* segment. Nor was it dealt with in most of the major press accounts. Yet, it is impossible to ignore the familiar racial power dynamics of this case.

Broadman's class bias was harder to conceal. The judge stated that his ruling was in no way related to Darlene Johnson's economic standing,[49] yet, at the original sentencing hearing, he asked her if she was on welfare. "He said it didn't have anything to do with whether she was poor or rich," Johnson's attorney said. "What does her welfare status have to do with anything if he's not concerned with her economic status?"[50] It was not the first time class entered into judicial deliberations on sterilization.

In 1965, the Superior Court of Santa Barbara granted Victoria Tapia, convicted of welfare fraud, a reduced sentence and probation after she agreed to be sterilized. In 1972, in *Cook v. State of Oregon,* the Oregon Court of Appeals ruled in favor of the sterilization of a seventeen-year-old girl on the grounds that she would be incapable of caring for children who would then become wards of the state.[51] In 1990, Tracy Wilder, a teenager convicted of smothering her baby, was given two years in prison and probation by a Florida court on the condition that she use birth control for ten years following her release. Talcott Camp of the ACLU noted the class bias of the Wilder case, contrasting it to the story of a fifteen-year-old girl in Connecticut who left her dead infant under her dorm bed at Miss Porter's Prep School. Her identity was never released to the press, which cited the unnamed student's "ordeal and trauma." She was not sentenced by the legal system, but briefly hospitalized and then sent home. "You can't give up fundamental right as a condition of probation," Camp said of Wilder's sterilization-for-probation deal. "You could not require a man convicted of slander to give up his right to free speech after he got out of jail."[52] A few weeks later, Broadman ordered Darlene Johnson not to speak publicly about her sentence.

The legality of Broadman's ruling was never resolved in California (Broadman recused himself and Johnson was arrested for violating her probation).[53] Nonetheless, the ruling struck a responsive chord across the country. In Washington State, medical staff lobbied to make Norplant mandatory for convicted women with crack-addicted babies. One doctor suggested paying pimps one hundred dollars for every prostitute they could coerce into accepting Norplant—removing women from the deliberations altogether.[54] In light of the fact that women in the sex trade are among the most consistent users of birth control, this proposition exposed the punitive impulses of some medical professionals engaged in Norplant policy debates. Colorado State Representative Bill Jerke called for "voluntary" sterilization of prisoners in return for ten days off their sentences. Betraying a belief in inheritability of felonious tendencies, he told the press, "we're talking about reducing the crime rate down the road."[55]

Douglas Besharov of the American Enterprise Institute has denied that these proposals carry any class prejudice. Quite the contrary, says Besharov, "If a woman is convicted of a crime and is given a choice to use something that the middle class is using, that's a signal that it's a lesser penalty.... If she doesn't want it, she can choose to go to jail."[56] But the middle class in question, presumably not standing before a judge, does not have a prison stretch hanging over their collective head.

The rash of punitive Norplant measures in the criminal justice system is motivated to a large degree by the same reasoning that has endowed birth control with the power to "fix" poverty. In this variation, Norplant is invested with the ability to curtail crime. At the same time, the use of Norplant as penalty serves to conflate socioeconomic status and criminal activity. It masks an underlying economic agenda: to attack welfare recipients. Soon after Broadman's ruling, the mask came off.

Incentives and Cutbacks: Norplant and Welfare Reform

It happened in an American city. A blind couple were the object of much sympathy, some little charity and a great deal of relief. They had six children, all born blind, and the community commiserated as Americans always do with profound misfortune. The idea that six young lives had been born into the world doomed never to see sunlight or trees or other human beings is profoundly shocking. Everyone in town felt sorry for the couple and for their children.

But the couple did not feel sorry for themselves. They had had their six children deliberately and planned to have more. They found that their relief checks and allowances for disability increased with each child, and only as a large family could they be comfortable.…

This is an extreme but not an isolated case of what happens in this country.[57]

Margaret Sanger's tale, spun in 1951, has deep resonance for current welfare debates. Like recent attempts to link welfare with Norplant, Sanger's anecdote reached out both to liberals—by emphasizing the need to alleviate poverty—and to conservatives—by alleging the depletion of public coffers by families on relief, going so far as to invoke the now familiar charge that the poor's primary motivation for having children is a bigger relief or welfare check.

Editorial pages trumpeted Norplant's entry onto the birth control stage. A *Washington Post* commentator proclaimed, "New Contraceptive Advances Freedom, Responsibility." The *Philadelphia Inquirer* wasted no time in suggesting that Norplant was particularly well-suited to alleviate African-American poverty. Not coincidentally, the editorial appeared at a time marked by a reinvigorated vilification of "welfare queens," almost uniformly coded as Black. Just one day after FDA approval, the *Inquirer* queried, "Poverty and Norplant: Can Contraception Reduce the Underclass?" The editors called for "an increased benefit for agreeing to use this new, safe, long-term contraceptive" for women "of any race," yet the entire piece centered on researchers' reports that nearly 50 percent of Black children in the United States were living in poverty. "The main reason more black children are living in poverty is that the people having the most children are the ones least capable of supporting them." The editors insisted that their interest was not in saving money, but in saving children.[58]

Kansas State Representative Kerry Patrick, on the other hand, was entirely up-front about his interest in saving money. In early 1991, just weeks after Darlene Johnson was sentenced in California, Patrick introduced a bill ideologically aligned with Sanger's sterilization "pension plan." The proposed legislation would have paid women on welfare five hundred dollars in exchange for their five-year acceptance of Norplant (already free in Kansas to AFDC recipients), plus an additional fifty dollars a year for medical follow-up. Women who opted for early removal would "rebate" a portion of the money.[59] Patrick, who described himself as "a pro-life Republic[an] Presbyterian," explained his reasoning:

> A cookie manufacturer gives you 25 cents off to buy their particular brand of chocolate-chip cookie. It's an incentive to try to get you to try their brand. You have to make the amount so that they'll respond.[60]

Patrick claimed his legislation would save the state the $205,000 per child it would otherwise spend on welfare from birth to adulthood. The bill's message, as Dick Kurtenbach of the ACLU noted, was "that for a certain class of women, the state would prefer they don't have children."[61]

Whereas the executive director of the NOW Legal Defense Fund had lambasted Broadman's sentencing just weeks before, NOW and Planned Parenthood of Kansas said there was "some merit" to Patrick's bill. Colleen Kelly Johnston, president of the Wichita chapter of NOW, stated, "If it is voluntary and if the state is going to pay for the procedure, I don't see any serious problems with it. . . . But I would prefer to see it extended to women in poverty, not just women on welfare."[62] Johnston's remarks exemplified the traditional liberal take on reproductive technologies and the limits of liberalism's "free choice" doctrine. Barbara Katz Rothman has written that the liberal stress on rational free choice does not factor psychological or economic forms of coercion into its analysis, the result being that "the 'choices' people make out of their poverty or need— choices individuals may experience as being coerced—liberals tend to see as freely chosen."[63] Liberalism makes no proviso for the relative nature of free choice.

Patrick's bill was voted down, but by early 1992, similar proposals to mandate Norplant for women on welfare—twenty in all—were introduced in thirteen state legislatures. The Texas House of Representatives considered an appropriations amendment that would have paid women a total of five hundred dollars for accepting Norplant (three hundred for receiving the implants, two hundred more for not removing them). In Louisiana, former grand wizard David Duke, running for governor, introduced a bill that would have provided one-hundred-dollar annual bonuses for women on welfare who used Norplant. He characterized the bill, which died in committee, as "tough love." In 1994, Connecticut legislators proposed giving AFDC recipients seven hundred dollars at implantation and an additional two hundred dollars annually. That same year, a Florida bill proposed a two-hundred-dollar bonus for women living below 125 percent of the poverty line who use contraception—Norplant or otherwise. Men were slated to receive four hundred dollars.[64]

All these bills failed, though some came very close to passing. However, as state after state terminates additional child allowances for welfare recipients, more and more women may be impelled to "voluntarily" accept Norplant or other measures. Arizona, Nebraska, Georgia, Wisconsin, California, and New Jersey have all passed child exclusion acts. The New Jersey "family development plan" denies benefits to children born more than ten months after their mother's application for welfare.[65] The California exclusion act exempts children who are born as a result of failed birth control, but only if the contraception in question is Norplant, an IUD, or sterilization—all long-term, semipermanent, and physician-controlled.[66]

Theresia Degener, writing in *Issues in Reproductive and Genetic Engineering*, observed, "Neo-eugenic population-control policies no longer principally rely on compulsory state intervention but on *voluntary eugenics* from below. The focus is now on the sovereign self-responsible individual with her/his own economic and social interests."[67] As welfare moves from federal entitlement to state block grants, and Aid to Families with Dependent Children becomes *Temporary* Assistance to Needy Families, explicit government directives or ultimatums vis-à-vis fertility will no longer be necessary. Women will have little recourse but to "choose" Norplant.

Norplant and the Pathologization of Youth

Before the FDA even gave Norplant the green light, Isabel Sawhill, a senior fellow at the Urban Institute, suggested that all girls should be encouraged to accept the implants as soon as they hit puberty. Two years

later, a "Norplant consortium" was assembled in Baltimore by the city's
health commissioner. Private doctors, public health administrators, and
representatives from hospitals and private foundations converged to
promote Norplant, "especially among the city's hard-to-reach 13- to 18-
year-old girls." The following month, clinics at Paquin Middle-Senior
High School (attended by girls who are pregnant and/or already mothers) began dispensing Norplant. Two additional city schools followed
suit in 1994.[68]

Much of the press coverage framed the Baltimore plan as a means to
improve the life chances of young girls. The only hitch was Norplant's
alleged potential to encourage "promiscuity." The policy's real drawbacks
were all but ignored. Neither health concerns nor the implicit denigration of youth, particularly low-income youth and youth of color, were
ever addressed as salient issues.[69]

FDA approval does not mean that Norplant was not experimental,
particularly for young women. It had never been tested on youth, but
the hysteria around teen pregnancy rendered this inconsequential.[70] Norplant contains the same hormone as Depo-Provera. According to three
studies cited by the National Women's Health Network (including one
produced by the World Health Organization), young women who used
the injectable for a number of years were at increased risk for breast
cancer. A Texas study found that young women on Norplant often stopped
using condoms. This should have been of grave concern to the consortium, given that Maryland, at the outset of the Baltimore Norplant program, had one of the highest rates of AIDS nationwide. Equally disturbing
was the revelation that the school clinics would not guarantee Norplant
removal. Consent forms in at least one Baltimore school stipulated that
the girls might have to go to a hospital or another clinic to have the implants removed.[71]

Nevertheless, the press was touting Baltimore's "Bold Attack on Teenage
Pregnancy."[72] Popular and media wisdom has long posited ever increasing rates of teenage pregnancy as the impediment to "breaking the cycle
of poverty," laying blame for economic disparity on girls who supposedly
cannot say "no." The truth is, adult *and teen* births had shown identical
trends for fifty years: a steady *decline* since 1960.[73] In 1999, the National
Center for Health Statistics reported a forty-year low in the birthrate of
fifteen- to seventeen-year-olds across all racial groups.[74] As Mike Males
wrote in *Scapegoat Generation,* "the view of roiling teenage biology de-

termining reckless teenage destiny clearly remains the mainstream view of the social and health scientists providing commentary to the media and to political authorities." Males issues the reminder that "teenage" in such rhetoric is often stand-in verbiage for "minority group" and/or "low-income group."[75] Like Adolph Reed, he discerns a strong resemblance between liberalism's position and Charles Murray's contention that childbearing by poor and low-income women is "dysgenic."

> Liberals who ascribe dire disadvantages to teenage motherhood, which is a surrogate for low-income motherhood, are making the same argument as Murray's without being as candid. If parents are poor, there is no good age, no healthy age, no age at which "social costs" will not result from childbearing.[76]

Rosetta Stith, principal of Paquin, told a reporter that dependence on the welfare system has left her students unwilling to think about the consequences of pregnancy. The article also noted that Norplant was "gaining favor as a policy option in cities where teenage pregnancy has become a way of life."[77] Actual numbers tell a different story. According to congressional studies produced before the 1996 Welfare Reform Act, before workfare pushed recipients off the welfare rolls and into sub-minimum wage jobs, the average teenage mother, despite being poorer before becoming pregnant than the average mother who gave birth in her twenties, remained on AFDC only one year longer than her adult counterpart. At the time of the report, half of all teenage mothers did not even receive welfare, and 40 percent of those that did were off the rolls within one year; 70 percent were off within four years.[78] Maryland Governor William Donald Schaeffer's 1993 State of the State suggestion to make Norplant mandatory to curb "runaway" welfare costs notwithstanding, the fact is that larger welfare payments have tended to correlate with lower, not higher, rates of teenage single motherhood in all fifty states and Washington, D.C.[79]

All this is not to say that girls should not be given access to contraception (the Baltimore school system had been dispensing birth control for two years before the Norplant program began) or that they should not be encouraged to delay pregnancy. However, a policy that induces girls to accept a method known to be hazardous—and is unresearched for their age group—coupled with an articulated focus on their welfare status and an equally intense unarticulated concentration on their race, is highly suspect. The girls in Baltimore seemed to be keenly aware of this. By May 1994, only thirty-six out of Paquin's seven hundred female students had had Norplant inserted. A total of two students had undergone the procedure at the additional two city schools where Norplant was available.

Robert Hill and J. Dennis Fortenberry, of the University of Oklahoma Health Sciences Center's Department of Pediatrics, have critiqued the construction of adolescence as "a developmental period defined by its problems." Decrying the view of adolescence as "a condition that is inherently pathological," they write, "Adolescence per se is seen as the inevitable 'risk' factor for these widespread problems as if the origin of these problems were innate to adolescents, rather than complex interactions of individual biology, personality, cultural preference, political expediency and social dysfunction."[80]

The idea that science will rescue America's youth from its own dysfunction has currency that goes beyond Norplant-related directives. In 1999, the Bureau of Alcohol, Tobacco, and Firearms began a joint venture with a "threat evaluation company" to pilot Mosaic-2000. The computer program, designed for school administrators, is supposed to evaluate "troubled" students and rate their potential for violence on a scale of 1 to 10. The ACLU has labeled it a "technological Band-Aid."[81]

The increasingly encrypted use of the technofix has strengthened its appeal to liberals who, consciously or not, join with conservatives in establishing race, class, and gender (augmented by youth and disability) as primary determinants in eugenics enterprises. The new technologies, in concert with public policy "innovations," have reinscribed old patterns wherein women and girls are first pathologized for membership in one or more specified demographic groups, and then pressured into accepting medically risky and constitutionally dubious "cures." Norplant and Depo-Provera fit into a longer history of eugenic sterilization mandated by state legislatures, enacted by individual doctors, and upheld by the judiciary as a practice integral to welfare "reform." The only change has been linguistic: *welfare status* and *teenage pregnancy* are cited by policy makers unwilling to utter *race* or *class*. *Eugenic* sterilization for the "feeble-minded" and "unfit" has evolved into *genetic* rationales for policing the fertility of women with disabilities.

Disability and Eugenics

The Constant Consensus

From the passage of the earliest sterilization statutes, warnings have been issued on the repercussions of eugenicists' assaults on the "feeble-minded." Eugenics opponents noted that women and people of color were frequently and erroneously so designated, and cautioned that endorsements of compulsory sterilization of the disabled would lead to an ever-widening circle of candidates among other reviled groups. Ironically, while physically and developmentally disabled women have historically been among the most prone to eugenic attack, their precarious position has rarely been viewed as anything other than an alarm, a call to safeguard the rights of nondisabled, though otherwise marginalized, individuals and groups.[1]

Robert Blank notes that while "strict reliance on a eugenic rationale seldom is used any longer... there continue to be some advocates of the genetic basis for nonconsensual sterilization."[2] The substitution of the word "genetic" for "eugenic" has ensured a continued consensus among physicians, policy makers, and commentators on an assumed imperative to prevent the birth of children with disabilities by monitoring the fertility of disabled women.

The defamation of people with disabilities has been a staple of eugenics oratory from the very beginning and their elimination a stated goal of almost every eugenics enactment. But recently, as with other intrusions into women's childbearing decisions, the issue has been couched in terms of expenditures and "the best interest of the children." Shortly after Norplant's debut, a columnist for the *Los Angeles Times* wrote:

It offers the possibility of reducing to some degree, the most expensive of all social ills: the annual arrival of thousands of unwanted, sometimes disabled children to parents who are unable to care for them.

In the past decade, California has inherited an army of these children. No one knows how they will be raised. In the case of the crack babies, no one knows if "raising" is a concept that applies.[3]

As in early twentieth-century eugenic treatises, those facing greatest jeopardy are constructed as the greatest menace; the gravest of all "social ills"; an "army" poised to destroy the strongest; the most secure, the least endangered. Margery Shaw, a lawyer and geneticist, went even further than the *L.A. Times:* women who decide to continue pregnancies after prenatal testing has detected a "seriously deformed or mentally defective" fetus, and parents who "knowingly and willfully transmit deleterious genes" should be held accountable to their children. Such parents, she explained, place additional burdens on society and on other family members and make their children suffer. "Society should decide to wipe out muscular dystrophy, Tay-Sachs disease, cystic fibrosis, and sickle cell anemia, just as small pox, polio and measles have been virtually eradicated."[4] Smallpox, polio, and measles, however, were suppressed by vaccines, not campaigns of coerced abortion or mandatory birth control. Furthermore, unlike Tay-Sachs or sickle cell anemia, they did not strike hardest at particular ethnic or racial groups.

The right to abort should be absolute, under any circumstances. At the same time, as Ruth Hubbard has stated, "'a woman must . . . have a right—and more than that, the opportunity—not to terminate . . . in the confidence that society will do what it can to allow her child to live a fulfilling life."[5] In actual practice, society has been unwilling to meet that obligation. Furthermore, women who undergo prenatal testing are generally not made aware of available support services or apprised of the possibilities open to a disabled child, only the limitations. They are counseled by *medical* personnel, using a *medical* model that presumes the undesirability of an "imperfect" offspring. Unless parents are themselves disabled or have had previous relationships in the disability community, they usually do not have all the information necessary to make a fully informed choice. Still more dissuasive is the fact that, currently, the decision to carry a pregnancy to term where disability is a high probability can subject women and their doctors to punitive legal action. Lisa Blumberg writes that "wrongful birth" and "wrongful life" litigation, based on allegations that "a person who is presently alive should not be here," have been banned in matters involving "healthy" children, because courts fear such suits would be damaging to their self-esteem. "However, judges have decided that it is a different matter when the child is 'defective.'" In wrongful birth and wrongful life cases, doctors are liable for damages if they did not recommend prenatal tests or inform parents of the likelihood of having a disabled child. While both types of cases are extremely troubling, there is an important distinction. Blumberg explains:

> in wrongful birth cases, it is the parents who are stating that they have been harmed by the child's birth, while in wrongful life cases, suit is being brought on "behalf" of the child on the premise that the child himself [sic] is harmed by his existence.... Thus, while wrongful birth suits at least have a certain logic to them, wrongful life suits, which have involved children with such varying disabilities as deafness, hemophilia, and mental retardation, engage the courts in absurd and philosophical speculations and have terrifying implications.

Blumberg has suggested that while all wrongful life suits, thus far, have been brought against health care providers, "given that the premise of such suits is that nonexistence is preferable to having a disability," it is not unforeseeable that a court-appointed guardian could sue a woman who refused an abortion or parents who declined prenatal screening.[6]

The irony, Blumberg writes, is that while society wants to ensure that the disabled are not born, it is unwilling to take measures that would prevent people from having disabilities. "Why is it that we collectively spend so much for ultrasounds when support for nutritional programs for infant[s] and nursing mothers is so lacking?" Blumberg contrasts the government's willingness to fund testing for the cystic fibrosis gene with its refusal to make existing new treatments accessible to people who currently have cystic fibrosis.[7] Attention to poverty, malnutrition, low birth weight, environmental racism, and exposure to toxins would go a long way toward lowering disability rates without targeting the disabled for eugenic curatives. Historically, however, these measures have not been as appealing as technological fixes.

The centrality of postconception, prenatal intervention is a relatively recent phenomenon. For most of this century the focus has been on preventing conception among people with real or imagined disabilities altogether. Early twentieth-century eugenic decrees targeted the developmentally delayed, the mentally ill, the epileptic, the physically disabled. A "Suggested Experimental Federal Law," proposed at the 1912 International Eugenics Conference, called for the sterilization of "inmates" of all federal hospitals, "defective immigrants and immigrants with defective heredity," among others.[8] That same year the president of the Michigan State Board of Health, resolute in his commitment to sterilize the disabled, declared:

> The State will not permit the reproduction of the weak-minded, the insane, the alcoholic and the criminal, and will deny parenthood to those suffering from diseases which cripple offspring. This prohibition will be enforced by segregation, sterilization, or by both.... The State has a right to protect its *honest* citizens against those that are *evil*, and no one can deny that the multiplication of the classes mentioned above and specified by the law is an evil.[9]

The American Eugenics Society, long active in restrictive immigration campaigns, actively lobbied for legislation mandating that individuals with disabilities or with disabled relatives be sterilized and/or post bond if they wanted to marry.[10] Sanger's *Birth Control Review* provided a platform for this kind of thinking. The banner on its June 1925 issue promised articles on "Eugenics and Birth Control." Inside was a reprint of P. W. Whiting's "Selection, the Only Way of Eugenics," originally delivered at the Sixth Annual Neo-Malthusian and Birth Control Conference. Whiting, a biologist and geneticist, proposed sterilization *and* segregation of the "unfit," all the while remaining attentive to fears of "situational" homosexuality:

> Moron communities have been suggested. Admission to such should be entirely voluntary but would in all cases involve sterilization. There seems to me no reason for isolation of the sexes; in fact such isolation would tend to generate sex perversions and dissatisfaction. Farm or other suitable work should be carried on under supervision and the colonies once started would be self-supporting or even productive. None would be compelled to enter nor to stay if once they entered.... Whole moron families might enter and the sterilization ensuing would act as a constant selective process for the elimination of the type. If they left the colonies, they would be racially innocuous at least and their places would be filled by others who would likely become racially innocuous. Very extensive colonies of this type might be established for the price of a few battleships.[11]

Whiting could afford to call for "voluntary" admission because surgery would render sterile even the briefest residents. His promise that the postoperative "racially innocuous" would surrender their places to others predated Justice Holmes's remark that Carrie Buck's sterilization might "open the asylum to others." Clearly, the purpose of Whiting's colonies would not have been to improve the quality of life of the disabled, but to neutralize their potential "contamination" of the gene pool.

Margaret Sanger was not so willing to make sterilization or other forms of contraception voluntary for those deemed physically or mentally unfit. Sanger, as previously noted, had long advocated sterilization for the "feeble-minded, the insane and the syphilitic," though she wrote in 1919 that such procedures were insufficient to cope with the "constantly growing stream of the unfit."[12] Several years later, she responded to a reader's query on the advisability of undergoing the operation:

> We do not think sterilization advisable for strong, healthy people, for
> they may change their minds about having children in four or five years
> time. However, if there is any taint of insanity or epilepsy in either
> husband's or wife's ancestry, sterilization is advisable.[13]

While a 1928 BCR editorial voiced the American Birth Control League's opposition to "those eugenists [sic] who would prefer compulsory [sterilization] laws for the unfit," Sanger personally took a harder line.[14] Bemoaning the "notorious fecundity of feeble-minded women," she wrote in The Pivot of Civilization, "Every feeble-minded girl or woman of the hereditary type, especially of the moron class, should be segregated during the reproductive period." Their male counterparts, she continued, were equally dangerous. But imposed isolation was not enough.

> Segregation carried out for one or two generations would give us only
> partial control of the problem. . . . we prefer the policy of immediate
> sterilization, of making sure that parenthood is absolutely prohibited to
> the feeble-minded. This, I say, is an emergency measure.[15]

Ellen Chesler has conceded that Sanger "refused to consider that the handicapped may also be worthy, that the rights of the individual, in any event, must reign supreme in a truly democratic society." Nevertheless, she framed Sanger's intention as benign. Sanger, she explained, favored enforced birth control for "the physically or mentally incompetent, who could not themselves understand the benefits of smaller families."[16] But this exoneration ignores the fact that "mentally incompetent" was a broad designation that encompassed many people who were more than able to make informed decisions. Additionally, it glosses over Sanger's assumption that the *physically* disabled were somehow cognitively impaired.

As the years passed, the disabled were persistently cast as insupportable and calls for intervention continued unabated. Ruth Hubbard reported on a 1942 article that appeared in the *Journal of the American Psychiatric Association*. The author, psychiatrist Foster Kennedy, advocated killing "retarded children up to the age of five."[17] Linus Pauling, the two-time Nobel prize recipient who called for premarriage exams for "defective genes," stated in 1968, "I have suggested that there should be tattooed on the forehead of every young person a symbol showing possession of the sickle-cell gene or whatever other similar gene." Despite the pretense that proposals such as Pauling's were disease-specific and not meant to single out marginalized groups, there is no escaping the disproportionately high rates of sickle cell anemia among African-Americans—a fact that made racist an already nefarious proposition.[18]

The heightened awareness of sterilization abuse in the 1970s (the HEW guidelines, the Relf hearings, the lawsuits) did not deter proponents of involuntary sterilization for the disabled and potential parents of the disabled any more than an awareness of Nazi eugenics hindered Pauling or Foster Kennedy in their calls for tattooing or mass murder. In *North Carolina Association for Retarded Children v. North Carolina* (1976), the North Carolina Supreme Court and a three-member federal district court upheld the compulsory sterilization of individuals deemed incapable of caring for children or likely to have children with "serious physical, mental, or nervous diseases or deficiencies."[19]

Rosalind Petchesky wrote that compulsory sterilization statutes rely on a set of erroneous and totalizing assumptions about mental retardation, including "that retardation is in most cases genetically determined; ... that most *mildly* retarded persons are 'incapable of managing temporary forms of birth control'; ... and that the capacity of retarded persons to understand and consent freely to contraceptive planning, including sterilization, is negligible or nonexistent."[20] This last supposition abandoned many developmentally delayed women to pharmaceutical experimentation. In 1973, a full twenty years before its FDA authorization as a birth control method, Depo-Provera was being administered at a Tennessee state facility for the mentally retarded. In 1984, after it had again been rejected by the FDA, a member of the administration's 1984 board of inquiry suggested that although Depo-Provera had not been proved safe for "*human subjects*," it could be used on *retarded women* and *drug addicts*.[21]

Buck v. Bell, which gave the constitutional stamp of approval to the involuntary sterilization of the "feeble-minded," has never been overturned. Ruth Macklin and Willard Gaylin attributed this to a move away from overt eugenics arguments toward debates that center on burden of care, rights of children, and the best interest of the developmentally disabled. The resulting cases are not constructed in such a way as to directly challenge the 1927 decision.[22] Yet even within this framework, eugenic ideology comes to the fore. A real concern with the "best interest" of the developmentally disabled would necessitate, as Petchesky wrote, a "commitment to programs involving sex counseling, body awareness, sex-integrated activities, private bedrooms, and many other basic conditions that are presently inaccessible to many retarded people."[23] In this regard, little has changed for people with developmental disabilities since Carrie Buck's day. If they are low-income or poor, they remain at increased risk for coerced sterilization or economically mandated, doctor-controlled contraception, especially as funding cuts eliminate other birth control options. Norplant, for example, is covered by Medicaid in every state. Other contraceptive methods are not. Advocates of externally imposed childlessness for the disabled, including "best interest" liberals, tell yet another class of people, "the state would prefer [you] do not have children."

Quinacrine, the Next Wave

Fast on the heels of Norplant and Depo-Provera came quinacrine, initially developed in the 1920s as a treatment for malaria, now a means of chemical sterilization.[1] As of this writing, no regulatory agency in the world has approved quinacrine as a birth control method. Nevertheless, tens of thousands of women in Indonesia, India, Pakistan, Egypt, Croatia, Chile, Bangladesh, Costa Rica, Pakistan, Iran, Venezuela, and Vietnam have all undergone insertions, including at least one hundred women on a Vietnamese rubber plantation who were told they were having their IUDs checked, and then inserted with quinacrine without their knowledge.[2]

While intrauterine use of quinacrine has been regarded as a possible sterilization method since the 1960s, no studies have been done to determine its link to cancer (though there has been at least one reported case of leiomyosarcoma, a rare uterine cancer). Research on primates has shown an impact on the liver and cardiovascular system. It is difficult, if not impossible, to reverse quinacrine-induced damage.[3] The potential risks are so great that the majority of population control groups have broken with their usual stance and come out against its use. But the drug still has champions.

The Dallas-based Leland Fikes Foundation has funded quinacrine studies by way of the Center for Research on Population and Security (in North Carolina). The Fikes Foundation has also played benefactor to the Federation of Americans for Immigration Reform, the driving force behind California's 1994 anti-immigrant initiative, Proposition 187. FAIR received $50,000 from the foundation in 1994. Two of quinacrine's chief promoters, Donald Collins and Sally Epstein, are FAIR board members.[4] This convergence of sterilization efforts and anti-immigrant lobbying typifies eugenic ideology and endeavor. It is neither incidental nor without precedent.

Henry Pratt Fairchild, vice president of Planned Parenthood for nine years (1939–1948), shared Margaret Sanger's belief that eugenics and population control were crucial for world peace. He was extremely active in the anti-immigration work of the American Eugenics Society and served for a time as its president. Guy Irving Burch, board member and contributor to the *Birth Control Review,* was equally vehement in his anti-immigrant stance. He too came from the ranks of the AES, where he served as secretary, and for a time he led the American Birth Control League.[5] In 1929, the *BCR* published "Immigration Control," in which Burch cautioned, "If the National Origins clause is repealed, justice and science will be defeated, and the pioneer stock that established this country's institutions and ideals might as well move out." Despite its appearance in this particular journal, Burch's article made but one reference to birth control:

> The two sources of population growth in this country are immigration and natural generation. If we are to have Birth Control, we should also have immigration control, and visa *[sic]* versa. The advocates of both types of population control should join forces to work for a just and rational population policy."[6]

Set against this backdrop, there is a certain consistency to Collins and Epstein's dual role as FAIR members and quinacrine hawkers, a historical logic to their 1995 appearance at the NGO forum at the United Nations Fourth World Conference on Women in Beijing, and their 1996 attendance at the Feminist Expo in Washington, D.C.[7] Though their efforts to peddle quinacrine subvert feminist goals of autonomy and self-determination and undermine women's physical well-being, Collins and Epstein have been able to operate under the guise of women's advocates. The door was held open for them by liberalism's long-standing compliance with and propulsion of eugenic/population control policies.

The history of eugenics in the United States is characterized by commonality, if not always cooperation, between reactionary and reformist agendas. Despite their areas of contention, in many instances liberalism furthered what conservatism and even out-and-out fascism could not.

Both liberals and conservatives assumed paternalistic postures with regard to women, the poor, the racialized, and the disabled. Both have failed to address systemic inequities. Both have supported population control projects in the United States and in "developing nations." This common cause was framed and fostered by one of liberal feminism's most cherished icons, Margaret Sanger, though certainly not by her alone.

Sanger's record as bridge builder between reformers and reactionaries and her recruitment of and by eugenicists and scientific racists is indisputable. The fallout from such efforts culminated in policies mandating the sterilization of women and girls both marginally and fully entangled in the welfare, health care, and/or criminal justice systems. Class, couched in the language of welfare status and expenditures, continues to be both the basis for selecting which women to monitor and a camouflager of racist motivations in those instances where vocal racism is to be avoided. The women and girls targeted, their advocates and their allies, have denounced eugenics practices through litigation, legislation, and popular education. Mainstream feminists, however, have responded by opposing federal regulations and/or failing to grant the issue of coerced sterilization priority status on par with abortion rights. It is a stance that has remained largely unchanged in the wake of new reproductive technologies, marketed as alleviatives to real or imagined "crises" (crime, welfare costs, teenage pregnancy, overpopulation). Liberal endorsements of Norplant "incentives" in the United States and overseas demand that some women forfeit the very claim to self-determination that liberalism extols.

It is crucial that we not allow ourselves to be manipulated into silence by a fear of giving ammunition to anti-choice proponents. It may be, as Asoka Bandarage has explained, that "when compared with right-wing fundamentalist opposition to abortion and women's reproductive freedom, this neo-Malthusian family-planning position seems liberal and consonant with feminist struggles for reproductive choice and human rights for women."[8] The elevation of Margaret Sanger to feminist hero, the homage she continues to be paid—uncompromised by revelations of her eugenicist bent—is a product of this reasoning. It is a tribute that demands interrogation.

In the current political climate, scrutinizing any method of birth control or questioning the motives of its providers may feel risky. This apprehension, accompanied by the preeminence accorded to middle-class and affluent women's health needs, has resulted, for example, in staunch support of RU486 despite its known and unknown health risks.[9] Similarly, an embrace of Norplant or Depo-Provera may seem logical, as right-wing groups like Concerned Women for America and the Christian Coalition (not to mention the Minnesota College Republicans) disingenuously posit themselves as defenders of the weak, pointing out health hazards and the threat of a eugenics revival.[10] The impulse to endorse anything the Christian Right opposes (and vice versa) will not work this time. A more nuanced response is required.

Conclusion

> Whereas inordinate individual wealth is damaging to society, and undesirable civic tendencies are transmissible by heredity, it is hereby enacted that each society for the improvement of the poor shall call in two philosophic anarchists and one socialist, who shall determine whether any person who shall have acquired inordinate wealth is by reason of the over development of his acquisitive greed a menace to the peace and welfare of the community, and if they so determine, they may cause to be performed upon him an operation for sterilization to prevent procreation, provided, in no event shall anarchists and socialist receive more than $3.00 for their consultation fee.
>
> —Charles A. Boston, "A Protest against Laws Authorizing the Sterilization of Criminals and Imbeciles"

While I was still in the early stages of my research, a friend gave me Charles Boston's send-up of sterilization statutes and it has remained tacked up over my desk ever since.[1] It offered a little comic relief, as I immersed myself deeper and deeper in the less than uplifting story of this country's eugenics past. More than that, it served as a reminder of the historical continuity of dissent. The necessary longevity of that challenge is itself a warning on the tenacity of eugenics.

It is tempting to relegate U.S. eugenics to its early twentieth-century heyday, or to ascribe to its components a linear and manageable history. In reality, the record of eugenics is neither linear nor manageable. It is not even merely historical, but ongoing. In truth, the distance eugenics has traveled since the word was first coined has been, at best, linguistic, but never paradigmatic. The premise that people with disabilities should, and could, be expunged/excised/eliminated by "controlled evolution" remains largely intact. This seems to be a point of concurrence, rather than contention, among liberals and conservatives. Gay gene research continues, threatened only by sensational reports on the gay germ. Lombroso's born criminal of the nineteenth century has given way to John Donohue and Steven Levitt's 1999 theory of "unborn criminals." According to the two economists, the drop in crime during the 1990s is the outcome of the 1973 *Roe v. Wade* decision and, they write, "appears to be attributable to higher rates of abortion by mothers whose children are most likely to be at risk for future crime." These mothers are identified by Donohue and Levitt as teenagers, single parents, and African-American women.[2]

George Mosse has described racism as a "scavenger ideology"—a "heightened nationalism" that posits the differences between peoples as immutable.[3] Eugenics, too, is a scavenger ideology, exploiting and reinforcing anxieties over race, gender, sexuality, and class and bringing them into the service of nationalism, white supremacy, and heterosexism—not for the first time, but under cover of a new phraseology. The verbiage of eugenics, the valor, neutrality, and redemptive power accorded science and its counterfeiters, has enabled it to extend itself not only to diverse demographic target groups, but to disparate political philosophies. The alliances that have resulted appear, at first glance, counterintuitive. The broad base of "social inadequates," the appeals to liberals and conservatives, the invocation of objective science by seemingly oppositional political camps, and the role of eugenics as both co-opter and supporter of a wide range of movements and ideologies may make these connections appear evasive. Indeed, it is this very elusiveness that has endowed eugenics and its permutations with such resilience. Yet commonalities persist.

The most tangible of these is the living causeway of individuals who participated in multiple campaigns. Their efforts lay bare the connections among eugenic assaults and between eugenics and other movements. Margaret Sanger's use of Yerkes and Terman's flawed IQ data; Harry Laughlin's 1914 declaration that "the compulsory sterilization of certain degenerates is . . . designed as a eugenical agency complementary to the segregation of the socially unfit classes and to the control of the immigration of those who carry defective germ plasm";[4] F. E. Daniel's expressed desire to castrate gay men and sterilize other impediments to "race improvement"; FAIR's involvement both in the passage of California's Proposition 187 and in the peddling of quinacrine—all demonstrate eugenicists' awareness of the interdependency of their efforts.

More enduring, if less overt, are the theoretical mechanisms shared by most, if not all, eugenic projects: the monitoring of national identity, the use of metaphor to shore up animosity toward already despised and/ or feared groups, and a subscription to technological antidotes to structural inequities. In combination, and with varying degrees of prominence, all eugenics endeavors have used claims of biological primacy to justify and further the disfranchisement of the most politically, socially, and/or economically vulnerable groups in the United States.

Each campaign saw reactionary practices and "progressive" rationales for turning to science for a quick fix, the merging of medical and judicial/legislative doctrine, and backlash against reviled and increasingly visible groups. The utility of eugenics to extreme (and not-so-extreme) right-wing objectives is clear. It offered biological justification for everything from economic apartheid to Jim Crow. Anti-immigrant agitators used whatever data they could manufacture to bemoan the "passing" of Anglo-Saxon dominance and promote restriction, exclusion, and deportation. For well over a century, a heterosexist medical establishment has intervened to punish, imprison, segregate, and experiment on queers, suspected and real. Scientific racists had not only the license and wherewithal to sterilize women against their will, but the ability and class animosity to withhold welfare and health care from women they did not want populating the country.

But eugenics has also been propelled by populist, reform, and liberal sentiments, whether artifice or genuinely felt. Ward's 1920 fingering of steamship companies and employers of "foreign-born hyphenates" was an attempt to garner the support of pro- (native) labor sympathizers. Here, eugenics posed as defender of the working man. In a more individuated science-as-savior approach, medical pronouncements on homosexuality have often centered on the humanity of healing the "afflicted," thereby enabling a normal life. More recently, such "benevolent" impulses have taken the form of calling biology to task and removing the guilt of personal choice and/or parental failure. Finally, sterilization, be it tubal ligation or Norplant or quinacrine, has promised liberal proponents everything from an end to teen pregnancy to the elimination of world hunger.

Eugenics enterprises emerge as inextricably linked, whether they originate from punitive conservatives or technofix liberals. Any embrace of biological determinism or technological palliatives as stand-ins for substantive challenges to institutionalized racism, xenophobia, classism, homophobia—no matter how congruent with liberalism it may have appeared—not only failed to deliver on promises of social redress, but threatened political alienation and physical assault.

The linguistic and material composition of eugenic policies relied, and continues to rely, on the interplay of legal, public policy, and medical discourses. The medicojuridical complex has been an indispensable and constant enabler of eugenics in the United States. The first and perhaps most obvious of its manifestations is the practice of direct state intervention, as when judicial ruling mandates a medical procedure (like sterilization or castration) as punishment, therapy, or prophylactic (to hinder or halt a "diseased" hereditary line). Legislative edicts such as the 1917 and 1924 Immigration Acts, which rested heavily on their eugenic and racist appeal, would also fall into this category, as would sterilization-for-probation deals, economically coerced acceptance of Norplant, and absorption of reproductive technologies by welfare policy.

The second medicojuridical front has asserted itself in the state's sanctioning of eugenics enterprise in a slightly more understated manner. This involves the housing of such research in government facilities (such as immigration stations, state or county hospitals, or federal research programs), the accreditation of such facilities and medical staff, and the federal funding (and by extension legitimacy) extended to such endeavors.

Lastly, and a little paradoxically, silence, too, becomes an enforcer of judicial, legislative, and medical consolidation. In a society where so much is regulated, an absence of involvement, or at least interest, on the part of the government is both striking and suspect. Failure to interrupt (or even note) Indian Health Service's mass sterilization policy or to ban electroshock "therapy" in the treatment of "childhood gender nonconformity" is conspicuous and constitutes state complicity with human rights abuses.

"Even Bogus Science Has Political Consequences"

The forces that shaped restrictionist eugenics are still in evidence: racism, xenophobia, the quest for (and construction of) a "pure" national core, the conflation of race and nationality, and the ability of eugenics to serve and absorb other ideologies.[5] An examination of the lobbying efforts of anti-immigration eugenicists and other exclusionists sympathetic or opportunistic enough to employ eugenics arguments reveal a century of scientific intrusion into the legislative process. This input helped to cement the connections between white supremacy, nationalism, and xenophobia by providing them with a legitimating and unifying discourse. Eugenicist number crunchers fed the flame by producing statistics to establish immigrants' innate propensity toward crime, disease, illiteracy, immorality, and insanity—all deemed manifestations of weak heredity and all costly to citizen taxpayers. While Congress no longer has a designated eugenics expert, many of the rhetorical devices employed by Laughlin and his colleagues continue to be trotted out more than seventy years after the passage of the Johnson-Reed Act: immigrant as malefactor (from Robert DeCourcey Ward's tirade against immigration laws that enabled the birth of "criminal children" to the California Coalition for Immigration Reform's proclamation that "open borders = more citizens murdered"), immigrant as contaminant and contaminated (from blame for the bubonic plague in 1936 to Alan Nelson's current claim that undocumented immigrants are responsible for spreading disease in the United States), immigrant as financial drain (from pre–World War I charges that Europe was sending the United States its "paupers" to the erroneous designation of immigrants as taxpayer burdens during the Proposition 187 campaign). Madison Grant's warnings of "national suicide" have yet to stop ringing in his countrymen's ears. Eugenics casts immigration restriction as national (self-)defense.

Like Laughlin, Grant, Ward, and Johnson, who played on apprehension about World War I, interracial unions, bolshevism, and labor "unrest," today's anti-immigrant lobbyists also draw from and nourish the political pressure points of their time: antiwelfare sentiment, the war on crime,[6] fears over a "population bomb," and even mainstream environmentalism. The tone of news coverage and the underlying rhetoric of FAIR, of contemporary Pioneer Fund beneficiaries, and of eugenicists and fellow travelers has not departed much from those early days. They, too, offer hysterical warnings about the implications of immigrant women's alleged hyperfertility and its fiscal and racial implications for the country, deflecting charges of racism by packaging their attacks as apolitical scientific findings. Increasingly, it seems, Robert Ward's "American race" is still the goal.

Like those leveled against immigrants, "scientific" assessments of lesbians, gays, bisexuals, and transgendered people have been widely regarded as value-free, even emancipatory (the chief difference being that, in the case of the former group, it was the rest of the nation that would be liberated from the grip of eugenic menace, whereas in the case of the latter, the menaces themselves would be freed from what made them threatening in the first place). Karoly Benkert's 1869 attestation that "Nature, in her sovereign mood, has endowed at birth certain male and female individuals with the homosexual urge, thus placing them in a sexual bondage,"[7] situated homosexuality as natural, but nevertheless tortuous. We seem to have held fast to this view. For the LGBT community, this has meant that we have frequently been coerced and manipulated into cooperating with our own victimization. From "orificial surgery" to Simon LeVay's autopsies, the conviction that medical experimentation can only enhance the lives of queers—either by "curing" us of homosexuality or trying to convince a hostile, straight world that it is not our fault—has eclipsed the inevitable cultural biases in researchers' equations.

A stalwart faith in the objectivity of scientific theories has proven costly. No matter what the intentions of theorists and practitioners, the historical record of the medicalization of homosexuality and the continuity of the resulting consequences cannot be denied. Causation theories of homosexuality have been additive and therefore have abetted, not confronted, homophobia. LeVay's hypothalamus hypothesis will not dislodge psychiatric diagnoses any more than the latter displaced morality-based condemnations. Like past endeavors, the current search for origins is severely flawed—conceptually and in execution. Adherence to it by some gay rights advocates is just as much an act of political desperation as the right's invocation of IQ tests to quash Head Start programs and affirmative action. In an era that has seen visible, militant, and well-organized responses by the queer community to AIDS and to homophobia, the rush to genetic explanations of homosexuality is neither incidental nor coincidental. Yet, despite the bankruptcy of biological models and their entanglement with the mechanisms of backlash, a turn to socialization arguments must also be resisted. Queers must opt out of the nature versus nurture paradigm altogether.

The validity of the biology/environment binary is seldom interrogated: What is driving the imperative to isolate, once and for all, the origins of homosexuality? Why the desire to affix blame—for it is blame that is being leveled—to parenting or heredity or both? To whose advantage is it and what issues are being obscured in the process?

Homosexuality is accorded a kind of eugenic primacy when sexuality is constructed along hereditary or evolutionary lines. Any glitch in a petri dish or on a slide can be attributed to queerness if that is the focal point of the research. Similarly, other eugenicist treatises have granted dominion to race or class or ethnicity, marking those as biological arbiters of all human behavior.

Human sexuality, as it has been documented at the laboratories of the Salk Institute and the National Institutes of Health, is reduced to a zero-sum equation when it comes to queers. It is treated by researchers and by some gay rights groups as a unitary construct, uncomplicated by the range of lived experience within the LGBT community. In fact, Donna Minkowitz of the *Village Voice*, who covered the Amendment 2 trial in its district court phase, reported that CLIP's attorneys seemed reluctant to even mention bisexuals for fear that they would jeopardize their claims to the biological constancy of homosexuality (thereby acquiescing to, instead of challenging, the erroneous assertion that social groups can only petition for civil rights on the basis of biological immutability).[8] Hamer wrote that, for his study, he had to measure sexual orientation in a way that was "scientifically sound, *quantitative,* and could be replicated by other investigators."[9] Basing their research on such a rigid monolith, scientists seek a gene, a hormone, a cluster to explain homosexuality, when they cannot even define it. Kate Bornstein has written that "A dominant culture tends to combine its subcultures into manageable units."[10] Both scientific and legislative pronouncements on homosexuals and homosexuality have had everything to do with making people manageable. Our response must be to remain as unruly as possible.

If any trajectory has served as a wake-up call, a notice to be wary of all eugenic maneuvers whatever their political origin, it is the long campaign of compulsory sterilization. It is a history not limited to the dangers of the right's scientific racism, but inclusive of the liberal fidelity to technological/surgical intervention as a means of ending privation. It goes a long way toward explaining, for example, why, in the midst of litigation and education drives by organizers fighting involuntary sterilization in the mid-1970s, NOW representatives joined with Zero Population Growth in opposing the regulation of sterilization in California.[11] It is important to restate, in these concluding pages, that eugenics has been championed largely by nationalists, heterosexists, and white supremacists possessed of enmity and paternalism toward women, the poor, and the disabled. Without glossing over the primacy of their culpability, liberal capitulation to, and propulsion of eugenics must be addressed.

A good place to start would be to end the deification of Margaret Sanger—not only because such idolization signifies blatant disregard for those deemed, by virtue of race or class, eugenic castoffs—but because Sanger erected the bridge between scientific race- and class-demonization and what should have been a truly liberatory birth control movement. Her role as consensus builder cannot be denied. Her coding of eugenics as sound and compassionate and as mutually dependent on birth control has been upheld by liberals, albeit with an overhauled vocabulary. Sanger's courtship of eugenicists, NOW's opposition to even the meager protections against sterilization abuse drafted by HEW, mainstream feminism's failure to challenge welfare-for-Norplant deals, the consensus of opinion evident in the derision and pathologization of youth and the disabled, Planned Parenthood's embrace of population control tactics, all underscore the simple fact that liberalism is not the polar opposite of conservatism.

Liberalism's investment in a body-by-body approach to economic inequity—an outlook that may align it more closely with the right wing than the left (despite the media's insistence that liberalism *is* the left)—assured, and continues to assure, its allegiance to eugenics. The history of eugenics is one of both overlapping and strategic affinities that contest preconceived, static notions of "left" and "right." Clearly, any project seeking to shatter the appeal of eugenics must call liberals and conservatives to task.

A close reading of U.S. eugenics brings us back to Balibar's estimation of national purification, a process through which the nation must "isolate within its bosom, before eliminating or expelling them, the 'false,' 'exogenous,' 'crossbred,' 'cosmopolitan' elements."[12] To these self-regulating efforts, eugenics brought avowals of hereditary superiority and biological fatalism, claims perfectly suited to political and cultural backlash against immigration, against increasing gay and lesbian visibility, against demands for civil rights and economic justice. A mere thirty years after *Brown v. Board of Education, The Bell Curve* informed us that affirmative action is a lost cause because African-Americans are genetically predisposed to intellectual inferiority. After two and a half decades of feminist theorists and activists struggling to force a sense of social and political culpability into the dialogue around rape and battering (indeed, struggling to get the very words "rape" and "battering" circulated), we saw the press jumping to print specious research findings suggesting that male aggression is not socially constructed at all, but genetically determined. All this has a potentially depoliticizing effect, as attempts to dismantle injustice are neutralized by unproven genetic hypotheses that naturalize disparity and suggest the biological futility of resistance.

A credible threat to the latest eugenics revival will only materialize if its various outcroppings are juxtaposed and viewed as mutually reinforcing. The preceding pages attempted such a decompartmentalization of what are frequently held to be isolated assaults, waged on the basis of race, gender, class, disability, immigration status, and/or sexuality. There are, of course, things missing from this list, critical issues in desperate need of attention: the gene mapping of the Human Genome Project; the "genetic prospecting" of seven hundred indigenous groups worldwide by the Human Genome Diversity Project; the harvesting, cloning, and patenting of genes from the peoples of Panama, Micronesia, the Solomon Islands, New Guinea, and East China by U.S. scientists; the implementation of probation-for-castration plea bargains; the ultimatums issued to women in industry to accept sterilization or face demotion; the search for a breast cancer gene and the absolute refusal to adequately address environmental suspects. Certainly, my own subject position influenced my decisions on what to include, though my political convictions cause me to dwell on what I know to be imperative but that which, for lack of space and time, remains absent. It is my hope that even what is missing will be illuminated in part by what is here.

Whatever guise eugenics arrives in, whatever promises its adherents make, its agenda is eerily transparent. It proffers "scientifically" sustained scapegoating and calls it sound public policy; it defends a violent status quo as a viable political strategy. There is no mystery to seeing through this subterfuge. All we have to do is ask ourselves, "Whose bodies will bear the brunt of the resulting enactments? Who has paid, and who will continue to pay, the physical and political consequences of these configurations?" Over a hundred years into eugenics practice, the answers remain unchanged. We ignore them at our own peril.

Notes

Introduction

1. Richard J. Herrnstein and Charles Murray, *The Bell Curve: Intelligence and Class Structure in American Life* (New York: Free Press, 1994), 360; *New York Times Book Review*, 16 October 1994, 3.

2. Dean Hamer and Peter Copeland, *The Science of Desire: The Search for the Gay Gene and the Biology of Behavior* (New York: Simon & Schuster, 1994); *New York Times Book Review*, 16 October 1994, 9. Given less prominence in the *New York Times Book Review* and elsewhere that year was *Race, Evolution, and Behavior: A Life History Perspective*, by J. Philippe Rushton (New Brunswick, N.J.: Transaction, 1994). Rushton, a psychology professor at the University of Western Ontario, published a paper in which he stated that the disproportionately high incidence of HIV among Blacks was due to genetically programmed sexual behavior that spread the virus (Adam Miller, "Professors of Hate," *Rolling Stone*, 20 October 1994, 110).

3. For an analysis of the treatment of "science as news," see Joan E. Bertin and Laurie R. Beck, "Of Headlines and Hypotheses: The Role of Gender in Popular Press Coverage of Women's Health and Biology," in *Man-Made Medicine: Women's Health, Public Policy, and Reform*, ed. Kary L. Moss (Durham, N.C.: Duke University Press, 1996): "Treating scientific developments as news stories in and of itself invites distortion. Science is a process by which hypotheses are formulated, tested, and replicated under various circumstances, and placed in context with other relevant information" (41). I would add that such "relevant information" would have to include not only other scientific data, but an accounting of political, social, and economic contexts as well.

4. My intention is not to produce an exhaustive account of the history of each U.S. eugenics campaign. The historians, activists, and writers cited in the pages that follow have excavated much of this past. My own work is deeply indebted to theirs and, I hope, complementary, but differs in significant ways. I am particularly beholden to Allan Chase, Carl Degler, Troy Duster, Sander Gilman, Stephen Jay Gould, David Greenberg, Betsy Hartmann, Antonia Hernandez, Ruth Hubbard, Jonathan

Ned Katz, Daniel Kelves, Nancy Leys Stepan, Barry Mehler, George Mosse, Philip Reilly, and Connie Uri, among others.

5. Charles Mills, *The Racial Contract* (Ithaca, N.Y.: Cornell University Press), 73.

6. Consider, for example, the findings of psychologist Mark Synderman and political scientist Stanley Rothman. As late as 1984, 53 percent of respondents in their survey who identified as liberal believed that genetic factors were partially responsible for the gap in IQ scores between whites and African-Americans (*Newsweek*, 24 October 1994, 59).

7. George Mosse, *Nationalism and Sexuality: Middle-Class Morality and Sexual Norms in Modern Europe* (Madison: University of Wisconsin Press, 1985), 9.

8. Étienne Balibar, "Racism and Nationalism," in Balibar and Immanuel Wallerstein, *Race, Nation, Class: Ambiguous Identities* (New York: Verso, 1991), 60.

9. Ibid.

10. Ibid., 39.

11. Barry Mehler's "The History of the American Eugenics Society, 1921–1940" (Ph.D. diss., University of Illinois at Urbana-Champaign, 1988) was extremely valuable in assessing eugenicists' claims. Similarly, Stephen Jay Gould's *The Mismeasure of Man* (New York: W. W. Norton, 1981) is a key background text for any discussion of pseudoscience and its history as anti-immigrant agent.

12. The Page Law barred entry to real or imagined Asian prostitutes, contract laborers, and felons. The Chinese Exclusion Act imposed a ten-year immigration suspension on Chinese laborers.

13. Thurman Rice, *Racial Hygiene: A Practical Discussion of Eugenics and Race Culture* (New York: Macmillan, 1929), xix.

14. Madison Grant, *The Passing of the Great Race* (New York: Charles Scribner's Sons, 1916), 14.

15. U.S. House Committee on Immigration and Naturalization, *Biological Aspects of Immigration: Hearings*, 16–17 April 1920, 18.

16. Ward, "Eugenic Immigration: The American Race of the Future and the Responsibility of the Southern States for Its Formation: The 'Survival of the Fittest,'" *American Breeders Magazine* 4, no. 2 (1913): 96–102. Ward's conflation of race and nationality came at the end of an article lamenting the number of South Asian and southern and eastern European immigrants entering the United States.

17. Madison Grant, *The Passing of the Great Race*, 81. Grant is perhaps the most notorious racial demographer in the history of U.S. eugenics. In addition to his treatise on the classification of European races, his often repeated comparison of two loathed and increasingly visible subject groups—Jews and African-Americans—epitomizes the technique of bringing disparate entities together in the racist imagination. Eugenics is, after all, an inherently comparative ideology. Years later, the district attorney of Chicago traced crime and the "degeneration" of the theater to Anglo-Saxon "intermixture with disharmonic and inferior races" (District Attorney Orebaugh, *Crime, Degeneracy, and Immigration*, reviewed in *Eugenical News* 16, no. 3 [March 1931]: 37).

18. Cited in Mehler, "The History of the American Eugenics Society," 216.

19. The McCarran-Walter Act severely limited immigration from countries considered "nonwhite," though it did grant the Japanese a small immigration quota and naturalization rights. Primarily, it eased the process of deportation and denaturalization of leftist union leaders and others deemed "subversive," in part by taking

such cases out of the courts and establishing its own boards, which were not bound by due process. Sucheng Chan, *Asian Americans: An Interpretive History* (Boston: Twayne Publishers, 1991), 42; Griffin Farielle, *Red Scare: Memories of an American Inquisition* (New York: Avon Books, 1995), 18–19; Philip S. Foner, *Organized Labor and the Black Worker, 1619–1981* (New York: International Publishers), 285.

20. *New York Times*, 19 October 1994, 78–79.

21. Richard J. Herrnstein and Charles Murray, *The Bell Curve*, 360.

22. Ibid.

23. Emphasis added; ibid., 549. The authors did, however, write that Chinese and Japanese immigrants were possessed of superior intellects.

24. Seymour W. Itzkoff's *The Decline of Intelligence in America: A Strategy for National Renewal* was also published that year. This author also favored a more restrictive immigration policy, calling for a ban on those immigrant families judged to be at risk for becoming permanently dependent on welfare. Malcome W. Browne, "What Is Intelligence, and Who Has It?" *New York Times Book Review*, 16 October 1994, 41.

25. "U.S. Deports Record Number of Illegals," *San Francisco Examiner*, 28 December 1995, A-2.

26. L. A. Kerr, "Bi-Sexuality in Mankind," *Sexology* 8, no. 4 (March 1941): 221–22.

27. Cited in Bruce Mirken, "Experts Debate Genetic Research," *Frontiers*, 25 March 1994, 34.

28. "Amazing Story... Read This," *San Francisco Bay Times*, 2 May 1996, 11.

29. LeVay himself has warned that a prenatal test for homosexuality would be "relatively routine" once the correct gene was identified. "What we're talking about here is do-it-yourself eugenics. It could have horrifying consequences." He has stated that he is "very optimistic that we can prevent people from aborting gay kids *[sic]* not by outlawing it, but by changing the world so that people don't want to change sexual orientation." Hamer believes that no such test will be developed, as long as the government can be persuaded to issue proper rules and regulations. However, as Mark Schoofs pointedly asked, can a Congress that is unwilling to pass gay rights legislation be trusted to ban genetic testing for homosexuality? (Schoofs, "Geneocide: Can Scientists 'Cure' Homosexuality by Altering DNA?" *Village Voice*, 1 July 1997, 40–42). Furthermore, as of this writing no laws on either a federal or state level safeguard the confidentiality of genetic test results or even ban discrimination by insurance companies or employers on this basis (Jan Plather, "Predicting the Potential for Disease: For Whose Benefit?" *Sojourner*, January 1997, 11). Related confidentiality and safety issues arose with the Centers for Disease Control's recent consideration of reporting the names of HIV-positive individuals to local health departments (Duncan Osborne, "Federal Gov't Considers Tracing Sex Partners of HIV Infected," *LGNY*, 8 June 1997, 1).

30. "Queer Anatomy" is beholden to the work of Lisa Duggan, Doug Futuyma, Sander Gilman, Richard Goldstein, R. C. Lewontin, Donna Minkowitz, and AIDS and breast cancer activists, whose work warns us away from biological determinism and an uncritical deference to the scientific establishment.

31. Siobhan Somerville, "Scientific Racism and the Emergence of the Homosexual Body," *Journal of the History of Homosexuality* 5, no. 2 (October 1994): 244.

32. G. Frank Lydston, *The Disease of Society (The Vice and Crime Problem)* (Philadelphia: J. B. Lippincott, 1904), 394.

33. Daniel made the former recommendation in an 1893 paper entitled, "Should Insane Criminals or Sexual Perverts Be Allowed to Procreate?" Daniel believed not only in the inheritability of sexual inclination and criminality, but also in the benevolence of an intersection of judicial mandate and medical procedure. He posited castration as prophylactic—more humane than existing sentencing laws in that it would obviate the need for capital punishment. In much the same way, later physicians and lawmakers would herald vasectomy as a kinder alternative to castration and, still later, cast Norplant as a more compassionate option than hysterectomy. In this way, Daniel and his ideological kin could cast themselves as progressive, advocating, as he wrote, an end to "the cruel and useless execution of criminals" and envisioning "a sanitary Utopia" (Harry H. Laughlin, *Eugenical Sterilization in the United States* [Chicago: Psychopathic Laboratory of the Municipal Court of Chicago, 1922], 351; Jonathan Katz, *Gay American History: Lesbians and Gay Men in the U.S.A.* [New York: Avon Books, 1976], 209, 110).

34. Michel Foucault, *The History of Sexuality,* vol. 1, *An Introduction* (New York: Vintage Books, 1990), 101.

35. Donna Minkowitz, "Trial by Science," *Village Voice,* 30 November 1993.

36. The attorneys who successfully argued against the amendment before the Supreme Court did not rely on a medical strategy. The high court upheld the ruling.

37. Hamer and Copeland, *The Science of Desire,* 218.

38. Ibid., 219.

39. The Lambda Legal Defense and Education Fund reports that, as of spring 2000, employment discrimination on the basis of sexual orientation remains legal in forty states.

40. Hamer and Copeland, *The Science of Desire,* 218. Data compiled by the Fight the Right Project of the National Gay and Lesbian Task Force.

41. 1999 interview with Laurie Andrews on *On the Line,* WNYC-AM. See also her *The Clone Age* (New York: Henry Holt, 1999).

42. Urvashi Vaid told a reporter from the *Village Voice* that her group, a National Gay and Lesbian Task Force–sponsored think tank, will start taking action on this. In contrast, the Gay and Lesbian Medical Association has taken no position at all. As Vaid pointed out, however, it is difficult to fault organizations for not grasping the urgency of the situation when the community is still besieged by hate-crime and AIDS crises. As for the gay scientists involved in this current round of genetic postulation, they clearly sense the danger, but have abdicated any responsibility. Dean Hamer was recently asked about the possibility of garnering signatures from leading scientists on a statement declaring that sexual orientation should not be tampered with and that all orientations are healthy. He replied, "That's a good idea. But no one is doing it" (Schoofs, "Gene-ocide," 43). The question is, why isn't Hamer, who extols the power of scientists to act as a check against eugenics, drafting such a document?

43. Margaret Sanger, "Address of Welcome" to the Sixth International Neo-Malthusian and Birth Control Conference, April 1925. Reprinted in *Birth Control Review* 9, no. 4: 100.

44. Jonathan Boyarin, *Storm from Paradise: The Politics of Jewish Memory* (Minnesota: University of Minnesota Press, 1992), 10. I borrow from Boyarin not to draw equivalencies between U.S. and German eugenics, but rather to signify the intellec-

tual support Nazi eugenicists received from American eugenicists and the latter's undaunted admiration for the former's policies. Boyarin was writing specifically about each country's history of genocide.

45. Robert H. Blank, *Fertility Control: New Techniques, New Policy Issues* (New York: Greenwood Press, 1991), 89.

46. Margaret Sanger, "Birth Control and Race Betterment," *Birth Control Review* 3, no. 2 (February 1919): 11, 12.

47. Thomas P. Bailey, superintendent of schools, Memphis, Tennessee, cited in Howard Odum, *Social and Mental Traits of the Negro: Research into the Conditions of the Negro Race in Southern Towns* (New York: Columbia University Press, 1910), 301.

48. Here, I was mindful of Barbara Katz Rothman's characterization of liberal philosophy as "an articulation of the values of technological society, with its basic themes of order, predictability, . . . rationalization of life, the systematizing and control of things and people as things, the reduction of all of us as component parts, and ultimately the vision of everything, including our very selves, as resources." Barbara Katz Rothman, *Recreating Motherhood: Ideology and Technology in a Patriarchal Society* (New York: W. W. Norton, 1989), 63.

49. Adolph Reed Jr., "The Content of Our Cardiovascular," *Village Voice*, 31 December 1996, 25.

ImagiNation

1. *Morning Edition,* National Public Radio, broadcast 25 October 1994; Barry Mehler, "In Genes We Trust," *Reform Judaism,* winter 1994, 14; Elizabeth Kadetsky, "Bashing Illegals in California," *Nation,* 17 October 1994, 419. FAIR's leadership has included politicians Richard Lamm, former governor of Colorado, and Eugene McCarthy, former U.S. senator from Minnesota ("Frustration over Immigration Growing in State," *San Francisco Chronicle,* 30 March 1994, A10).

2. Lost in the preelection deliberations was the fact that undocumented immigrants pay $3 billion a year in taxes, three times as much as they receive in benefits (Connie Montoya, "The Anti-Immigrant Backlash" session participant, *Dangerous Intersections: Feminist Perspectives on Population, Immigration and the Environment,* transcript of a conference held by the Committee on Women, Population, and the Environment, New York, 25–26 October 1996, 43. Many thanks to Heidi Dorow for making this report available to me. Kitty Calavita argues that the effort to deprive undocumented immigrants of social services and education represents an "almost single-minded focus on the fiscal burden," though she also writes of the symbolic import that backers accorded the measure (Calavita, "The New Politics of Immigration: 'Balanced-Budget Conservatism' and the Symbolism of Proposition 187," *Social Problems* 43, no. 3 [August 1996]: 286). In this, the current anti-immigrant fervor mirrors the attack on welfare and welfare recipients. Aid to Families with Dependent Children (now Temporary Assistance to Needy Families) accounts for just 1 percent of the federal budget, yet the hype surrounding need-based allotments exploits the popular misconception that the percentage is much higher. Likewise, undocumented workers constitute just 13 percent of all immigrants in the country and a mere 1 percent of the U.S. population (Mimi Abramovitz, *Under Attack, Fighting*

Back: Women and Welfare in the United States [New York: Monthly Review Press, 1996], 25; David Cole, "Five Myths about Immigration," *Nation* 259, no. 12 [17 December 1994]: 410).

3. Charles Davenport, cited in Albert P. Van Dusen, "Birth Control as Viewed by a Sociologist," *Birth Control Review* 8, no. 5 (May 1924): 134.

4. Étienne Balibar, "Racism and Nationalism," 39.

5. Remarks of John T. Morgan, 25 April 1882, *Congressional Record,* 47th Cong., 1st sess., 1882, 8: 3270.

6. Balibar, "Racism and Nationalism," 50.

7. Ibid., 60.

8. George William Curtis, cited in Robert DeCourcey Ward, "Our Immigration Laws from the Viewpoint of Eugenics," *American Breeders Magazine* 111, no. 1 (1912): 26.

9. *Eugenical News* 17, no. 1 (January–February 1932): 31.

10. Ward, "Our Immigration Laws from the Viewpoint of Eugenics," 20. Ward was probably not speaking of Georgia, whose first European settlers were convicts.

11. Ward, "Eugenic Immigration: The American Race of the Future and the Responsibility of the Southern States for Its Formation: The 'Survival of the Fittest,'" *American Breeders Magazine* 4, no. 2 (1913): 99, 101. Ward was a regular contributor to the *American Breeders Magazine,* which published its debut issue in 1910, six years after the founding of its parent organization, the American Breeders Association. Initially, the association was created to enable "animal and plant breeders to realize that they are working under laws of heredity common alike to animals and plants." There were articles on plowing, feed, and crop cultivation. As eugenics and scientific racism blossomed, however, more and more pages were devoted to their champions. An article entitled "Eugenics and Immigration," for example, might appear directly after a piece on "Breeding Sugar Beets." The magazine published and reprinted the written work of the eugenics heavyweights of the day: Madison Grant, Charles Davenport, Harry Laughlin, H. H. Goddard, C. W. Saleeby, and of course, Francis Galton himself. In 1914, the magazine was rechristened the *Journal of Heredity.* The American Breeders Association became the American Genetic Association and declared the journal "A Monthly Publication Devoted To Plant Breeding, Animal Breeding and Eugenics." Immigrant bashers like Ward found an ideological blank check in the journal's stated mission and were thus able to construct xenophobia as the logical extension of the thinking breeder's concerns.

12. Balibar, "Racism and Crisis," in Balibar and Immanuel Wallerstein, *Race, Nation, Class: Ambiguous Identities* (New York: Verso, 1991), 223.

13. Balibar, "Racism and Nationalism," 49, 54.

14. Ward, "The American Race of the Future," 101; and "Our Immigration Laws from the Viewpoint of Eugenics," 25. Daniel Kelves has also noted this tendency among eugenicists (*In the Name of Eugenics: Genetics and the Uses of Human Heredity* [New York: Knopf, 1985]).

15. Ward, "Our Immigration Laws from the Viewpoint of Eugenics," 21, 22.

16. Ibid., 21.

17. "Race Genetics Problems" (editorial), *American Breeders Magazine* 11, no. 3 (1911): 230.

18. Editorial, *American Breeders Magazine* 11, no. 3 (1911): 232.

Calculating Hysteria

1. George Mosse, *Nationalism and Sexuality: Middle-Class Morality and Sexual Norms in Modern Europe* (Madison: University of Wisconsin Press, 1985), 133, 134. An insistence, I would add, that extended to other categories.

2. George Stocking, *Bones, Bodies, Behavior: Essays on Biological Anthropology,* (Madison: University of Wisconsin Press, 1988), 8.

3. Cited in Ward, "The Immigration Problem of Today," *Journal of Heredity* 11, no. 7 (September–October 1920): 326.

4. Ibid. Similar rationales have been used in recent years to pass and uphold English-only legislation currently on the books in twenty-three states (Cathi Tactaquin, "The Anti-Immigrant Backlash" session participant, *Dangerous Intersections,* 44).

5. The IRL, founded by a handful of Harvard graduates, had the ear of Senator Henry Cabot Lodge. They were, Kelves reports, effective lobbyists in Washington and bombarded hundreds of daily newspapers with their literature. In Ward's 1931 obituary, the *Eugenical News* characterized the IRL's aims as "primarily the conservation of the essential race-structure of the American people." Kelves, *In the Name of Eugenics,* 24; Mehler, "The History of the American Eugenics Society," 191; *Eugenical News* 16, no. 11 (November 1931).

6. "The Tide of Immigration," book review, *Journal of Heredity* 7, no. 12 (December 1916): 554, 545. It is unclear if Warne found literacy in any language acceptable, or if he drew the line at reading and writing English. Frequently, "illiteracy" was a charge leveled at any immigrant without reading and writing knowledge of English, whether the individual in question was literate in their first language or not.

7. Sidney Gulick, "An Immigration Policy," *Journal of Heredity* 7, no. 12 (December 1916): 547, 552. At the same time, Gulick advocated an immigration policy that would "be applied equally to every land, and thus avoid differential race treatment." The policy he envisioned was, nevertheless, equally if not more biased against Chinese and Japanese immigrants as legislation already in place. He emphasized this, along with his plan's hindrance to eastern and southern European immigrants (particularly Italians), but it did not assuage his detractors. In 1919, Robert Ward said this plan to repeal the Chinese Exclusion Act and the Gentlemen's Agreement, to put Chinese and Japanese immigration under a quota system (a percentage "of the American born children of foreign parents of that people plus the number of those from that same people who have already become naturalized citizens"), and to offer Asian immigrants naturalization "proposes to put the thin edge of a wedge under the door which our national policy has built against Oriental immigration. It is a thin wedge now, but it is a wedge." "The Immigration of Orientals," *Journal of Heredity* 10, no. 3 (March 1919).

8. Craniometry is a good example. In use since the early nineteenth century, it had established a foothold long before immigrants were being asked to count backward or memorize objects in a given picture. For a description of intelligence tests administered by the Public Health Service at Ellis Island, see Howard A. Knox, "Tests for Mental Defects," *Journal of Heredity* 5, no. 3 (March 1914): 122–30. Binet, who died in 1911, did not believe, as did IQ examiners in the United States, that intelligence

could be gleaned or measured by a single number. For more on his philosophy and intent, see Gould, *The Mismeasure of Man*.

9. Mehler, "The History of the American Eugenics Society," 190, 191.

10. Allan Chase, *The Legacy of Malthus: The Social Costs of the New Scientific Racism* (New York: Knopf, 1977), xix. Eight decades later, Herrnstein and Murray's claims betrayed their selective reading of history. *The Bell Curve* asserts that, while the mean IQ of immigrants to the United States in the 1980s "works out to about 95," Asians and Ashkenazi Jews are the highest test scorers (thereby effectively splitting two groups from the body of ethnic immigrants, which, following the authors' logic, disproves their theory). Herrnstein and Murray further explained that IQ is race-based and remains more or less fixed from generation to generation (Herrnstein and Murray, *The Bell Curve*, 275; Malcome Browne, "What Is Intelligence and Who Has It?" *New York Times Book Review*, 16 October, 1994, 3). What they did not explain was how then to reconcile the low scores of Ashkenazi Jews in the 1910s with their high scores seventy years later. Moreover, if Charles Murray truly believes in the intellectual superiority of Ashkenazim, he must consider that the "group intelligence" of whites might be artificially inflated by the presence of these intellectual ringers. Following his own specious reasoning to its conclusion, if Jews were factored out of the white IQ pool, the gap he perceives between Black and white might be significantly diminished. The Jewish case should have complicated Herrnstein and Murray's analysis. That it didn't only reveals predetermined results dictated by social categorization.

Jews occupied a peculiar position in the hearts and minds of eugenicists, alternatively singled out for vehement derision or praised for keeping a "pure" bloodline (the impossibility of which has still not been fully owned up to). English eugenicist C. W. Saleeby, who claimed that Jews "show a larger proportion of physical and mental defectives than any other civilized religious, social, or ethnic group of people," also maintained that thousands of years of persecution had weeded out the "weaklings" and the "fools" from the Jewish people, "the one human race of which we know assuredly that it has persisted unimpaired, [has] been the most continuously and stringently selected of any race that can be named" (Saleeby, cited in "Eugenics in Jewish Life," *Journal of Heredity* 8, no. 12 [December 1917]: 543). There is nothing unique in such inconsistencies. For example, Jews were simultaneously charged with "capitalism" and "bolshevism." An examination of the contradictory nature of eugenic assessment of Jews reveals not only the contradictions, hypocrisies, and leaps of logic necessary to maintain a pro-eugenics stance, but demonstrates the inherent pitfalls in racializing any people.

11. Gould, *The Mismeasure of Man*, 165, 166. Gould framed the results this way: "For the evident reason, consider a group of frightened men and women who speak no English and have just endured an oceanic voyage in steerage. Most are poor and have never gone to school; many have never held a pencil or pen in their hand. They march off of the boat; one of Goddard's intuitive women takes them aside shortly thereafter, sits them down, hands them a pencil, and asks them to reproduce on paper a figure shown to them a moment ago, now withdrawn from their sight. Could their failure be a result of testing conditions, of weakness, fear, or confusion, rather than of innate stupidity? Goddard considered the possibility, but rejected it" (166).

12. Cited in ibid., 168.

13. Knox, "Tests for Mental Defects," 125.

14. Ibid., 122, 123, 126. Ward shared the assistant surgeon's concerns. Prior to Goddard's Ellis Island excursions, he had suggested that immigrants be "stripped to the skin" and given rigorous mental and physical exams upon landing. In addition, he wanted "immigrant inspectors and our own surgeons on board of all immigrant-carrying vessels" who would be charged with "mingling with the immigrants on the voyage over" with an eye toward better detection of "defects" (Ward, "Our Immigration Laws from the Viewpoint of Eugenics," 23, 24).

15. "The Tide of Immigration," 542.

16. L. E. Cofer, "Eugenics and Immigration," *Journal of Heredity* 6, no. 4 (April 1915): 173.

17. Ibid., 171. While charging Europe with deliberately inundating the United States with "unfit" immigrants, Cofer reported that the U.S. Public Health Service (PHS) was trying to stay on top of the situation. PHS issued a "Book of Instructions for the Medical Inspection of Aliens," listing four classes of "diseases:" Class A-1 included "idiocy, imbecility, feeblemindedness, epilepsy, insanity, and tuberculosis." Class A-2 was composed of "loathsome, contagious or dangerously contagious diseases," including "ringworm of scalp . . . leprosy and venereal diseases . . . trachoma . . . hookworm." Class B listed "those defects which affect the ability on the part of the immigrant to earn a living," such as hernia, "defective nutrition," and anemia. Children with disabilities requiring "unusual care" were also placed in this category. Lastly, class C was the catchall for conditions that were "reportable" but not "deportable" (172). Cofer's accusation of "dumping" has held fast to anti-immigrant doctrine for the better part of the century, though blame has shifted from European countries to those in Latin America and Southeast Asia. Of course, in later years the U.S. media was twice lucky in being able to level similar charges at Cuba—a country both Latin/Caribbean and communist.

18. Kelves, *In the Name of Eugenics*, 23.

19. Ibid., 23, 24.

20. W. C. Billings (chief medical officer, Immigration Service, Angel Island), "Oriental Immigration," *Journal of Heredity* 6, no. 10 (October 1915): 464.

21. "The Tide of Immigration," 544.

22. *U.S. Statutes at Large* 39 (1917): 874–98. As evidenced by debate recorded in the *Congressional Record*, President Wilson's veto had less to do with humanitarian impulses than with concerns over response from the Japanese government.

23. Lombroso's "born criminal" has given way to John Donohue and Steven Levitt's "unborn criminal." The two economists have suggested that the drop in crime between 1985 and 1997 is tied to the increase in abortions following the 1973 *Roe v. Wade* decision.

24. Stephen Steinberg, *The Ethnic Myth: Race, Ethnicity, and Class in America* (Boston: Beacon Press, 1989), 116.

25. Theodore Bingham, "Foreign Criminals in New York," *North American Review,* September 1908, 383.

26. Cofer, "Eugenics and Immigration," 174.

27. "The Tide of Immigration," 543.

28. Ward, "Our Immigration Laws from the Viewpoint of Eugenics," 24.

29. Ward, "Eugenic Immigration: The American Race of the Future," 96, 97, 98. On one level or another, proponents of the 1917 legislation were all concerned with the working class. The act called for more thorough inspection of steerage quarters.

Those coming to the United States who could afford to travel first class had, no doubt, already displayed their eugenic mettle by their ability to finance better accommodations (Ward et. al., "War, Immigration, Eugenics: Third Report of the Committee on Immigration, American Genetic Association, Prescott F. Hall, Chairman," *Journal of Heredity* 7, no. 6 [June 1916]: 241). In later years Ward complained that "many eugenically and otherwise undesirable aliens are trying to escape rigid medical examination by traveling second, or even first class" ("Immigration and Eugenics: Second Report of the Sub-Committee on Selective Immigration of the Eugenics Committee of the United States of America," *Journal of Heredity* 16, no. 8 [August 1925]: 288).

30. Edward Ross, who alarmed white workers with claims of "Chinese race vitality," remains the standard-bearer. There were important exceptions. The Industrial Workers of the World, until their demise, and the Congress of Industrial Organizations before it was purged and absorbed by the AFL, resisted divisive tactics that pitted immigrant workers against the native born.

31. S. L. Gulick, "An Immigration Policy," *Journal of Heredity* 7, no. 12 (December 1916): 546.

32. Balibar, "Racism and Crisis," 219.

33. Ward, "Immigration after the War," *Journal of Heredity* 8, no. 4 (April 1917): 149.

34. Ibid.

35. Ward et. al., "War, Immigration, Eugenics," 243, 247.

36. "The Tide of Immigration," 514.

37. Ward, "Immigration after the War," *Journal of Heredity* 7, no. 3 (March 1916): 134. Ward suggested heavier fines be levied against steamship companies that attempted to bring "mentally defective aliens" to the United States.

38. Ward et. al, "War, Immigration, Eugenics," 244, 243.

39. Ward, "Immigration after the War" (1917), 147.

40. Grant, cited in Mehler, "The History of the American Eugenics Society," 196 n.31.

41. Ward et. al., "War, Immigration, Eugenics," 248.

42. *U.S. Statutes at Large* 39 (1917): 875.

43. Knox, "Tests for Mental Defects," 126.

44. Ward, "Immigration after the War" (1917), 151.

45. "The Tide of Immigration," 542.

46. Not everyone was satisfied with the five-year period. Ward continued to argue that immigrants should be deportable if they became a member of an excludable class or a public charge, no matter how long they had lived in the United States (Ward, "Immigration and Eugenics," 289, first issued on 9 March 1925). The 1925 Deportation Bill would have allowed for such an open-ended policy, but, although it passed the House, it failed to become law. Over seventy years later, noncitizen immigrants face a one-strike-and-you're-out approach to crime, regardless of their years of residency or the time elapsed since they were charged or imprisoned.

47. *U.S. Statutes at Large* 39 (1917): 876.

48. *United States v. Balsara* (1910) and *United States v. Ajkoy Kumar Mazumdar* (1913). These rulings were reversed in *United States v. Bhagat Singh Thind*, in which the U.S. Supreme Court ruled that South Asians were ineligible for naturalization on the grounds that they may be Caucasian, but they were not "white."

49. Charles Mills, *The Racial Contract* (Ithaca, N.Y.: Cornell University Press), 81.

50. *Congressional Record,* 64th Cong., 2d sess., 1917, 44: 2618. Of course, most of the immigrants the senator was referring to were Sikh, not Hindu, but such distinctions were lost in the racist amalgamation of eugenic legislation.

51. Ronald Takaki, *Strangers from a Different Shore: A History of Asian Americans* (New York: Penguin Books, 1989), 297.

52. *Congressional Record,* 64th Cong., 2d sess., 1917, 44: 2620.

53. Reed, quoting Woodrow Wilson, ibid., 2618.

54. Ibid., 2621.

55. Ward et al., "War, Immigration, Eugenics," 246, 248. The AGA was formerly the American Breeders Association, publisher of the *American Breeders Magazine,* which in turn became the *Journal of Heredity.*

56. Ward, "Immigration after the War" (1917), 151, 152.

57. "Some Present Aspects of Immigration: Fourth Report of the Committee on Immigration of the American Genetics Society," *Journal of Heredity* 10, no. 2 (February 1919): 69, 70.

58. Commissioner F. A. Wallis, cited in Ward, "The Immigration Problem Today," *Journal of Heredity* 11, no. 7 (September–October 1920): 323.

59. Ibid., 324.

60. Similarly, the targeting of Bhagat Singh Thind and the resulting Supreme Court decision was partly prompted by Thind's advocacy of Indian independence (Sucheng Chan, *Asian Americans: An Interpretive History* [Boston: Twayne, 1991], 94).

61. This persisted throughout postwar red-baiting and well into the 1920s. Thurman Rice, for example, a bacteriologist at the University of Indiana and the author of *Racial Hygiene,* wrote that eastern and southern European immigrants "do not mix well with our stock...and if they do cross with us their dominant traits submerge our recessive traits; they are often radicals and anarchists causing no end of trouble; they have very low standards of living; they disturb the labor problems of the day; they are tremendously prolific" (Thurman B. Rice, *Racial Hygiene: A Practical Discussion of Eugenics and Race Culture* [New York: Macmillan, 1929], 302–3).

62. Frederick Adams Woods, "The Racial Limitation of Bolshevism," *Journal of Heredity* 10, no. 4 (April 1919): 190.

63. Ezra Bowen, "Battling over Birth Policy," *Time,* 24 August 1987, 31.

64. Ward, "The Immigration Problem Today," 327. In another of restrictionists' many contradictions, immigrant workers were viewed as both labor agitators—foreign radicals spreading propaganda and discontent—and tools of capitalists who, lacking any sense of solidarity, allowed themselves to displace higher-paid, U.S.-born, white workers.

65. My use of the term "populism" implies a tactical and orchestrated use of appeals to labor and others by eugenicists and exclusionists. Clearly, eugenics was a top-down ideology, and its proponents, by and large, were men of position and power, "pillars of society." Furthermore, eugenics thrived, in part, by its ability to simultaneously level blame at capitalist interests while filling eugenicists' coffers with industrialists' money. Harry Laughlin's Eugenics Record Office, for example, was partially underwritten by John D. Rockefeller.

66. Ward, "Immigration and the Three Per Cent Restrictive Law," *Journal of Heredity* 12, no. 7 (August–September 1921): 320–21. Similarly, automakers, airlines,

phone companies, and oil companies who donate to immigrant groups are held in contempt by contemporary anti-immigrant groups such as Citizens for Responsible Immigration. "Frustration over Immigration Growing in State," *San Francisco Chronicle*, 30 March 1994, A11.

67. Galton Society members included eugenics heavyweights Madison Grant, Harry Laughlin, Carl C. Brigham, Lothrop Stoddard, and Charles Davenport, among others. In 1927, Johnson asked the organization for suggestions on eugenic uses of the upcoming 1930 census. Laughlin proposed that it become "a permanent and complete pedigree record of the American people as individuals, and [as such] also would enable the nation to measure its racial trend" (*Eugenical News* 12, no. 12 [1927]: 172).

68. Mehler, "The History of the American Eugenics Society," 196, 197.

69. Ibid., 200–201.

70. Ward, "Immigration and the Three Per Cent Restrictive Law," 318. Ward wanted the "percentage limitation principle" made permanent, except when it might supersede the Chinese Exclusion Act and the Gentlemen's Agreement. Japanese and Chinese immigrants, Ward argued, should not be admitted on any percentage basis (325). The year 1910 was selected simply because it had the most recent available census data (Kelves, *In the Name of Eugenics*, 65).

71. *Congressional Record*, 64th Cong., 2d sess., 1917, 44: 2619.

72. Mehler argues that the restriction laws of this era should be situated within a larger body of "anti-foreign statutes" that began with the 1917 Espionage Act. State laws enacted during the 1917–1920 period barred "aliens" from practicing medicine, pharmacy, chiropractic, architecture, surgery, surveying, engineering, operating a motor bus, and executing wills (Mehler, "The History of the American Eugenics Society," 181).

73. Ibid., 391, 428, 197–98.

74. Grant, *The Passing of the Great Race*, 227; Chase, *The Legacy of Malthus*, 174.

75. Mehler's "The History of the American Eugenics Society" is a highly informative guide to the intricacies of the AES and includes an invaluable who's who of the organization's membership and leadership.

76. Gould, *The Mismeasure of Man*, 194. See also Kelves, *In the Name of Eugenics;* Chase, *The Legacy of Malthus;* and Troy Duster, *Back Door to Eugenics* (New York: Routledge, 1990). Recruits given the Alpha test were asked to unscramble sentences, complete analogies, and add the missing numbers to given sequences. The Beta test was picture-based, but by and large was thematically the same. For more on the content of the exams, see Gould, *The Mismeasure of Man*, 199–200.

77. Gould, *The Mismeasure of Man*, 196, 197. The tests were interpreted by E. G. Boring, then a captain in the army and Yerkes's assistant.

78. Ibid., 197.

79. Cited in Rhett S. Jones, "Proving Blacks Inferior" (1965), reprinted in *The Death of White Sociology*, ed. Joyce A. Lander (New York: Random House, 1973), 132.

80. Gould, *The Mismeasure of Man*, 199, 224. Gould has noted that the army tests were the "first written IQ tests to gain respect, and they provided essential technology for implementing the hereditarian ideology that advocated . . . the testing and ranking of all children" (230). Indeed, one writer credited the army tests with increasing military efficiency and declared that they "demonstrated the applicability of the group method of measuring intelligence to educational and industrial

work. The army methods, although not adapted to the usual educational or industrial requirements, can readily be modified or used as a basis for the development of similar procedures" ("Army Tests Reveal the Vast Differences in Mankind," *Journal of Heredity* 10, no. 4 [April 1919]: 190). Generations of U.S. schoolchildren who had to endure such testing made these remarks prophetic.

81. Brigham, *A Study of American Intelligence* (Princeton, N.J.: Princeton University Press, 1923), xx.

82. Ibid., 205.

83. Ibid., 190, 189, 207–8.

84. Ibid., 190. In 1929, Lewis Terman wrote that the Jewish population filled its quota of "gifted children" twice over. Edward East, "Genetics of the Gifted: A Review of *Genetic Studies of Genius,* vol. 1, *Mental and Physical Traits of a Thousand Gifted Children,"* *Birth Control Review* 13, no. 7 (July 1929): 191. As elsewhere, this claim was mitigated by assertions of low uniformity among Jews. Maurice Fishberg, a physician at New York's Montefiore Hospital and author of *The Jews,* believed there were "two diametrically opposed extremes among the modern Jews. . . . On the one hand we have a very high proportion of feeble-minded, idiots, imbeciles and insane and also physical defectives, weakly and decrepit people; and on the other hand . . . we have also an amazingly high proportion of persons of marked ability in nearly all walks of life." The reason for this, Fishberg maintained, was different ethnic and racial elements among Jews. Assimilation and eugenics merged as Fishberg explained the recent decrease in Jewish variability. As Jews adapted the habits of their non-Jewish neighbors, "these peculiarities are gradually being effaced. Whether the loss sustained in the number of capable is compensated by the decrease in the number of defectives depends on the point of view." Perhaps trying to cast Jews as eugenically conscientious, he went so far as to claim that "the Rabbinical teachings are teeming with positive eugenic suggestions and one is inclined to say that the rabbis anticipated Galton by about sixteen hundred years." Even matchmakers, he wrote, "worked along eugenic lines" (Fishberg, "Eugenics in Jewish Life," *Journal of Heredity* 8, no. 12 [December 1917]: 544, 549, 545, 546).

85. Brigham, *A Study of American Intelligence,* 158, 159, 170.

86. Ibid., 210.

87. Paul Popenoe, "In the Melting Pot," *Journal of Heredity* 14, no. 5 (August 1923): 223.

88. Yerkes, foreword to Brigham, *A Study of American Intelligence.*

89. Gould, *The Mismeasure of Man,* 232.

90. Terman, cited in Mehler, "The History of the American Eugenics Society," 195. Terman was a member of the Eugenics Record Association and the American Eugenics Society. In later years, he turned his attention to quantifying what he perceived as differences in mental ability between women and men (430).

91. Jennifer Terry, "Anxious Slippages between 'Us' and 'Them': A Brief History of the Scientific Search for Homosexual Bodies," in *Deviant Bodies: Critical Perspectives on Difference in Science and Popular Culture,* ed. Terry and Jacqueline Urla (Bloomington: Indiana University Press, 1995), 138–48.

92. Mehler speculates that Madison Grant was instrumental in Johnson's evolution as a eugenicist, introducing him to the key players in New York (Mehler, "The History of the American Eugenics Society," 196). As for Laughlin, in late 1923 and early 1924, under the auspices of the Department of Labor, he investigated the

possibility of overseas examinations of potential immigrants in eleven European countries ("Third Report of the Sub-Committee on Selective Immigration of the Eugenics Committee of the United States of America," *Journal of Heredity* 16, no. 8 (August 1925): 293.

93. House Committee, *Biological Aspects of Immigration*, 17.

94. Ibid., 10. A few years before Laughlin's death, it became apparent that he himself had a form of epilepsy (Mehler, "The History of the American Eugenics Society," 391).

95. House Committee, *Biological Aspects of Immigration*, 7.

96. Ibid., 15, 18. As is often the case today, blame for this overrepresentation was laid squarely and solely on the incarcerated, and not on discriminatory sentencing or placement practices.

97. Ibid., 8, 7.

98. J. David Smith and K. Ray Nelson, *The Sterilization of Carrie Buck* (Far Hills, N.J.: New Horizon Press, 1989), 183.

99. Cited in Thomas Shapiro, *Population Control Politics: Women, Sterilization, and Reproductive Choice* (Philadelphia: Temple University Press, 1985), 40.

100. "Doctor Laughlin Honored," *Eugenical News* 21, no. 4 (July–August 1936).

101. Cited in Popenoe, "In the Melting Pot," *Journal of Heredity* 14, no. 5 (August 1923): 23, 24, 25. See also U.S. House Committee on Immigration and Naturalization, *Analysis of America's Melting Pot: Hearings*, 67th Cong., 3d sess., 21 November 1921.

102. Cited in Mehler, "The History of the American Eugenics Society," 185.

103. Ibid., 197, 198, 206. Mehler notes that there was some dissent in the ranks. Biologist and zoologist Herbert Jennings, a member of the American Eugenics Society, was so dismayed by Laughlin's skewed statistics that he left the organization and became a critic of eugenics (208, 202).

104. Ibid., 203.

The Immigrant Within

1. "Four Prizes for Eugenics Essays," *Journal of Heredity* 19, no. 9 (September 1928): 424.

2. *Birth Control Review* 13, no. 10 (October 1929): 287; *Journal of Heredity* 20, no. 11 (November 1929): 542.

3. *Eugenical News* 18, no. 2 (March–April 1933).

4. "Meeting of the Board of Directors of the AES," *Eugenical News* 13, no. 3 (March 1928): 38; "American Eugenics Society: Committee on Cooperation with Clergymen," *Eugenical News* 13, no. 4 (April 1928): 53. This attention to the clergy is particularly interesting in light of Galton's disdain for Christianity and his desire to see it supplanted by eugenics. Galton, who coined the term "eugenics," blamed Christianity for the downfall of the Roman Empire because of the precept "the meek shall inherit the Earth." According to Galton, "Man has already furthered evolution very considerably, half unconsciously, and for his own personal advantages, but he has not yet risen to the conviction that it [is] his religious duty to do so deliberately and systematically." More recently, Raymond Cattell, a former psychology professor at the University of Illinois, introduced "Beyondism," the tenets of which state that in order to avert the propagation of the unfit and the subsequent demise

of civilization, the richest and the most intelligent should inherit the earth. "The Reason for Eugenics" (an excerpt from Galton's 1883 *Inquiries into Human Faculty*), *Journal of Heredity* 5, no. 5 (May 1914): 221; Mosse, *Nationalism and Sexuality,* 10; Mehler, "In Genes We Trust," 77.

5. Laughlin, Grant, Fairchild, H. F. Osborn, Charles Davenport, and C. C. Little, among others. The AES had nearly one thousand charter members in forty-five states, Washington, D.C., Hawaii, Canada, England, Germany, Cuba, Italy, Puerto Rico, and Switzerland. Contributors to the AES included John D. Rockefeller Jr., who bestowed a total of $10,000 on the AES in 1925 and 1926 (Mehler, "The History of the American Eugenics Society," 81, 82, 83). In later years, Rockefeller would become a contributor to Margaret Sanger's *Birth Control Review.* Along with Winston Churchill, Alexander Graham Bell, and former Harvard president Charles Eliot, he attended the first International Eugenics Congress in 1912. Their involvement is significant, for it underscores the esteem in which eugenicists were held and belies the myth that scientific racists were a mere fringe element. At this gathering, a "Suggested Experimental Federal Law" was proposed, calling for the sterilization of "defective immigrants and immigrants with defective heredity" (*Problems in Eugenics,* First International Eugenics Conference [Adelphi: Eugenics Education Society, 1912], 466). Of course, eugenicists were willing to claim anyone as their own, including William Randolph Hearst. In the platform of principles that the Hearst papers endorsed, the magnate's *New York American* included "Selective immigration to admit only those suited for citizenship and American needs," to which one eugenicist responded, "It will be interesting to watch the evolution of this ideal to see whether, in a few years, these same papers which now make 'suitability for citizenship' the standard change to the 'breeding stock standard'" (*Eugenical News* 16, no. 1 [January 1931]: 5).

6. Mehler, "The History of the American Eugenics Society," 217.

7. "Third Report of the Sub-Committee on Selective Immigration of the Eugenics Committee of the United States of America," 298, 296.

8. Mehler, "The History of the American Eugenics Society," 216, 215.

9. "Memorial on Immigration Quotas: To the President, the Senate and the House of Representatives," *Eugenical News* 12, no. 3 (March 1927): 27.

10. "Immigration Control: Statements on Immigration by Representative John C. Box of Texas, Member of the House Immigration Committee. Statement Number 3," *Eugenical News* 12, no. 11 (November 1927): 163, 164. Like all statistical "evidence" introduced by eugenicists and their fellow travelers, these are highly suspect. Immigrant families, owing to a variety of factors, were more likely to be noticed by child welfare workers. Places that used to belong to Mexico, e.g., Galveston, Los Angeles, and Fort Worth, would of course have a significant Mexican and Mexican-American (it is doubtful Box made the distinction) population that would claim a proportionate amount of available services.

11. House Committee, *Biological Aspects of Immigration,* 15. Laughlin's perceptions resonated through the decades that followed. In a statement before the Senate Subcommittee on Labor and Public Welfare in 1973, Dr. Arlene Parsons testified that in the United States 70 to 80 percent of African-American and Latino children classified as either "Educable Mentally Retarded," "Borderline Retarded," "Dull Normal," or "Socially Maladjusted" had been categorized erroneously (U.S. Senate Subcommittee on Health, Committee on Labor and Public Welfare, *Quality of*

Health Care—Human Experimentation: Hearings, 93d Cong., 30 April, 28, 29 June, and 10 July 1973, 1614).

12. "American History in Terms of Human Migration," *Eugenical News* 13, no. 8 (August 1928): 122.

13. Mehler, "The History of the American Eugenics Society," 216.

14. "Immigration Commission of California," *Eugenical News* 13, no. 2 (February 1928): 24. Unlike the 1990s discourse on illegal entry, the northern border was also considered a site for some vigilance. The commission announced that the average "Habitant" of Quebec was "eugenically low-powered." In the main, however, eugenicists were unconcerned with Canadians slipping across the border.

15. Ward, "Immigration and Eugenics: Second Report of the Sub-Committee on Selective Immigration," 289.

16. Princeton League of Women Voters, *Heredity and Twelve Social Problems* (Princeton, N.J.: Department of Social Hygiene of the Princeton League of Women Voters, 1935/36), 31. Leon F. Whitney of the AES wrote that the Depression "has done an immeasurable amount of good for eugenics. . . . In this country a man has to feel the pinch on his pocketbook before he begins to realize that anything is wrong, and this depression has made him feel that pinch" (Whitney, "Neither Dead nor Sleeping," *Journal of Heredity* 24, no. 4 [April 1933]: 150).

17. Princeton League of Women Voters, *Heredity and Twelve Social Problems,* 31. This was one of the few eugenic tracts that did not place African-Americans at the bottom of the mental-racial hierarchy. However, in its tirade against the "foreign-born" who "contribute inmates to institutions for the insane at a rate twice that of the whole population," the writers did declare that "Negroes are in far excess of the ratio to which they are entitled by their number in the general population" (23).

18. Ibid., 31. Likewise, eugenicists worried that the cutoff of cheap labor from Europe was resulting in an increased migration of southern Blacks to northern industrial centers. "The census of 1930 shows that the negro population of the north increased nearly 1,000,000 between 1920 and 1930, while during the same period in the south it increased less than one-half that number" ("Negro Migration to the North," *Eugenical News* 16, no. 9 [September 1931]: 150). Of course, before the restrictive legislation in the 1920s, the Great Migration and the large immigration of southern and eastern European Jews overlapped considerably.

19. Osborn, cited in Mehler, "The History of the American Eugenics Society," 219.

20. C. M. Goethe, "Patriotism and Racial Standards," *Eugenical News* 21, no. 4 (July–August 1936): 66, 68.

21. "Japanese Immigration Quota: Pro and Con," *Eugenical News* 16, no. 9 (September 1931): 144. The writer was arguing against putting China and Japan on the quota system. Due to previous exclusionary legislation, he could not cite statistics on high numbers of immigrants from these countries in an effort to conjure an encroaching Asian threat. Instead, he invoked the "particularly striking" rise in the Filipino population of Hawaii.

22. Princeton League of Women Voters, *Heredity and Twelve Social Problems,* 31.

23. "Puerto Rico as State," *Eugenical News* 20, no. 4 (July–August 1935): 58, 59.

24. Ibid., 59.

25. Cited in Grant, *The Passing of the Great Race,* 68.

26. Princeton League of Women Voters, *Heredity and Twelve Social Problems,* 31.

27. George Stocking, *Race, Culture, and Evolution: Essays in the History of Anthropology* (Chicago: University of Chicago Press, 1982), 48, 49.

28. Madison Grant, "Further Notes on 'The Racial Elements of European History,'" *Eugenical News* 13, no. 9 (September 1928): 120.

29. Brigham, *A Study of American Intelligence*, 208.

30. Siobhan Somerville, "Scientific Racism and the Emergence of the Homosexual Body," 265.

31. *American Breeders Magazine* 2, no. 3 (1911): 232. Of course, southern and eastern European immigrants were neither legally barred nor routinely physically punished for marrying "whiter" partners, as were Asians, African-Americans, and Mexicans.

32. Mehler, "The History of the American Eugenics Society," 92, 94.

33. Ibid., 92, 94.

34. *Eugenical News* 16, no. 3 (March 1931): 37.

35. C. G. Campbell, "The Present Position of Eugenics," *Journal of Heredity* (April 1943): 147. Campbell praised the United States for having "the distinction of being the first nation in modern times to enact laws for a eugenic purpose. Since the first [sterilization] law in Indiana in 1905..."

36. This "history" project relied not only on the erasure of millions of people, but on eugenicists' selective group recall. In conjuring the vision of a glorious, eugenic U.S. past, they forgot their own panicked antimiscegenation calls and the calls of their nineteenth-century ideological predecessors against both African-Americans and earlier waves of immigrants. They forgot, too, the children born to Black women raped by white men during and after slavery.

37. Stephen Steinberg, *Turning Back* (Boston: Beacon Press, 1995), 25, 32, 33. See also George Stocking, *Bones, Bodies, Behavior* and *Race, Culture, Evolution*.

38. K. Holler, "The Nordic Movement in Germany," *Eugenical News* 17, no. 5 (September–October 1932): 117.

39. "Doctor Laughlin Honored," *Eugenical News* 21, no. 4 (July–August 1936).

40. "Eugenical Sterilization In Germany," *Eugenical News* 18, no. 5 (September–October 1933): 89.

41. C. M. Goethe, "Patriotism and Racial Standards," *Eugenical News* 21, no. 4 (July–August 1936): 65–66. Goethe was extremely active in anti-Mexican immigration restriction and he used his brief presidency of the ERA to drum up support for Nazi policies in the United States (Mehler, "The History of the American Eugenics Society," 356).

42. Holler, "The Nordic Movement in Germany."

43. Cited in Gould, *The Mismeasure of Man*, 195.

44. "Eugenical Sterilization in Germany," *Eugenical News* 18, no. 5 (September–October 1933): 90. With no sense of irony, the writer reassured readers, "To one acquainted with English and American law, it is difficult to see how the new German Sterilization Law could, as some have suggested... be made an 'instrument of tyranny' for the sterilization of non-Nordic races. Of the more than 16,000 cases of legal sterilization under the recent American state sterilization statutes, no one has ever suggested that in any single case... any racial, religious or political prejudice had ever prompted any single operation under these laws" (90). While this may or may not have been the case in 1933, it did not remain so for long. By 1935, the Black press in the United States was making just such accusations—and very effectively.

45. "Race Hygiene (Eugenics) in Germany," *Eugenical News* 19, no. 5 (September–October 1934): 136.

46. "Hitler and Race Pride," *Eugenical News*, 17, no. 2 (March–April 1932): 61–62.

47. "Jewish Refugees from Germany," *Eugenical News* 19, no. 2 (March–April 1934) 44. Actually, having decried the presence of Jews in the United States for decades, this was merely a rhetorical question for the journal. In the same issue, the editors printed a letter Madison Grant had received from the Count de Lapouge, bemoaning the presence of Syrians, Poles, Moroccans, Spaniards, and Italians in France. "And to complete this, it is 'raining' German Jews" ("A French View," ibid., 39).

48. "A Letter From Dr. Ploetz" and "Jewish Physicians in Berlin (from *Rassenpolitische Auslands-Korrespondenz*, no. 2, 1934), *Eugenical News* 16, no. 5 (September–October 1934): 129, 126.

49. Even among eugenicists, there were early critics of Nazi rhetoric and policies. However, their criticisms were muted. The most strident merely blamed "men like Hitler" for discrediting eugenics, an otherwise sound philosophy in their eyes. J. H. Kempton, seeing the glass half full, wrote that while Hitler and his ilk "subject Eugenics to further ridicule, their antics stimulate a demand for information on the subject of human heredity that may, in the end, prove beneficial" ("Bricks without Straw," *Journal of Heredity* 24, no. 12 [December 1933]: 463).

50. C. G. Campbell, "The German Racial Policy," *Eugenical News* 21, no. 2 (March–April 1936): 25.

51. Ibid., 29.

52. "Must We Raise Our Birth Rate?" *Sexology* 8, no. 4 (March 1941): 230.

The Pioneer Fund

1. Certificate of Amendment of the Certificate of Incorporation of the Pioneer Fund, Inc., under Section 803 of the Not-For-Profit Corporation Law, 30 April 1985, State of New York. Thanks to Rochelle Ordover for tracking down this document. See also Mehler, "The History of the American Eugenics Society," 115–116; and Adam Miller, "Professors of Hate," *Rolling Stone*, 20 October 1994, 112.

2. Mehler, "In Genes We Trust," 14.

3. Certificate of Amendment of the Certificate of Incorporation of the Pioneer Fund. It is difficult to ascertain the fund's full range of activities. As a nonprofit, tax-exempt organization, they are required to file changes in their charter, but because they receive no money from the government, there are certain disclosures they are not mandated to make to the public. According to federal law, the Pioneer Fund can retain its tax-exempt status so long as it refrains from "carrying on propaganda or otherwise attempting to influence legislation." As this restriction does not apply to grantees, Pioneer Fund recipient Ralph Scott, who studied "forced busing and its relationship to genetic aspects of educability," used part of his award for antibusing/anti–school integration seminars in Louisville and Boston ("Fund Backs Controversial Study of 'Race Betterment,'" *New York Times*, 11 December 1977, 76).

4. Miller, "Professors of Hate," 112; Mehler, "In Genes We Trust," 14.

5. Mehler, "The History of the American Eugenics Society," 116.

6. Bonnie Squires, "'Bell Curve' Given Too Much Attention," *Philadelphia Tribune*, 28 February 1995, 7A; Mehler, "The History of the American Eugenics Society," 116 n.70.

7. Steven J. Rosenthal, "The Pioneer Fund: Financier of Fascist Research," *American Behavioral Scientist* 39, no. 1 (September–October 1995): 50; and Squires, "'Bell Curve' Given Too Much Attention."

8. Senator James Eastland was also a member of the board in the 1950s (Rosenthal, "The Pioneer Fund," 50; Griffin Fariello, *Red Scare: Memories of the American Inquisition* [New York: Avon Books, 1995], 18–19, 471). The Pioneer Fund began publishing *Mankind Quarterly* in 1960. Ottmar von Verschner, Nazi "scientist" and mentor to Joseph Mengele, was on the editorial board (Rosenthal, "The Pioneer Fund," 51).

9. U.S. House Subcommittee No. 1, Committee on the Judiciary, *To Amend the Immigration and Nationality Act, and for Other Purposes: Hearings on H.R. 2580*, 89th Cong., 1st sess., 20 May 1965, 251, 237. In 1942, the American Coalition of Patriotic Societies (ACPS) was named in a Justice Department sedition indictment. The ACPS called for the release of Nazi war criminals in the 1960s and was proudly pro-apartheid. Other Pioneer Fund directors include Thomas F. Ellis, who backed Ronald Reagan's 1976 presidential run, and Jesse Helms's 1972 campaign manager ("Fund Backs Controversial Study," 76; Miller, "Professors of Hate," 112, 113).

10. House Committee, *To Amend the Immigration and Nationality Act*, 240, 239.

11. Ibid., 241.

12. Ibid., 245, 253.

13. Some individuals have managed to collect huge sums from the Pioneer Fund over time. Arthur Jensen, of the University of California at Berkeley, received over a million dollars through the years (Mehler, "In Genes We Trust," 12, 14; Miller, "Professors of Hate," 114).

14. Miller, "Professors of Hate," 114. Many Californian academics have proved true to the Pioneer Fund's credo and have been amply rewarded, Jensen and the late William Shockley topping the list. Their grants were dispensed through the University of California at Berkeley and Stanford, respectively ("Fund Backs Controversial Study," 76). While this mode of disbursement is not uncommon, it does betray a certain measure of institutional complicity. It also attests to the historical continuity of the top-down racism provided by eugenic tabulators.

15. Rosenthal, "The Pioneer Fund," 51.

16. Elizabeth Kadetsky puts the total award at $800,000 (Kadetsky, "Bashing Illegals in California," 419; Rosenthal, "The Pioneer Fund," 54). The Pioneer Fund granted FAIR $100,500 in 1994, $108,500 in 1993, and $150,000 in 1992. A second anti-immigrant recipient of recent years is the American Immigration Control Foundation, though their $10,000 award from the Fund is paltry in comparison (*Dangerous Intersections*, appendix).

17. Special issue, *New Republic*, 31 October 1994; and Kadetsky, "Bashing Illegals in California," 419.

18. "Frustration over Immigration Growing in State," *San Francisco Chronicle*, 30 March 1994, A10; Pamela Burdman, "Grass-Roots Anger Takes Hold in California," *San Francisco Chronicle*, 30 March 1994, A10.

19. Rosenthal, "The Pioneer Fund," 54.

20. Malthusianism holds that resources grow arithmetically, while population grows exponentially.

21. Domestic warnings frequently rely on the construction of a hyperbolic fertile Other: "Statisticians say the population of Southern California since 1940 has been

growing at twice the rate of Bangladesh" (*San Francisco Chronicle*, 14 May 1995, Sunday sec., 8). Popular culture, too, has tapped into fear of a non-U.S. population "explosion." A Certs commercial warns, "More people on earth means more bad breath to pollute our air."

22. Cited in Aaron G. Lehmer, "The Greening of Hate," *San Francisco Bay Guardian*, 19 February 1997, 23.

23. Cited by Nikki Bas, "State of the Environmental Justice Movement" session participant, *Dangerous Intersections*, 33.

24. Betsy Hartmann, keynote speech, *Dangerous Intersections*, 3.

25. Pamela Burdman, "The 'I-Word' Creates Tense Environment at Sierra Club," *San Francisco Chronicle*, 30 March 1994.

26. Cathi Tactaquin, Saturday plenary session participant, *Dangerous Intersections*, 22.

27. A discussion on the acceptability of "neutrality" in an age of backlash and scapegoating is beyond the scope of this chapter.

"Indiscriminate Kindness" and "Maudlin Sentimentalism"

1. "The Most Dangerous Conservative," *New York Times Magazine*, 9 October 1994, 48.

2. E. J. Lidbetter, "Heredity, Disease, and Pauperism," *Birth Control Review* 13, no. 7 (July 1929): 192.

3. Ibid., 193.

4. Grant, *The Passing of the Great Race*, 228.

5. Ward, "Immigration after the War," *Journal of Heredity* 8, no. 4 (April 1917): 147–49.

6. Fishberg, "Eugenics in Jewish Life," 547.

7. Cited in Chase, *The Legacy of Malthus*, 174. In addition to the obvious similarity in their views of the immigrant "menace," Chase contends that Coolidge's invocation of sentimentality's power of destruction may have been entirely borrowed from Madison Grant. Coolidge was an avid reader of Grant's work. John Trevor's appeal to Congress forty-four years later echoed the statements of Coolidge and Ward. Trevor's use of Radzinski included his ascription of juvenile delinquency, treason, and "excessive homicide" to America's "*reckless generosity* to the people of other lands" (emphasis added). Radzinski, himself an immigrant, had written, "The U.S.A. can no longer afford to be the foster home for the unfortunates of the world. Biologically, there are already present here so many human types that further additions can hardly enhance the genetic end product. But such additions will tend to postpone indefinitely the salutary fusion necessary for a harmonious society" (House Committee, *To Amend the Immigration and Nationality Act*, testimony of John Trevor Jr., 241).

8. Eugenicists and their sympathizers were so vested in their image as rational, science-over-sentiment crusaders, that the American Eugenics Society could not bear to advocate for a child labor law based on concern for the well-being of minors. The society chose instead to include the law in its 1928 proposed legislative program "not as an euthenic measure . . . but as a definitely eugenic measure since large families are encouraged in the very poor in regions where child labor is easily

exploited" ("Legislative Program of the American Eugenics Society," *Eugenical News* 13, no. 3 [March 1928]: 39). Also included in this platform was a reworking of the parole system to allow for consideration of "possible social and hereditary menace," extension of the quota system to North and South America, extension of deportation provisions, and the "registration of all aliens." In July 1994, in a comparable refusal to bow to tenderheartedness, shoppers at two supermarkets in San Anselmo and San Rafael in California threatened to boycott and/or picket United Market stores if they did not discontinue use of shopping bags with the face and immigration story of a fourth-grader from El Salvador who entered the United States without documentation. Despite the fact that bag supporters seemed to outnumber detractors, United Market capitulated, dumping 40,000 bags that were part of a Public Art Works program for children. Rick Oltman, chairman of the Proposition 187 campaign, received phone calls from several people complaining that the ten-year-old's appearance on the bag encouraged illegal immigration. Heartened by the response, he told a reporter, "We always like to see our base broadening" ("Marin Markets Dump Bags after Protest," *San Francisco Chronicle,* 21 July 1994, 1, 13).

The Abiding Panic

1. "Frustration over Immigration Growing in State," *San Francisco Chronicle,* 30 March 1994, A10.

2. "Latinos Attack VA Bill to Require Reporting of Illegal Immigrants," *Washington Post,* 27 February 1994, B1, B3, B2.

3. Alan C. Nelson, "Open Forum: Something Must Be Done," opinion piece, *San Francisco Chronicle,* 17 August 1994, 19. Raising the specter of disease may become a self-fulfilling prophesy. AIDS activists and care providers have warned of the devastating ripple effect welfare reform will have on immigrants, particularly in the Latino community, which has been disproportionately affected by the pandemic and will now receive even fewer services ("AIDS Activists State Fears of Welfare Reform," *Bay Windows* 14, no. 37 [5–11 September 1996]: 3, 21).

4. Mosse, *Nationalism and Sexuality,* 190. Mosse was speaking of racism in Europe, but this was certainly an apt description of what has transpired in the United States.

5. "Frustration over Immigration Growing in State," A11.

6. Ibid.

7. Mosse, *Nationalism and Sexuality,* 9.

8. J. H. Landman, *Human Sterilization* (New York: Macmillan Company, 1932), 302–4.

9. R. C. Lewontin, *Biology as Ideology: The Doctrine of DNA* (New York: Harper Collins, 1991), 23.

Science as Savior

1. *Prevention* magazine's 1968 charge that fluoride in the drinking water causes homosexuality was reported by Ralph Blair, *Etiological and Treatment Literature on Homosexuality* (New York: National Task Force on Student Personnel Services and

Homosexuality, 1972), 1; "Crash Made Him Gay, Jury Made Him Rich," *Oakland Tribune*, 15 February 1976.

2. Cited in Donna Minkowitz, "Trial by Science," *Village Voice*, 30 November 1993.

3. Dean Hamer and Peter Copeland, *The Science of Desire: The Search for the Gay Gene and the Biology of Behavior* (New York: Simon & Schuster, 1994), 212.

4. Cited in Ruth Hubbard, "False Genetic Markers," *New York Times*, 2 August 1993, A15.

5. Emphasis added; "What Makes Folks Gay?" *Bay Area Reporter*, 10 November 1994, 44.

6. Caleb Crain, "Did a Germ Make You Gay?" *Out*, August 1999, 48.

7. Neenyah Ostrom, "'Gay Gene' Research Doesn't Hold Up under Scrutiny," *New York Native* 638 (10 June 1995): 25; Ruth Hubbard, "False Genetic Markers," A15. Geneticist Richard Lewontin of Harvard has offered some critical insight into reading twin studies, some of which certainly applies to Hamer's study on gay brothers: Studies that advertise for twin subjects will attract twins who are most alike and who are certainly closer to each other. This self-selection is rarely, if ever, acknowledged by researchers designing or executing the study (Lewontin, *Biology as Ideology: The Doctrine of DNA* [New York: Harper Collins, 1991], 33).

8. Hamer and Copeland, *The Science of Desire*, 49, 108, 50, 51, 99, 102. This, despite his profession that he holds "the Baconian concept that scientific models should be based on observation rather than on what a scientist already is convinced is true" (28).

9. "Study Links Brain Path to Sexual Orientation," *Oakland Tribune*, 18 November 1994, A3; Natalie Angier, "Study Links Brain to Transsexuality," *New York Times*, 2 November 1995.

10. Radio interview with Dean Hamer on "Technation: Americans and Technology," KALW-FM, San Francisco, airdate 17 January 1995.

11. Simon LeVay, "A Difference in the Hypothalamic Structure between Heterosexual and Homosexual Men," *Science* 253 (30 August 1991): 1034, 1035.

12. "Born or Bred?" *Newsweek*, 24 February 1992, 48.

13. Don Jackson, review of *The Stereotaxic Treatment of Homosexuality*, by F. Roeder and D. Muller, *Bay Area Reporter*, 28 June 1972; Jackson, "Psychosurgery," *Gay Sunshine* 13 (June 1972): 1. Jackson's report did not enumerate the charges that landed these men in prison or suggest that their sentences were augmented by knowledge of their homosexuality.

14. Dean Gengle and Norman C. Murphy, "Revolutionary Extinction? An Emerging Model of the Origin of Sexuality," *Advocate* 253 (1 November 1978): 17.

15. Cited in Don Jackson, "Legislature Acts to Ban Surgical Cures," *Bay Area Reporter*, 28 June 1972.

16. "News for Neurosurgeons: Brain Cells Scattered," *Advocate* 4, no. 24 (20 January–2 February 1971): 4.

17. Hamer and Copeland, *The Science of Desire*, 48.

18. Marcia Barinaga, "Is Homosexuality Biological?" *Science* 253 (30 August 1991): 956, 957.

19. Ibid., 956; William Byne and Bruce Parsons, "Human Sexual Orientation: The Biologic Theories Reappraised," *Archives of General Psychiatry* 50, no. 3 (March 1993): 235.

20. "Born or Bred," 48, 52, 50. I need hardly add that Hamer's research implicates biology *and* mothers.

21. Doug Futuyma, "Is There a Gay Gene? Does It Matter?" *Science for the People*, January–February 1980, 15.

22. Wendell Ricketts, "Biological Research on Homosexuality: Ansell's Cow or Occam's Razor?" *Journal of Homosexuality* 9, no. 4 (1984): 88.

Delineating Deviance

1. George Chauncey, *Gay New York: Gender, Urban Culture, and the Making of the Gay Male World, 1890–1940* (New York: Basic Books, 1994), 60, 14.

2. Ibid., 122.

3. Jennifer Terry, "Anxious Slippages between 'Us' and 'Them,'" 134.

4. Cited in Siobhan Somerville, "Scientific Racism and the Emergence of the Homosexual Body," *Journal of the History of Homosexuality* 5, no. 2 (October 1994): 259.

5. Michel Foucault, *The History of Sexuality*, vol. 1, *An Introduction* (New York: Random House, 1994), 118, 54.

6. Cited in Havelock Ellis, *Studies in the Psychology of Sex*, vol. 2, *Sexual Inversion* (1901; reprint, Philadelphia: F. A. Davis, 1928), 68.

7. Cited in David F. Greenberg, *The Construction of Homosexuality* (Chicago: University of Chicago Press, 1988), 414. Krafft-Ebing's *Psychopathia Sexualis* was translated into English in 1892.

8. James Foster Scott, *The Sexual Instinct: Its Use and Dangers as Affecting Heredity and Morals: Essentials to the Welfare of the Individual and the Future of the Race*, (New York: E. B. Treat, 1907), 421, 427. Scott defined "onanism" as "a term of comprehensive meaning, applicable in a broad sense to all forms of sexual stimulation employed by either sex, singly or mutually, to produce orgasm in unnatural ways— i.e., otherwise than by coitus" (419).

9. Ibid., 105.

10. Ibid., 47, 105, 425.

11. Ibid., 42, 104. Justice Holmes would later lead the Supreme Court in upholding the states' right to sterilize without consent.

12. Cited in Vern L. Bullough, "Homosexuality and the Medical Model," *Journal of Homosexuality* 1 (fall 1974): 106, 107.

13. Ibid., 107.

14. Ibid.

15. Greenburg, *The Construction of Homosexuality*, 403. By way of illustration, Greenberg cites the naming of "alcoholism," termed "drunkenness" prior to reclassification by Swedish physician Magnus Huss in 1852. Like homosexuality, it now had the status of disease conferred upon it. Over a hundred and forty years later, Newt Gingrich would classify homosexuality and alcoholism together as well, this time not explicitly as diseases, but as "orientations."

16. Ibid., 415–16.

17. Ibid., 412, 413.

18. Terry, "Anxious Slippages between 'Us' and 'Them,'" 132.

19. Chauncey, *Gay New York*, 132.

20. Emphasis added; G. Frank Lydston, *Addresses and Essays* (Louisville, Ky.: Renz & Henry, 1892), 244; originally published in the *Philadelphia Medical and Surgical Reporter,* 7 September 1889. Havelock Ellis cited Lydston's work excessively in *Sexual Inversion,* particularly the doctor's use of hypnotic suggestion.

21. Lydston, *Addresses and Essays,* 346.

22. Lydston, *The Diseases of Society (The Vice and Crime Problem)* (Philadelphia: J. B. Lippincott, 1904), 15.

23. Some years later Ellis would write of "latent congenital inversion," which manifests later in life, perhaps after childbearing/siring years (*Sexual Inversion,* 84).

24. Lydston, *Addresses and Essays,* 248.

25. Jonathan Katz, *Gay American History: Lesbians and Gay Men in the U.S.A.* (New York: Avon Books, 1976), 197; Vern Bullough and Martha Voght, "Homosexuality and Its Confusion with the 'Secret Sin,'" *Journal of the History of Medicine and Allied Sciences* 38, no. 2 (April 1973): 152.

26. *Boston Medical and Surgical Journal* 61, no. 8 (22 September 1859): 165.

27. "The Orificial Philosophy," *Journal of Orificial Surgery* 1, no. 8 (February 1893): 536.

28. E. P. Miller, "Sensuality as a Cause of Disease," *Journal of Orificial Surgery* 4, no. 9 (March 1896): 401.

29. Lydston, *The Diseases of Society,* 564.

30. E. H. Pratt, editorial, *Journal of Orificial Surgery* 2, no. 6 (December 1893): 281.

31. Wayne R. Dynes, ed., *Encyclopedia of Homosexuality* (New York: Garland Publishing, 1990), 740. Vern Bullough wrote that Lombroso advocated internment not in prisons, but in asylums (Bullough, "Homosexuality and the Medical Model," 107).

32. Greenberg, *The Construction of Homosexuality,* 415; Wayne Dynes, *Encyclopedia of Homosexuality,* 740.

33. Lombroso, cited in Greenberg, *The Construction of Homosexuality,* 418.

34. Katz, *Gay American History,* 209. Later doctors and policy makers would speak similarly of the humanity of vasectomy over castration and the supposedly lesser invasiveness of Norplant over tubal ligation.

35. Ibid.

36. Ibid., 210.

37. J. H. Landman, *Human Sterilization: The History of the Sexual Sterilization Movement* (New York: Macmillan, 1932), 312–13.

38. Ibid., 57–58.

39. Ibid., 61, 62, 63, 304, 306. During the mid- to late twenties, gays fell within the scope of sterilization laws passed by Idaho, North Dakota, and Michigan. Within three years of the statute's passage, Michigan had sterilized 629 people, the overwhelming majority (476) of whom were women (73).

40. Ibid., 76, 77.

41. Ibid., 75.

42. Emphasis added; *Warden Davis v. Walton,* Supreme Court of Utah, 9 April 1929; and Landman, *Human Sterilization,* 103.

43. The judges were not even wholly convinced that the alleged "sexually criminal offense" had, in fact, occurred (*Warden Davis v. Walton,* Supreme Court of Utah, 9 April 1929).

44. *U.S. Statutes at Large* 39, part 1 (1917): 875. U.S. immigration law, however, has never clearly defined "moral turpitude," leaving it to the discretion of the courts and the Immigration and Naturalization Service. For an excellent discussion on homosexuality and U.S. immigration law, including the uses of the 1952 McCarran-Walter Act and the Immigration Act of 1990, see Shannon Minter, "Sodomy and Public Morality Offenses under U.S. Immigration Law: Penalizing Lesbian and Gay Identity," *Cornell International Law Journal* 26 (1993): 771.

45. Michel Foucault, *Discipline and Punish: The Birth of the Prison* (New York: Vintage Books, 1979), 22.

Biological Apologists

1. "Born or Bred," 48. Shilts's remark was perhaps an unconscious reference to "batting lefty"—a metaphor for homosexuality.

2. Barinaga, "Is Homosexuality Biological?" 957.

3. Deeg, "Gay Genes Found on Queer Butts," *Ultra Violet* (newsletter of Lesbians and Gays against Intervention) 6, no. 1 (summer 1995): 10.

4. Cited in Gunter Schmidt, "Allies and Persecutors: Science and Medicine in the Homosexuality Issue," *Journal of Homosexuality* 10, no. 3/4 (winter 1984): 128.

5. Ibid., 129.

6. Ibid., 132.

7. Despite Hirschfeld's embrace of the procedure, Ellis noted some years earlier that the German sexologist "concludes that from a eugenic standpoint the marriage of a homosexual person is always very risky" (*Sexual Inversion*, 335).

8. Schmidt, "Allies and Persecutors," 133, 134.

9. Fifty years later, such operations to sever the part of the brain believed to control "homosexual drives" were indeed carried out in Germany. Ibid., 134–35.

10. Greenberg, *The Construction of Homosexuality*, 418.

11. Ellis, *Sexual Inversion*, 179.

12. Ibid., 265.

13. Ibid., 139.

14. Greenberg, *The Construction of Homosexuality*, 419.

15. Ellis, *Sexual Inversion*, 335.

16. James P. Winsco, "The Real Homosexual," *Sexology* 2, no. 10 (June 1935): 639.

17. Ellis, *Sexual Inversion*, 327.

18. Ibid., 338.

Gender, Race, and the Strategy of Metaphor

1. Perkins's comment would have more aptly described the omission of women from AIDS drug trials.

2. "Born or Bred?" 48.

3. Ellis, *Sexual Inversion*, 310 n.2.

4. Lynda I. A. Birke, "Is Homosexuality Hormonally Determined?" *Journal of Homosexuality* 6, no. 4 (summer 1981): 41.

5. The fact that they were making genetic claims without being geneticists did not undermine their credibility with the press. It was not even remarked on by a media only too willing to embrace unsubstantiated findings. This is not an uncommon phenomenon. More recently, two economists stepped out of their area of expertise to assert that *Roe v. Wade* is responsible for the drop in crime during the 1990s.

6. "Genetic Clues to Female Homosexuality," *Science News* 142 (22 August 1992): 117; J. Michael Bailey and Richard C. Pillard, "A Genetic Study of Male Sexual Orientation," *Archives of General Psychiatry* 48, no. 12 (December 1991). "Bisexual" was subsumed under "lesbian" or "gay" in each study. For accounts of further studies on twins and sexuality, see Wayne Dynes and Stephen Donaldson, *Homosexuality and Medicine, Health, and Science* (New York: Garland Publishing, 1992); and Ralph Blair, *Etiological and Treatment Literature on Homosexuality* (New York: National Task Force of Student Personnel Services and Homosexuality, 1972). Twin studies have long been a favorite of researchers determined to root out a cause of homosexuality. More generally, Sir Francis Galton, who coined the term "eugenics" in 1883, was, according to Mehler, the first to call for twin studies to differentiate between what nature impacted and what nurture impacted (Mehler, "In Genes We Trust," 77).

7. Their results, however, could very well be skewed by their self-selected recruitment strategy (they advertised for subjects in the queer and feminist press). It may well be that lesbians who had a lesbian twin were more likely to be out and to feel safe/supported enough to enroll themselves and their siblings in such a study.

8. "Study Cites Genetic Basis for Lesbianism," *San Francisco Examiner*, 11 March 1993, A1, A8. The potential ramifications of using "sexual orientation" and "fetus" in the same sentence are chilling and deserving of much more attention than they garnered in the press coverage.

9. Ibid.

10. Duncan Osborn, "The Escalating Drive to 'Cure' Queer Kids," *LGNY* 121 (16 December 1999): 4.

11. Here, Bailey, Pillard, et. al. (Bailey and Pillard, "A Genetic Study of Male Sexual Orientation"; Bailey, Pillard, Michael C. Neale, and Yvonne Agyei, "Heritable Factors Influence Sexual Orientation in Women," *Archives of General Psychiatry* 50, no. 3 [March 1993]) concluded that gay men had been less interested than straight men in sports as children. This lack of interest was said to indicate "effeminacy in male subjects. Harassment from other children was apparently not considered to be a deterrent to participation.

12. Bailey, Pillard, et al., "Heritable Factors Influence Sexual Orientation in Women," 219, 222; Bailey and Pillard, "A Genetic Study of Male Sexual Orientation," 1090, 1095, 1094.

13. Somerville, "Scientific Racism and the Emergence of the Homosexual Body," 247.

14. Albert Moll, *Perversions of the Sex Instinct* (Newark: Julian Press, 1931), 235. Moll's work was translated into English and cited extensively by British and American writers, Havelock Ellis being among the most notable.

15. Terry, "Anxious Slippages between 'Us' and 'Them,'" 131.

16. Lydston, *The Diseases of Society,* 49, 178, 263.

17. Ibid., 263.

18. Ibid.

19. C. A. Weirick, editorial, *Journal of Orificial Surgery* 7, no. 10 (April 1899): 479, 477. Weirick was an associate editor of the journal.

20. Cited in Katz, *Gay American History*, 209, 210, 211.

21. Emphasis added; Pratt, editorial, *Journal of Orificial Surgery*, 280.

22. "The Orificial Philosophy," 527.

23. Bullough and Voght, "Homosexuality and Its Confusion with the 'Secret Sin,'" 152.

24. Weirick, editorial, *Journal of Orificial Surgery*, 479. In his elevation of parental blame to research imperative, Weirick anticipated the *Newsweek* coverage of causation research by almost a hundred years.

25. Bullough and Voght, "Homosexuality and Its Confusion with the 'Secret Sin,'" 150.

26. Julia Holmes Smith, "Three Disappointing Cases," *Journal of Orificial Surgery* 7, no. 9 (March 1899): 397, 398.

27. Ellis, *Sexual Inversion*, 226, 253, 254.

28. Ibid., 255, 256.

29. This particular myth appears to have had some currency over a period of time. Ellis wrote that Ulrichs was the first to identify the deficiency in men, and that Hirschfeld had found male inverts unable to whistle in 23 percent of the cases (ibid., 291). A 1940 article in *Sexology* magazine also attributes an inability to whistle to homosexuality in boys ("Can Homosexuality Be Cured?" *Sexology* 7, no. 12 [November 1940]: 827). In one of Ellis's case studies, a gay man, perhaps attempting to offset the stereotype or distance himself from other male inverts, takes care to mention that he is "perhaps a better whistler than most men" (Ellis, *Sexual Inversion*, 191).

30. Greenberg notes that Franz Joseph Gall, founder of the phrenologist school, posited "adhesiveness" as a brain function responsible for a friendship instinct. It was as susceptible as any other process to "pathology," such as homosexuality (Greenberg, *The Construction of Homosexuality*, 404).

31. Cited in Katz, *Gay American History*, 206.

32. Ellis, *Sexual Inversion*, 201.

33. George W. Henry, ed., *Sex Variants: A Study of Homosexual Patterns* (New York: Paul B. Hoeber, 1941), 1080, 1082–99. How Dickinson determined this is anybody's guess, but an intense focus on erection, then or now, can hardly be called a rarity. Dickinson sat on the American Eugenics Society's Committee on the Eugenic and Dysgenic Effects of Birth Control (Mehler, "The History of the American Eugenics Society," 331). See also Jennifer Terry's essay on the Committee for the Study of Sex Variants, "Anxious Slippages between 'Us" and 'Them'"; and chapters 6 and 7 of her *An American Obsession: Science, Medicine, and Homosexuality in Modern Society* (Chicago: University of Chicago Press, 1999).

34. Maurice Chideckel, *Female Sex Perversion: The Sexually Aberrated Woman as She Is* (New York: Eugenics Publishing, 1935), 125.

35. George Chauncey Jr., "Christian Brotherhood or Sexual Perversion: Homosexual Identities and Sexual Boundaries in the World War I Era," in *Hidden from History: Reclaiming the Gay and Lesbian Past* (Marham, Ontario: Penguin/New American Library Books, 1989), 546 n.74.

36. August Forel, *The Sexual Question* (New York: Rebman Company, 1908), cited in Katherine Bement Davis, *Factors in the Sex Life of Twenty-Two Hundred Women*,

Publication of the Bureau of Social Hygiene (New York: Harper and Brothers, 1929), 241.

37. This sort of distinction was not limited to cataloging women, nor was it unique to the age. In Louis-Rene Villerme's 1824 report of French prisons, the doctor not only decreed which gay men were playing "male" and "female" roles, but labeled the former's homosexuality circumstantial and the latter's "instinctual," perhaps indicating heritability (Greenberg, *The Construction of Homosexuality*, 404).

38. Ricketts, "Biological Research on Homosexuality," 84.

39. Muriel Wilson Perkins, "Female Homosexuality and Body Build," *Archives of Sexual Behavior* 10, no. 4 (1981): 340, 344, 341. Implicit in this hierarchy is the construction of thinness as normative.

40. Of late, some gay rights advocates have also turned to metaphor as a civil rights strategy, leveraging claims of biological immutability to elide differences between racism and homophobia.

41. Cited in Blair, *Biological and Treatment Literature on Homosexuality*, 3.

42. Chideckel, *Female Sex Perversion*, 136, 137.

43. Ibid., 163, 136.

44. Ibid., 299.

45. Ibid., 303–5; Somerville, "Scientific Racism and the Emergence of the Homosexual Body," 262.

46. Moll, *Perversions of the Sex Instinct*, 61; Ellis, *Sexual Inversion*, 21.

47. Ellis, *Sexual Inversion*, 257, 258.

48. Cited in Moll, *Perversions of the Sex Instinct*, 20.

49. Ibid., 222, 172.

50. Paul Moreau, *Des aberrations du sens genetique*, cited in Bullough, "Homosexuality and the Medical Model," 106.

51. Scott, *The Sexual Instinct*, 48.

52. Indeed, travel journals continue to rhapsodize about the unhindered sexuality of "native children."

53. Glen Wadsworth, "Treatment of Masturbation in Children," *Sexology* 2, no. 18 (April 1935): 507.

54. "Premature Sex Development," *Sexology* 1, no. 5 (January 1934): 313.

55. Davis, *Factors in the Sex Life of Twenty-Two Hundred Women*, 239; Moll, *Perversions of the Sex Instinct*, 63. Moll did not believe that gays and lesbians were underrepresented among Jews. He wrote that he himself knew a few personally and noted that "Jewish Uranists [male] bear the names of Jewish women, such as Rebecca, Sarah, etc." (86).

56. Chideckel, *Female Sex Perversion*, 323.

57. Scott, *The Sexual Instinct*, 48. Likewise, assumptions about the "purity" of Jewish blood contributed to notions of sinlessness. Scott also wrote that "Jewesses, who belong to an unmixed people, menstruate at about the same age in all latitudes, i.e., at fourteen or fifteen years of age," whereas "half-castes" reach puberty at an earlier age (48). Scott did not elaborate on the supernatural phenomenon that would enable a diasporic people who had been conquered, enslaved, raped, absorbed (to a degree), and expelled, over and over again, to keep a pure bloodline dating back to the three patriarchs.

58. "The Orificial Philosophy," 527, 534.

59. Sander Gilman, *The Jew's Body* (New York: Routledge, 1991), 5, 126.

60. George Mosse, *Nationalism and Sexuality: Middle-Class Morality and Sexual Norms in Modern Europe* (Madison: University of Wisconsin Press, 1985), 140, 142, 158.

61. Nancy Leys Stepan, "Race and Gender: The Role of Analogy in Science," in *Anatomy of Racism*, ed. David Theo Goldberg (Minneapolis: University of Minnesota Press, 1990), 44, 48, 47.

62. Into this cultural context comes Simon LeVay's pronouncement that a region of the hypothalamus in gay men and (presumably heterosexual) women is less than half the size of the same region in straight men (LeVay, "A Difference in the Hypothalamic Structure between Heterosexual and Homosexual Men," 1034). Once again, size equals sexuality, and once again women and gay men fail to measure up as hard science confirms the associative link between them. LeVay certainly has ideological ancestors, in terms of both method and analogy. In 1931 Albert Moll reported on French writers Magnan and Gley, who theorized on a "feminine brain in Uranists." Moll also noted autopsies performed by Recklinhausen to determine the roots of homosexuality (Moll, *Perversions of the Sex Instinct*, 166). Perhaps the most extremist of subscribers to the gay-man-as-true-woman theory was Theo Lang. Beginning in 1934, Lang studied 1,015 gay men whose names and addresses were obtained from the Munich and Hamburg police. The men had a total of 1,734 brothers and 1,532 sisters. Lang concluded that this ratio of 100 females to 121 males must mean that some of the male subjects were genetic females in men's bodies. The compilation of lists of known and suspected gay men by the police under the Third Reich comes as no surprise, yet Lang's assertion that research findings would "contribute to the final solution of homosexuality" is chilling, especially given the fact that his theory was not quashed until the 1950s when it became possible to determine chromosomal sex. Theo Lang, "Studies on the Genetic Determination of Homosexuality," *Journal of Nervous and Mental Disease* 92, no. 1 (July 1940): 61, 56, 63; C. M. B. Pare, "Etiology of Homosexuality: Genetic and Chromosomal Aspects," in *Sexual Inversion*, ed. Judd Marmor (New York: Basic Books, 1965), 72.

Thanks in large part to the many years science has spent establishing the physical, emotional, and intellectual inferiority of women, "woman" has become the worst possible insult to hurl at a man, while behavior categorized as "feminine" translates to "gay marker" when engaged in by men or boys. I am problematizing the equation of gay men and gay male sexuality with straight women and heterosexual female sexuality. Both women and gay men have been pathologized at different times and have seen the defamation of their sexuality used to normalize male heterosexuality. They have become metaphoric foils for the denigration of one another. I am not speaking of male to female transgendered people, heterosexual or homosexual, for whom such equivalences have a varied implication. None of these researchers were or are interested in explicating a fluidity of gender or sexuality.

Homosexuality and the Bio/Psych Merge

1. Psychologists' "treatment" regimens are still fraught with peril for queers. The discipline's early years were no different. In an article titled "The Gynecology of Homosexuality," Robert Latou Dickinson matter-of-factly relayed the case of Kathleen M.: "A psychologist advised relations with men; he slept with her, after which she had vaginal bleeding for two months" (in Henry, ed., *Sex Variants*, 1092).

2. Hamer and Copeland, *The Science of Desire,* 73.

3. Greenberg, *The Construction of Homosexuality,* 422, 425, 423, 26; Ellis, *Sexual Inversion,* 83.

4. George B. Lake, "Sex 'Inversion,'" *Sexology* 1, no. 7 (March 1934): 419. Demanding that gays be treated at least as well as other groups who were systematically maligned and scorned was not asking for much, for either population. I am indebted to Susan Stryker for alerting me to *Sexology*'s store of information.

5. Kermit Reidner, "Cure for Homosexuals?" *Sexology* 1, no. 18 (April 1934): 490, 492.

6. Myron D. Jacoby, "Homosexuality—Another Letter," *Sexology* 8, no. 6 (May 1941): 351.

7. J. W. P., "Can Homosexuality Be Cured?" *Sexology* 7, no. 12 (November 1940): 828.

8. Dynes, *Encyclopedia of Homosexuality,* 826.

9. Moll, *Perversions of the Sex Instinct,* 180, 65, 64.

10. Ibid., 168, 148, 149, 160.

11. Dynes, *Encyclopedia of Homosexuality,* 826.

12. Moll, *Perversions of the Sex Instinct,* 187.

13. Ibid., 147.

14. Ellis, *Sexual Inversion,* 286, 330, 329.

15. Max, cited in Katz, *Gay American History,* 252.

16. Notes from 16 December 1937, meeting with doctors. Pauli Murray Papers, collection #MC412, box 4, file 71, Schlesinger Library, Cambridge, Massachusetts. My exercise in paraphrasing Murray's notes is no match for the original in impact or eloquence.

17. Ibid., notes from meeting on 17 December 1937.

18. *New York World-Telegram,* 3 November 1939, 1.

19. "Sex Tablets Stir Medics," *New York Amsterdam News,* 11 November 1939, 1.

20. Letter dated 4 November 1939, Pauli Murray Papers.

21. "Summary of Symptoms of Upset," 8 March 1940, Pauli Murray papers. Because the discourse around homosexuality at the time was far from nuanced, Murray's repeated references to wanting to be male can be interpreted in a number of ways (which are in no way mutually exclusive): a belief that she belonged in a man's body or that an attraction toward women and a desire to live and work in "a man's world" must translate into wanting to be a man in the strict, anatomical sense.

22. "The Invert Personality (Part Three)," *Sexology* 6, no. 9 (May 1939): 585, 586.

23. D. H. Keller, "A New Theory of Homosexuality," *Sexology* 10, no. 4 (November 1943): 208.

24. Editor, *Sexology* 7, no. 5 (March 1940): 329.

25. "Endocrimes," *Sexology* 10, no. 2 (September 1943): 128.

26. *Sexology* 2, no. 8 (April 1935): 5538.

27. Stilbesterol is a synthetic form of estrogen. Estriol is an estrogen derivative.

28. The U.S. Public Health Service kept the Tuskegee experiments going from 1932 to 1972. For forty years, over four hundred African-American men were the unwitting subjects of a government study on untreated syphilis. They were told they were receiving treatment, subjected to spinal taps, and given what amounts to a placebo in lieu of actual medical attention.

29. Ricketts, "Biological Research on Homosexuality," 72.

30. Early articles that cast doubt on hormone "therapy" include Hyamn S. Barahal, "Testosterone in Psychotic Male Homosexuals," *Psychiatric Quarterly* 14, no. 2 (1940): 319–30; Abraham Myerson and Rudolph Neustadt, "Bisexuality and Male Homosexuality: Their Biologic and Medical Aspects," *Clinics* 1, no. 4 (1942): 932–57; and Myerson and Neustadt, "Essential Male Homosexuality and Results of Treatment," *Archives of Neurology and Psychiatry* 55, no. 3 (1956): 291–93.

31. "Castration of a Male Homosexual," *British Medical Journal* 4894 (1954): 1001.

32. Charles Bery and Clifford Allen, *The Problem of Homosexuality* (New York: Citadel Press, 1958), 73.

33. Ibid., 32.

34. D. Srnec and K. Freund, "Treatment of Male Homosexuality through Conditioning," *International Journal of Sexology* 7 (1953), cited in Blair, *Biological and Treatment Literature on Homosexuality*, 35. One purveyor of 35 mm "stimulus slides" was the Farrall Instrument Company of Grand Island, Nebraska. Their inventory, "For use with visual shockers, manual shockers and systematic desensitization," included: "Dating scenes, movies, picnics. COUPLES"; "Simulated gang rape, sadistic with belts and whips. Faces not shown. MALE"; "Nude, in bed, glamour poses. MALE (Bruce) 27, long hair, mustache"; "Nude, bed scenes. FEMALE (Annette) 21, long brown hair, excellent figure."

35. James Basil, "Case of Homosexuality Treated by Aversion Therapy," *British Medical Journal* 5280 (1962): 768–70.

36. M. J. MacCulloch and M. P. Feldman, "Aversion Therapy in Management of 43 Homosexuals," *British Medical Journal* 5552 (1967): 594–97.

37. Terry, "Anxious Slippages between 'Us' and 'Them,'" 158, 159.

38. "New Homosexuality Study Supports Physical Origin," *Advocate* 58 (28 April–11 May 1971): 2. The hormones-as-destiny frenzy continued as the *New England Journal of Medicine* published the findings of a group of doctors who measured lower sperm counts and levels of testosterone in thirty gay college students. Of course, many things can cause a decrease in testosterone, but, as none of those were the object of analysis, they were disregarded. Doctors Kolodny, Masters, and Toro wrote, "Whether the defect is testicular, pituitary or hypothalamic awaits further investigation" (cited in "Baker Says Researchers Neglect Hormone Study," *Advocate* 77 [19 January 1972]: 7).

39. "Report on Study Triggers Demonstrations in LA," *Advocate* 58 (28 April–11 May 1971): 2.

40. Cited in Gerald T. Fitzgerald, "Improve Human Heritage Means 'Kill the Queers,'" (editorial), *Advocate* 53 (17 February–2 March 1971): 25, 26. Glass was retiring president of the American Association for the Advancement of Science.

41. "Born or Bred?" 49; Greenberg, *The Construction of Homosexuality*, 430.

42. Don Jackson, "Psychosurgery," *Gay Sunshine: A Newsletter of Gay Liberation* 13 (June 1972): 1.

43. David L. Aiken, "Ervin Committee Report Bares 'Clockwork Orange' Horrors," *Advocate* 154 (1 January 1975): 5.

44. Bryant, Florida orange juice hawker and former beauty queen, led the 1977 charge to repeal Dade County's gay rights ordinance. After her success in Florida, she took her Save the Children campaign national.

45. Dean Gengle and Norman Murphy, "Revolutionary Extinction? An Emerging Model of the Origin of Sexualities," *Advocate* 253 (1 November 1978): 15.

46. Ibid., 16.

47. Ibid., 19.

48. Ibid., 18.

49. Ibid., 18; Dean Gengle, "An Interview with Revolutionary Psychologist Norman C. Murphy," *Advocate* 253 (1 November 1978): 22.

50. Gengle, "An Interview with Revolutionary Psychologist Norman C. Murphy," 24.

51. Lake, "Sex 'Inversion,'" 419.

52. John A. W. Kirsch and James Eric Rodman, "The Natural History of Homosexuality," *Yale Scientific Magazine* 51, no. 3 (winter 1977): 7, 8, 11. Dean Hamer has fleshed out this argument somewhat, positing a genetic increase in fertility rates among heterosexual female relatives of gay men. These women, he reasons, may have more children than average, thus ensuring the continued presence of a gay gene in humans (cited in Hamer and Copeland, *The Science of Desire*, 183). Hamer has further proposed that if a gay gene has a beneficial side effect, other than making the bearer homosexual, heterosexuals might even want to introduce it into their progeny (186). Bailey and Pillard have in fact suggested that a gene coded for homosexuality might yield such benefits as immunity against certain diseases (J. Michael Bailey and Richard C. Pillard, "A Genetic Study of Male Sexual Orientation," 1095).

53. Gengle, "An Interview with Revolutionary Psychologist Norman C. Murphy," 23, 22.

54. G. Dörner, "Hormonal Induction and Prevention of Female Homosexuality," *Journal of Endocrinology* (Great Britain) 42 (1968): 163–64.

55. Birke, "Is Homosexuality Hormonally Determined?" 46.

56. G. Dörner et. al., "Prenatal Stress as Possible Aetiogenetic Factor of Homosexuality in Human Males," *Endokrinologie* 75, no. 3 (1980): 365. Perinatal stress would include malnutrition or an absentee father.

57. Ibid., 365–68.

58. Also implicated in compromising German virility was the Allied bombing. Perhaps, if the prevention of war bears the promise of the prevention of homosexuality, then Dörner's theory may one day make pacifism safe for manhood everywhere (or at least as good as war and nationalism for cultivating masculinity).

59. Gunter Schmidt, "Allies and Persecutors: Science and Medicine in the Homosexuality Issue," *Journal of Homosexuality* 10, no. 3/4 (winter 1984): 136.

60. Hamer and Copeland, *The Science of Desire*, 30. Hamer has also linked his work to LeVay's by postulating that Xq28 (the region of the X chromosome he claims holds the gay gene) makes a protein that "is directly involved in the growth or death of neurons in INAH-3. Alternatively, the gene could encode a protein that influences the regulation of this region by hormones" (163).

61. Gengle and Murphy, "Revolutionary Extinction?" 21.

AIDS, Backlash, and the Myth of Liberatory Biologism

1. Sadly, the liberal response to this latest incarnation of social Darwinism was to maintain that, far from eliminating "only" the weakest links in the evolutionary chain—queers and IV drug users—everyone was at risk, even strong, highly evolved specimens like heterosexual athletes—hardly a courageous rebuttal.

2. Cindy Patton, *Sex and Germs: The Politics of AIDS* (Boston: South End Press, 1985), 19.

3. Richard Goldstein, "AIDS and the Social Contract," in *Taking Liberties: AIDS and Cultural Politics,* ed. Erica Carter and Simon Watney (London: Serpent's Tail, 1989), 81.

4. Cited in Hamer and Copeland, *The Science of Desire,* 212.

5. Lisa Duggan, "Queering the State," in *Sex Wars: Sexual Dissent and Political Culture,* Lisa Duggan and Nan D. Hunter (New York: Routledge, 1995), 189, 190. Many thanks to Rachel Rosenbloom for calling my attention to this essay.

6. Blair, *Biological and Treatment Literature on Homosexuality,* 1.

7. Tony Kushner, "Copious, Gigantic, and Sane," in *Thinking about the Long-standing Problems of Virtue and Happiness* (New York: Theater Communications Group, 1995), 52.

8. Somerville, "Scientific Racism and the Emergence of the Homosexual Body," 266.

Liberal Loopholes

1. "U. of Minn. Republicans Oppose Sanger as Racist," *Chronicle of Higher Education,* 8 November 1996, 8.

2. Janice Raymond, *Women as Wombs: Reproductive Technologies and the Battle Over Women's Freedom* (San Francisco: Harper, 1993), 41.

3. The use of "technofix" here is borrowed from Robert Blank's discussion of technological fixes that serve to divert attention from substantive challenges to institutionalized inequity (Robert H. Blank, *Fertility Control: New Techniques, New Policy Issues* [New York: Greenwood Press, 1991]).

4. David Theo Goldberg, *Racist Culture: Philosophy and the Politics of Meaning* (Cambridge, Mass.: Blackwell, 1993), 5, 213.

5. Blank, *Fertility Control,* 89, 119.

6. Loren R. Graham, "Political Ideology and Genetic Theory: Russia and Germany in the 1920's," *Hastings Center Report 7,* no. 5 (October 1977): 35. Graham was writing of eugenics in Germany and Russia, though this would be an equally apt observation in a U.S. context. His specific analyses of German and Soviet eugenics movements are worth noting here. Graham wrote that, with some exceptions, Mendelian eugenics were associated with the right and Lamarckian eugenics (which dictate that acquired traits can be hereditary) with the left. In the USSR, there were arguments over which of the two was more counterrevolutionary. In Weimar Germany, the more right-wing eugenicists used the term *Rassenhygiene,* while those further to the left employed *Eugenik* (31, 34).

7. Cited in Steven J. Rosenthal, "The Pioneer Fund: Financier of Fascist Research," *American Behavioral Scientist* 39, no. 1 (September–October 1995): 55. Of course, the two are not mutually exclusive, and plenty of liberal policy directives easily incorporate what Noel Ignatiev has called "attempts to splice genetics and sociology" (Noel Ignatiev, *How the Irish Became White* [New York: Routledge, 1995], 1).

8. In fact, the compulsory nature of eugenic prescriptions necessitates a distinction between current meanings of "birth control" and Sanger's usage, for in courting eugenicists and embracing eugenic ideology, the voluntary nature of birth control was supplanted by what would later be termed population control. Popula-

tion control used, and continues to use, the same basic criteria (race and income) as domestic eugenic campaigns that targeted specific groups for externally imposed fertility caps.

9. Stephen Steinberg, *Turning Back: The Retreat from Racial Justice in American Thought and Policy* (Boston: Beacon Press, 1995), 135.

Buck v. Bell and Before

1. Carl Degler, *In Search of Human Nature: The Decline and Revival of Darwinism in American Social Thought* (New York: Oxford University Press, 1991), 45.

2. Cited in Harry H. Laughlin, *Eugenical Sterilization in the United States* (Chicago: Psychopathic Laboratory of the Municipal Court of Chicago, 1922), 351.

3. J. H. Landman, *Human Sterilization: The History of the Sexual Sterilization Movement* (New York: Macmillan, 1932), 52, 54.

4. Nancy Leys Stepan, *The Hour of Eugenics: Race, Gender, and Nation in Latin America* (Ithaca, N.Y.: Cornell University Press, 1991), 31.

5. House Committee, *Biological Aspects of Immigration*, 7.

6. Laughlin, "Legal Status of Eugenical Sterilization," *Birth Control Review* 12, no. 3 (March 1928): 78. Laughlin's article was part of a "Sterilization Symposium" published in *Birth Control Review*. By 1928, the year Laughlin elucidated this vision, 75 percent of U.S. colleges were teaching courses on eugenics (Garland Allen, "Genetics, Eugenics, and Class Struggle," *Genetics* 79, supplement [June 1975]: 33; cited in Carole R. McCann, *Birth Control Politics in the United States, 1916–1945* [Ithaca, N.Y.: Cornell University Press, 1994], 102).

7. Degler, *In Search of Human Nature*, 151.

8. Landman, *Human Sterilization*, 276, 277.

9. David Smith and K. Ray Nelson, *The Sterilization of Carrie Buck* (Far Hills, N.J.: New Horizon Press, 1989), 226.

10. Landman, *Human Sterilization*, 84.

11. *Buck v. Bell* (argued 22 April 1927; decided 2 May 1927), *U.S. Reports* 274 (1928): 207–8. The only dissenter was Justice Butler, who did not write an opinion, though some scholars have posited that he cast his vote out of religious conviction. Voting with the majority were Louis Brandeis, William Howard Taft, and Harlen Stone (Blank, *Fertility Control*, 60).

12. In 1942, compulsory sterilization was challenged in *Skinner v. Oklahoma*. Oklahoma permitted the practice upon an individual's third felony conviction (Skinner had been found guilty on three separate occasions, once for chicken thieving and twice for armed robbery). The case was argued on several points, including Fourteenth Amendment grounds: some felonies (embezzlement, political offenses, violations of prohibition laws and revenue acts) were expressly excluded by the statute. The Supreme Court found that the law did indeed violate equal protection standards. Justice Douglas wrote, "When the law lays an unequal hand on those who have committed intrinsically the same quality of offense and sterilizes one and not the other, it has made as invidious a discrimination as if it had selected a particular race or nationality for oppressive treatment. . . . Oklahoma makes no attempt to say that he who commits larceny by trespass or trick or fraud has biologically inheritable traits which he who commits embezzlement lacks." Chief Justice Stone

concurred with the ruling, but laid emphasis on due process: "Although petitioner here was given a hearing to ascertain whether sterilization would be detrimental to his health, he was given none to discover whether his criminal tendencies are of an inheritable type." Finally, Justice Jackson wrote, "There are limits to the extent to which a legislatively represented majority may conduct biological experiments at the expense and dignity and personality and natural powers of a minority—even those who have been guilty of what the majority define as crimes" (*Skinner v. Oklahoma*, no. 782 [argued 6 May 1942; decided 1 June 1942]). This ruling did not, however, overturn *Buck v. Bell* or intimate an opinion on whether Oklahoma statute constituted cruel and unusual punishment. Troy Duster commented: "The unmistakable implication of this ruling is that if a general association can be demonstrated between criminal behavior and genetic make-up, and if it did not so blatantly exempt one class of criminal offender then sterilization by the state is permissible" (Duster, *Backdoor to Eugenics*, 30–31).

13. Popenoe, author of *Sterilization for Human Betterment*, was an active member of the American Eugenics Society and a host of other eugenics organizations.

14. That's 5,069 females versus 2,482 males (Philip R. Reilly, *The Surgical Solution: A History of Involuntary Sterilization in the United States* [Baltimore: The Johns Hopkins University Press, 1991], 98).

15. E. H. Pratt, "Circumcision of Girls," *Journal of Orificial Surgery* 6, no. 8 (February 1898): 385; H. E. Beebe, "The Clitoris," *Journal of Orificial Surgery* 6, no. 1 (July 1897): 9 (first delivered before the Homeopathic Medical Society of Ohio at Akron); Cora Smith Eaton, "Circumcision for Headaches," *Journal of Orificial Surgery* 8, no. 8 (February 1900): 369; "Removal of the Ovaries as a Cure for Insanity" (from the Tenth Annual Report of the Committee on Lunacy of the Board of Public Charities of Pennsylvania), *American Journal of Insanity* 49 (January 1893): 397; Pratt, editorial, *Journal of Orificial Surgery*, 279. There were dissenters. Thomas W. Barlow, member of the Philadelphia Committee on Lunacy, stated, "I am of the opinion that the operation of oophorectomy [the removal of one or both ovaries] upon insane women . . . unless necessary to save lives, is not only illegal, but, in view of its experimental character, it is brutal and inhuman and not excusable on any reasonable ground. . . . The zeal of the gynecologist is being carried to an unusual extent when it proposes to use a State Hospital for the Insane as an experimental station, where lunatic women are to be subjected to doubtful operations for supposed cure." Thomas G. Morton, surgeon and fellow committee member, concurred, calling the operations "unwarrantable and indefensible" and questioning the moral or legal right of a relative or guardian to give consent for hysterectomies (Morton, "Removal of the Ovaries as a Cure for Insanity," 399–401.

16. Harry H. Laughlin, *Eugenical Sterilization in the United States* (Chicago: Psychopathic Laboratory of the Municipal Court of Chicago, 1922), 440–41. Laughlin compared the "unprotected females of the socially unfit classes" to "females of mongrel strains of domestic animals," which, he emphasized, were reduced in number "through the destruction or unsexing of the female" (*Proceedings of the First National Conference on Race Betterment*, 8–12 January 1914 [Battle Creek, Mich.: Race Betterment Foundation, 1914], 484).

17. Many progressives, for example, came to eugenics by way of neo-Malthusian belief in a resource pool that grew arithmetically and a population that grew exponentially.

18. Ralph Bevan, "God's Call to Birth Control," *Birth Control Review* 8, no. 9 (1924): 252.

Margaret Sanger and the Eugenic Compact

1. Linda Gordon, *Woman's Body, Woman's Right: Birth Control in America* (New York: Penguin Books, 1990); Angela Davis, *Woman, Race, and Class* (New York: Vintage Books/Random House, 1983).

2. Goldberg, *Racist Culture*, 5.

3. Charles Mills, *The Racial Contract* (Ithaca, N.Y.: Cornell University Press, 1997), 56.

4. McCann, *Birth Control Politics in the United States*, 19, 125.

5. Ellen Chesler, *Woman of Valor: Margaret Sanger and the Birth Control Movement in America* (New York: Simon and Schuster, 1992), 123.

6. McCann, *Birth Control Politics in the United States*, 123.

7. Chesler, *Woman of Valor*, 216.

8. Ibid., 215.

9. McCann, *Birth Control Politics in the United States*, 101, 19.

10. Sanger, *The Pivot of Civilization* (New York: Brentanos: 1922), 240. For Terman and Yerkes citations, see 241 n.3 and 263.

11. Betsy Hartmann, *Reproductive Rights and Wrongs: The Global Politics of Population Control* (Boston: South End Press, 1995), 97.

12. Martha C. Ward, *Poor Women, Powerful Men: America's Great Experiment in Family Planning* (Boulder, Colo.: Westview Press, 1986), 8.

13. "Address of Welcome" to the Sixth International Neo-Malthusian and Birth Control Conference, reprinted in *Birth Control Review* 9, no. 4 (April 1925): 100; emphasis in original.

14. McCann, *Birth Control Politics in the United States*, 101.

15. Emphasis added; Sanger, *Pivot of Civilization*, 78.

16. Ibid., 70. In 1950, Sanger told those gathered to honor her that the Holocaust was the result of "population pressure." She stated, "Human beings are herded into concentration camps, into vast slave labour prisons. Whole nations are made homeless and displaced. *These manifestations are symptoms of a complete lack of population policies* and of political foresight as to the value and meaning of dignified humans living on this earth" (emphasis added). The "Lasker Award Address," by Margaret Sanger, was read at the Thirtieth Annual Luncheon of the Planned Parenthood Federation of America, 25 October 1950, and reprinted in a supplement to *Malthusian* (January 1951) (Margaret Sanger Papers, Collected Document Series [hereafter MSP-CDS], folder 1950–1951).

17. Sanger, *The Pivot of Civilization*, 78.

18. Ibid., 82.

19. See Sanger's editorial in *Birth Control Review* 9, no. 6 (June 1925): 163, 164.

20. Sanger, *The Pivot of Civilization*, 187.

21. Ibid., 279. This appears in the book's appendix, under "Principles and Aims of the American Birth Control League" (founded by Sanger in 1921).

22. Sanger, "The Eugenic Value of Birth Control Propaganda," *Birth Control Review* 5, no. 10 (October 1921): 5. MSP-CDS.

23. Sanger, *The Pivot of Civilization*, 104; "The Eugenic Value of Birth Control Propaganda," *Birth Control Review* 5, no. 10 (October 1921): 5.

24. Sanger to Mrs. William P. Driscoll, 28 November 1928, MSP-CDS.

25. McCann, *Birth Control Politics in the United States*, 16–17.

26. Mehler, "The History of the American Eugenics Society," 416, 417; McCann, *Birth Control Politics in the United States*, 108, 109.

27. McCann, *Birth Control Politics in the United States*, 119. McCann fails to comment on why the ABCL extended this invitation to Davenport in the first place. The offer of such a prominent conference role to Davenport suggests a much tighter link between Sangerists and eugenicists than McCann acknowledges. According to Mehler, Davenport was a founding member of the Eugenics Committee, an editorial committee member of *Eugenical News*, president of the Galton Society, honorary president of the Eugenics Record Association, chairman of the American Eugenics Society's Committee on Research Problems in Eugenics, a member of the AES board of directors, president of the Third International Congress of Eugenics, director of the Station for Experimental Evolution and Eugenics Record Office, president of the International Federation of Eugenics Organizations, and the author of various eugenic texts, including *Heredity in Relation to Eugenics* (Mehler, "The History of the American Eugenics Society," 329–30).

28. Gunther, cited in "Notes and News," *Eugenical News* 15, no. 2 (December 1930): 178. Grant had earlier reviewed Gunther's book in the September 1928 edition of *Eugenical News* 13, no. 9. Sixty years later, Richard J. Herrnstein would reiterate Gunther and Grant's concern over the falling birthrate of the "fit," and their belief in women's culpability in the resulting calamity. In a 1989 article, Herrnstein bemoaned the "fact" that high IQ, "well-educated women" were having fewer and fewer children. While attempting to reconcile the sovereignty of the individual with nationalist duty, he wrote, "Nothing is more private than the decision to bear children, yet society has a vital interest in the aggregate effects of those decisions" (Herrnstein, "IQ and Falling Birth Rates," *Atlantic Monthly*, May 1989, 73–79).

29. Albert E. Wiggam, "Will the Good or the Bad Inherit the Earth?" *Birth Control Review* 13, no. 12 (December 1929): 347; emphasis in original. This appeared after Sanger's tenure as the journal's editor-in-chief.

30. Mehler, "The History of the American Eugenics Society," 439.

31. Wiggam, "Will the Good or the Bad Inherit the Earth?" 349.

32. Sanger, "The Eugenic Value of Birth Control Propaganda," 5.

33. Ibid.

34. Ibid.

35. Sanger, *The Pivot of Civilization*, 189. Ellis wrote of the relationship between eugenics and birth control: "Here we touch on the highest ground and are concerned with our best hopes for the future of the world. For there can be no doubt that Birth Control is not only a precious but indispensable instrument in moulding the coming man to the measure of our developing ideals" (Ellis, "Birth Control in Relation to Morality and Eugenics," *Birth Control Review* 3, no. 2 [February 1919]: 8).

36. Hartmann, *Reproductive Rights and Wrongs*, 97.

37. Gordon, *Woman's Body, Woman's Right*, 222.

38. Little, "Will Birth Control Promote Race Improvement?" *Birth Control Review* 13, no. 12 (December 1929): 343; Laughlin, "Legal Status of Eugenic Sterilization," *Birth Control Review* 12, no. 3 (March 1928): 78.

39. Mehler, "The History of the American Eugenics Society," 393, 394, 392.

40. *Eugenic Reform*, reviewed in *Birth Control Review* 12, no. 8 (August 1928): 234, lamented the "many human cuckoos living in our midst." A review of *What Is Eugenics* appeared the following year (*Birth Control Review* 13, no. 8 [August 1929]: 226). Ellis review cited in Gordon, *Woman's Body, Woman's Right*, 278.

41. In Gordon, *Woman's Body, Woman's Right*, 182.

42. McCann, *Birth Control Politics in the United States*, 133, 134.

43. Chesler, *Woman of Valor*, 15. Chesler also cites instances of Sanger's anti-Semitism and then maintains she "outgrew it" (50, 51).

44. Raymond Pearl, "The Differential Birth Rate," *Birth Control Review* 9, no. 10 (October 1925): 278. Pearl explained these numbers, in part, by claiming that not only did the relatively wealthier have a broader knowledge about contraception, but they also had "more varied intellectual interests, and generally wider outlets for nervous energy" (279).

45. The ongoing eugenic assault on the disabled is historicized in a later chapter, but it is important to raise the issue in the context of heredity-based attacks on the poor as well.

46. Pearl, "The Differential Birth Rate," 301.

47. John C. Duvall, "The Purpose of Eugenics," *Birth Control Review* 8, no. 12 (December 1924): 345.

48. "The Story of a Subsidized Family, or How to Populate the Earth with the Unfit," *Birth Control Review* 9, no. 7 (July 1925): 200, 201.

49. "Unprofitable Children," *Birth Control Review* 8, no. 5 (May 1924): 144.

50. Sanger, "Address of Welcome," Sixth International Neo-Malthusian Conference, 100; emphasis in original.

51. Ibid. The linking of the developmentally disabled—exposed to political and personal abuse—with prisoners who were, and continue to be, defined almost exclusively as victimizers, typifies the way in which society's most vulnerable groups are recast as its greatest villains. Sanger went on to complain, "This is a free country, a democratic country of universal suffrage. We can all vote, even the mentally arrested. And so it is no surprise to find the moron's vote as good as the vote of the genius. The outlook is not a cheerful one" (ibid.).

52. Ibid.

53. Gordon, *Woman's Body, Woman's Right*, 300, 304.

54. Sanger, cited in ibid., 310.

55. Chesler, *Woman of Valor*, 298.

56. Cited in Thomas Shapiro, *Population Control Politics: Women, Sterilization, and Reproductive Choice* (Philadelphia: Temple University Press, 1985), 49, 50.

57. "With our relief and other social expenditures as high as they are," the PBCF literature asked, "is it not time that a program of sterilization of the unfit be considered anew?" (ibid., 49).

58. Reporting that over 50 percent of the women dropped out of the study, Chesler noted only that they "expressed dissatisfaction" (*Woman of Valor*, 378, 379). Robert Latou Dickinson is renowned for his 1941 study, "The Gynecology of Homosexuality." He had a long association with Gamble, who funded his National Committee

on Maternal Health. At one point, Dickinson served on the AES Committee on the Eugenic and Dysgenic Effects of Birth Control (Mehler, "The History of the American Eugenics Society," 331).

59. Gamble to Sanger, 29 June 1937, Margaret Sanger Papers, Smith College Collection (hereafter MSP-SCC), S13:0261. Gamble had supervised testing of the foam-powder-sponge method a year earlier in Florida. In later endeavors, he did not restrict his use of captive and quasi-captive acceptors to women enmeshed in health or welfare systems. In 1961, for example, he wrote to Sanger that he had finally located a site for the research he had been trying to organize. There had been complaints of "discomfort" from women using foam tablets, and Gamble had been trying for two to three years to find volunteers to test different brands. Now it seemed the search was over. "An Israel doctor said the problem was easily solved: women in the army would do it. Gamble had samples of sixteen brands shipped to him (Gamble to Sanger, 16 April 1961, MSP-SCC, S58:0617).

60. Quoted by E. A. Ross, sociologist, professor of economics, and coiner of the term "race suicide," in "Reason IX" of "Ten Good Reasons for Birth Control," *Birth Control Review* 12, no. 12 (December 1928). Speaking at the Chicago Birth Control Conference several years earlier, Ross stated that whereas the prospect of "race suicide" in the United States had once frightened him, he was now more scared of overpopulation. Rev. Albert P. Van Dusen, "Birth Control as Viewed by a Sociologist," *Birth Control Review* 8, no. 5 (May 1924): 135 (first presented at the Birth Control Conference at Syracuse, New York, 29 June 1924).

61. Hartmann, *Reproductive Rights and Wrongs*, 247; Yamila Azize-Vargas, "The Roots of Puerto-Rican Feminism: The Struggle for Universal Suffrage," *Radical America* 23, no. 1 (June 1990): 72.

62. Gordon, *Woman's Body, Woman's Right*, 333.

63. Shapiro, *Population Control Politics*, 53. This, despite the fact that Bill 64, legalizing birth control in Puerto Rico, was not signed into law until 1937 (Gamble to Sanger, 2 May 1937, MSP-SCC, S13:0016).

64. Gamble to Sanger, 6 February 1938, MSP-SCC, S14:0380; Sanger to Gamble, 12 February 1938, MSP-SCC, S14:0450.

65. Shapiro, *Population Control Politics*, 54, 53.

66. Ibid., 53. Hartmann notes that, as tax-free investment incentives and underpaid labor began to attract U.S. manufacturing industries to Puerto Rico in the 1940s, women there became increasingly vulnerable to sterilization abuse. Sterilization was seen as a way "to help 'free' them for employment, as opposed to, for example, providing good child-care facilities." The government of Puerto Rico, along with International Planned Parenthood and other private agencies, with U.S. government funds, pushed sterilization on women, offering it at little or no monetary cost. In 1968, Puerto Rican women had the highest sterilization percentage in the world (Hartmann, *Reproductive Rights and Wrongs*, 247).

67. Shapiro, *Population Control Politics*, 53.

68. Sanger to Gamble, 15 April 1954, MSP-SCC, S43:0571.

69. Gamble to Sanger, 6 December 1947, MSP-SCC, S27:0791.

70. This included repeated references to Joseph Sunnen of St. Louis who pledged $100,000 per year for three years to the Asociación Puertorriqueña Pro Bienestar de la Familia (Gamble to Sanger, 5 August 1956, MSP-SCC, S50:0463; 4 September 1956, MSP-SCC, S50:0723; and 13 March 1957, MSP-SCC, S51:0651 [this last was from a

general mailing and addresses "Dear Collaborators"]; Gamble to Sanger, 18 January 1957, MSP-SCC, S51:0285).

71. Sanger to Gamble, 7 March 1957, MSP-SCC, S51:609; Gamble to Sanger, 9 March 1957, MSP-SCC, S51:0623.

72. Gamble to Sanger, "Dear Collaborators" letter, 13 March 1957, MSP-SCC, S51:0651. Hartmann notes that the "ethical and scientific standards of the early pill studies in Puerto Rico...left much to be desired." She cites one study in which a woman developed pulmonary tuberculosis and another died of congestive heart failure. Researchers were unwilling to attribute either the death or the tuberculosis to the women's participation in the drug trial. It certainly was not atypical for medical science, in an effort to assert expertise, to minimize women's symptoms. In the context of the Puerto Rico experiments, Hartmann reports, "One study even blamed most occurrences of side effects such as nausea, vomiting, and dizziness on psychological factors" (Hartmann, *Reproductive Rights and Wrongs*, 190).

73. Gamble to Sanger, 13 March 1957, MSP-SCC, S51:0651; and 9 March 1957, MSP-SCC, S51:0623. Tests were also being conducted on women in San Juan.

74. Gamble to Sanger 19 March 1957, MSP-SCC, S51:0675; 5 April 1957, MSP-SCC, S51:0855; 2 September 1957, MSP-SCC, S52:0643; 19 March 1958, MSP-SCC, S53:0659; 5 June 1959, MSP-SCC, S55:0514; 20 June 1959, MSP-SCC, S55:0547; 12 February 1962, MSP-SCC, S59:0446; Gamble to Joseph Sunnen, 1 February 1960, MSP-SCC, S56:0576.

75. Cited in Gordon, *Woman's Body, Woman's Right*, 328.

76. Jessie Rodrique, "The Black Community and the Birth Control Movement," in *Unequal Sisters: A Multicultural Reader in U.S. History*, ed. Ellen DuBois and Vicki Ruiz (New York: Routledge, 1990), 27.

77. "Sterilization," *Pittsburgh Courier*, 30 March 1935, 10.

78. Ibid.

79. W. E. B. DuBois, "Forum of Fact and Opinion," *Pittsburgh Courier*, 27 June 1936, sec. 2, 1. This watchfulness in no way undermined the Black community's support for accessible contraception. In 1941, the National Council of Negro Women became the first national women's organization to officially endorse birth control (McCann, *Birth Control Politics in the United States*, 218). For more on this, see Rodrique, "The Black Community and the Birth Control Movement." Also see Nancy Leys Stepan and Sander Gilman, "Appropriating the Idioms of Science: The Rejection of Scientific Racism," in *The Bounds of Race: Perspectives of Hegemony and Resistance*, ed. Dominick Lacapra (Ithaca, N.Y.: Cornell University Press, 1991).

80. Emphasis added; Gamble to Florence Rose, 26 November 1939, MSP-CDS; Sanger to Gamble, 26 November 1939, MSP-CDS.

81. Reilly, *The Surgical Solution*, 134.

82. Gamble to Sanger, 6 December 1947, MSP-SCC, S27:0791. Sanger's severe dislike for Catholics, owing ostensibly to the Church's opposition to birth control, is well-documented. For more on this, see Chesler, *Woman of Valor*.

83. Reilly, *The Surgical Solution*, 134.

84. Margaret Sanger, "Lasker Award Address," read at the Thirtieth Annual Luncheon of the Planned Parenthood Federation of America, 25 October 1950, reprinted in a supplement to the *Malthusian* (January 1951). MSP-CDS, folder 1950–1951, Lasker Award Address.

85. Ibid.

86. Ibid.

87. Sanger, "Sterilization: A Modern Program for Human Health and Welfare," June 1951, MSP-CDS, subseries 4, folder 1950–1951, 4.

88. Gamble to Sanger, 20 November 1950, MSP-SCC, S33:0414; Sanger to Gamble, 28 November 1950, MSP-SCC, S33:0456. Sanger anticipated William Shockley's Volunteer Sterilization Bonus Program by twenty years. "Bonuses will be offered for sterilization. Income-tax payers get nothing. Bonuses for all others, regardless of sex, race, or welfare status, would depend on best scientific estimates of hereditary factors in disadvantages such as diabetes, epilepsy, heroin addiction, arthritis, etc.... A motivation boost might be to permit those sterilized to be employed at sub-minimum wages without loss of a welfare floor income. Could this provide opportunity for those newly unemployable?" (Shockley, "Dysgenics—A Social Problem Evaded by the Illusion of Infinite Plasticity of Human Intelligence?" in *Shockley on Eugenics and Race: The Application of Science to the Solution of Human Problems,* ed. Roger Pearson [1971; reprint, Washington, D.C.: Scott-Townsend Publishers, 1992], 181). While claiming to favor welfare programs, Shockley nonetheless stated that they had "anti-evolutionary effects" (Shockley, "Racial Aspects of the Environment— Heredity Uncertainty," proposal read before the National Academy of Science, 24 March 1968, cited in ibid., 103).

89. Sanger, *Sterilization: A Modern Medical Program for Human Health and Welfare,* 1951, MSP-CDS.

90. Robert Latou Dickinson and Clarence Gamble both served on the organization's Medical and Scientific Committee when it was the Human Betterment Association of America.

91. Gamble to Sanger, 5 December 1950, MSP-SCC, S33:0505.

92. Sanger, "Sterilization (Address to Be Read in Absentia)," MSP-CDS, February 1951, 2.

93. Ibid. Sanger's remarks predated "wrongful life" suits by years.

94. Ruth Proskuaer Smith to Margaret Sanger, 20 July 1960, MSP-CDS. In 1962, Dr. Mary S. Calderone, medical director of Planned Parenthood Federation of America– World Population Emergency Campaign, told workshop participants, "Sterilization has never been a part of the Planned Parenthood program.... We have a committee authorized to refer such requests to the Human Betterment Association" ("Birth Control Services in Tax-Supported Institutions," digest of a workshop sponsored by the PPFA in San Francisco, 16 May 1962 [New York: PPFA-World Population Emergency Campaign, 1962], 21, MSP-CDS).

95. Smith to Sanger (telegram), 4 January 1961; Sanger to Smith, 5 January 1961, MSP-CDS.

96. H. Curtis Wood Jr., February 1961, MSP-CDS. Sanger herself made a ten-dollar donation.

97. McCann, *Birth Control Politics in the United States,* 173 and throughout book.

98. As for Gamble, Sanger not only maintained more than cordial ties with the Pennsylvania physician, but praised him as "a splendid organizer who initiates trial and error experiments here and there" (Sanger to Irene Headley Armes [executive director of the HBA of America], 16 January 1951, MSP-CDS).

99. MSP-SCC, reels S36, 37, 47, 56. Chesler's report that Sanger repudiated but did not altogether alienate Gamble after 1955 would thus seem to be less than completely accurate (Chesler, *Woman of Valor,* 438). In fact, Sanger returned the favor in 1958, submitting Gamble's name, among others, on a list of candidates to succeed

her as president of the International Planned Parenthood Federation (IPPF) (Gamble to Sanger, 11 September 1958, MSP-SCC, S54:0441).

Physical Fallout

1. "The Right to Reproduce," editorial, *New Leader*, 3 March 1945, 16; *[sic!]* in original.

2. Alice Tanabe Nehira's testimony was reprinted in "The Commission on Wartime Relocation and Internment of Civilians: Selected Testimonies from the Los Angeles and San Francisco Hearings," *Amerasia Journal* 8, no. 2 (fall–winter 1981): 92.

3. Emphasis added; Nils P. Larsen, "Post-Partum Sterilization: Reduction of Infant and Maternal Mortality in Hawaii" (excerpts from an article in *Human Fertility*, March 1944, MSP-CDS).

4. John Stuart Mill, British philosopher, cited in Mills, *The Racial Contract*, 149 n.57.

5. Gamble to Sanger, 29 August 1949, MSP-SCC, S30:0989. That Gamble, just four years after the war, could describe German sterilization policies as "arbitrary" (policies that were modeled on U.S. eugenicists' proposals) reveals a profound misreading of Nazi eugenics.

6. Moya Woodside, *Sterilization in North Carolina: A Sociological and Psychological Study* (Chapel Hill: University of North Carolina Press, 1950), 161, 78–79, 24. My argument here is not that U.S. eugenics constituted the practical or ideological equivalent of Nazi eugenics, merely that Gamble and Woodside were so ready to dismiss concerns about racism, coercion, and white supremacy as irrational, even in the immediate wake of the full enactment of eugenic ideals in Europe. Twenty years later, William Shockley would declare, also without irony, "Only the most anti-Teutonic racist can believe that the German people are such an evil breed of man that they would have tolerated the concentration camps and gas chambers if a working First Amendment had permitted exposure and discussion of Hitler's 'final solution.'... The First Amendment makes it safe for us in the United States to find humane eugenic measures" (Shockley, "Dysgenics—A Social Problem Evaded by the Illusion of Infinite Plasticity of Human Intelligence?" 180).

7. Among the most brutal were the surgeries performed by the "Father of American Gynecology," J. Marion Sims. Between 1845 and 1849, Sims operated on enslaved women (at least one of whom he bought explicitly for experimentation purposes) in his makeshift hospital in Alabama. The surgeries were performed without anesthesia, often before an audience. In the process, he became one of the wealthiest doctors in the United States. Sims is credited with introducing the speculum to North American medical practice. For a critical account of the surgeries and their implications, see Terri Kapsalis, *Public Privates: Performing Gynecology from Both Sides of the Speculum* (Durham, N.C.: Duke University Press, 1997).

8. Nancy Krieger and Elizabeth Fee, "Man-Made Medicine and Women's Health: The Biopolitics of Sex/Gender and Race/Ethnicity," in *Man-Made Medicine*, ed. Moss, 17. Cartwright pathologized his subjects' very resistance, announcing his "discoveries" of "dysesthesia," a condition that resulted in slaves sabotaging their masters' orders, and "drapetomania," a mental disturbance characterized by escape attempts (Mike A. Males, *The Scapegoat Generation: America's War on Adolescents*

[Monroe, Maine: Common Courage Press, 1996], 243; Adolph Reed Jr., "The Content of Our Cardiovascular," *Village Voice*, 31 December 1996, 25).

9. Krieger and Fee, "Man-Made Medicine and Women's Health," 18. According to a 1913 text, the chief statistician of the Prudential Insurance Company of America declared syphilis and tuberculosis to be inevitable consequences of Blacks' "immense immorality, which is a race trait" (Edward Eggelston, *The Ultimate Solution of the American Negro Problem* [Boston: Gorham Press, 1913], 226).

10. Charles McCord, *The American Negro as a Dependent, Defective, and Delinquent* (Nashville: Benson Printing, 1914), 168, 169, 98, 99. McCord related the story of "Bob," charged with killing his wife, suspected of killing his mother (*after* emancipation, McCord was careful to note), and hanged. "Suppose Bob's mother had been unsexed at the proper time; at least three violent deaths would have been prevented, and the expense of prosecution, detention, and execution saved." McCord pronounced Bob's mother feebleminded and blamed her for her son's hanging, her daughter-in-law's murder, and her own death. Such women, he wrote, "spread immorality and disease and breed degenerates" (307). Bob's father, a white slave owner, was barely mentioned.

11. G. Frank Lydston, *The Disease of Society (The Vice and Crime Problem)* (Philadelphia: J. B. Lippincott, 1904), 394.

12. Woodside, *Sterilization in North Carolina*, 16, 7, 33.

13. Ibid., 5, 138.

14. Ibid., 6.

15. Ibid., 83. Woodside's pronouncement was but a variation on a theme. Charles McCord, for example, charted the "estimated Annual Cost of Sickness and Death among Negroes to the States of the South" in 1914, maintaining that whites became poor because they were unfortunate, incapable, or of "pauper stock," whereas "the Negro becomes a pauper more often because he is of a child race, undeveloped and . . . defective" (McCord, *The American Negro as a Dependent, Defective, and Delinquent*, 151, 152).

16. Woodside, *Sterilization in North Carolina*, 8.

17. Ibid., 21. An excerpt of Woodside's study on North Carolina, bemoaning the failure of contraception education and use among women "of borderline or subnormal mentality," was found in Margaret Sanger's papers. It read: "Workers in the contraceptive service are unanimous about the difficulty of persuading such persons to attend clinics, of instructing them in procedures, and ensuring that advice is carried out." In the margin, in Sanger's hand, is written "They should be steriliz[ed]" (MSP-CDS, folder 1–18 July 1960, *Sterilization in North Carolina* excerpt).

18. Woodside, *Sterilization in North Carolina*, 204–5.

19. Eugene Harris, "The Physical Condition of the Race; Whether Dependent on Social Conditions or the Environment," *Atlanta University Publications*, no. 2 (1987): 25.

20. Reilly, *The Surgical Solution*, 134.

21. Emphasis added; cited in Adelaida R. Del Castillo, "Sterilization: An Overview," in *Mexican Women in the United States: Struggles Past and Present*, ed. Magdalena Mora and Adelaida R. Del Castillo (Los Angeles: Chicano Studies Research Center Publications, University of California, Los Angeles, 1980), 73. Similar bills were debated across the country. In Illinois, a member of the Public Aid Commission asked that state law be amended to allow for the mandatory sterilization of women convicted of prostitution. This would have included "any person siring or

giving birth to more than one illegitimate child" or undergoing an abortion, all of which fell under the definition of "prostitute."

22. *Biennial Report of the Eugenics Board of North Carolina,* 1 July 1939 to 30 June 1940; and Del Castillo, "Sterilization," 67. As in other eugenic crusades, this increase is also indicative of a backlash against civil rights gains.

23. Claudia Dreifus, "Sterilizing the Poor," *Progressive,* 14 December 1975, 18.

24. Del Castillo, "Sterilization," 68. These type of surveys inevitably raise questions about who constitutes canvassable "average Americans."

25. Ibid., 67. Del Castillo reported that in this survey of physicians' attitudes, obstetrician-gynecologists were the most punitive.

26. Ibid., 68, 69.

27. Hartmann, *Reproductive Rights and Wrongs,* 248.

28. It is not insignificant that this was the year the Moynihan Report was issued.

29. *ACLU News,* submitted to the Senate Subcommittee on Health, Committee on Labor and Public Welfare, *Quality of Health Care—Human Experimentation,* 1586. Nial Ruth Cox's case revealed that in the forty years since 1933, 5,000 out of 7,686 sterilizations performed under the auspices of the North Carolina Eugenics Commission were carried out on African-Americans (Angela Davis, *Women, Race, and Class* [New York: Vintage Books, 1983], 217). Cox herself did not learn of the irreversibility of the procedure until five years after the operation. She suffered migraines, backaches, painful menstruation, inability to control her bladder, and weakness in her legs (*ACLU News,* Senate Subcommittee, *Quality of Health Care,* 1596). All these symptoms dated back to the operation, and some are consistent with documented side effects of sterilization.

30. Antonia Hernandez, "Chicanas and the Issue of Involuntary Sterilization: Reforms Needed to Protect Informed Consent," *Chicano Law Review* 3 (1976): 41.

31. Davis, *Women, Race, and Class,* 217; *[sic]* in original.

32. Hernandez, "Chicanas and the Issue of Involuntary Sterilization," 22.

33. Dreifus, "Sterilizing the Poor," 18. Wood articulated the position of many physicians who believed they had a responsibility to the general public to sterilize welfare recipients and other "parasites." "People pollute, and too many people crowded too close together cause many of our social and economic problems. . . . As physicians, we have obligations to the society of which we are a part. The welfare mess, as it has been called, cries out for solutions, one of which is fertility control" (cited in Carlos Velez, "Se Me Acabó La Canción: An Ethnography of Non-Consenting Sterilization among Mexican Women in Los Angeles," in Mora and Del Castillo, *Mexican Women in the United States,* 77).

34. Hernandez, "Chicanas and the Issue of Involuntary Sterilization," 18.

35. Cited in Susan Pleck, "Voluntary Sterilization: Attitudes and Legislation," *Hastings Center Report* 4, no. 3 (June 1974): 8.

36. Davis, *Women, Race, and Class,* 216. The fact that Depo-Provera was administered to humans before lab tests were concluded did not register as significant to most reporters covering the Relf story.

37. Senate Subcommittee, *Quality of Health Care,* 1502, 1501.

38. Shapiro, *Population Control Politics,* 90.

39. Senate Subcommittee, *Quality of Health Care,* 1497.

40. Ibid., 1496.

41. Shapiro, *Population Control Politics,* 90.

42. Hernandez, "Chicanas and the Issue of Involuntary Sterilization," 9, 10. In 1973, the assistant secretary of HEW attributed the absence of such guidelines to the department's belief "that professional judgment and ethics should govern the practice of medicine in health service projects" (Senate Subcommittee, *Quality of Health Care,* 1571). This belief did not extend to abortion, however, which was unavailable at HEW-funded facilities.

43. Hernandez, "Chicanas and the Issue of Involuntary Sterilization," 17.

44. Senate Subcommittee, *Quality of Health Care,* 1498.

45. Allan Chase, *The Legacy of Malthus: The Social Costs of the New Scientific Racism* (New York: Knopf, 1977), 17; Shapiro, *Population Control Politics,* 91.

46. Reilly, *The Surgical Solution,* 151–52.

47. Boston Women's Health Book Collective, *The New Our Bodies/Ourselves* (New York: Simon and Schuster, 1984), 257.

48. Jeannie I. Rosoff, "Sterilization: The Montgomery Case and Its Aftermath," *Hastings Center Report* 3, no. 4 (September 1973): 6.

49. Robert E. McGarrah Jr., "Sterilization without Consent: Teaching Hospital Violations of HEW Regulations," Public Citizen's Health Research Group Report, 21 January 1975, 3, 4.

50. Ibid., 1, 2, 5. McGarrah, staff attorney at Public Citizen, surveyed teaching hospitals, where great importance had long been attached to cutting. McGarrah related two illustrative incidents. In the first, a student who asked why a woman was getting a hysterectomy rather than a tubal ligation was told by a resident, "we like to do a hysterectomy, it's more of a challenge. . . . you know a well-trained chimpanzee can do a tubal ligation . . . and it's good experience for the junior resident . . . good training." In the second exchange, a doctor told his colleague, "I want to congratulate the resident staff for 11 postpartum tubal ligations a week ago." The other physician replied, "They did an even better job last week, they did 15." McGarrah also noted, "The early 'rewards' for doing more operations on the poor and disadvantaged in the form of residency certification and specialty board qualification are translated, after training, into financial rewards wherein, the more you cut, the more money you make." Interns who resisted hard-sell tactics to "persuade" poor patients to accept sterilization were pressured by medical staff, the result being "systematic violations of fundamental constitutional patients' rights" by obstetrics and gynecology interns and residents at Boston City Hospital, Baltimore City Hospitals and the Women's Hospital of the Los Angeles County Medical Center (McGarrah, "Voluntary Female Sterilization: Abuses, Risks, and Guidelines," *Hastings Center Report* 4, no. 3 [June 1974]: 5, 6).

51. American Indian Inter-Agency Council, *Drumbeat of the Bay Area: News and Opinions of the American Indian Community* 1, no. 4 (May 1977): 4.

52. Andrea Carmen, "Native American Growing Fight against Sterilizations of Women," *Akwesane Notes* (late winter 1979).

53. Rosalind Pollack Petchesky, "Reproduction, Ethics, and Public Policy: The Federal Sterilization Regulations," *Hastings Center Report* 9, no. 5 (October 1979): 32; "American Indian Women Sterilized without Informed Consent," *Hastings Center Report* 7, no. 1 (February 1977): 3; Richard Louv, "The Sterilization of American Indian Women," *Playgirl* 4, no. 12 (May 1977): 43.

54. On-air interview, *Woman: Concerns of American Indian Women,* WNED, Western New York Educational Television Association, 1977. Uri's own investigation

was prompted by a visit from a woman requesting a "womb transplant" (Shapiro, *Population Control Politics,* 187).

55. Louv, "Sterilization of American Indian Women," 43.

56. Cited in Andrea Carmen, "Native American Growing Fight against Sterilizations of Women," *Akwesane Notes* (late winter 1979).

57. Louv, "Sterilization of American Indian Women," 53.

58. Ibid., 100. According to one report, there was some evidence that full-blooded Native Americans were being targeted. "Native sources report that there is one tribe in Oklahoma where there are no full-blooded women who have not been sterilized" (Carmen, "Native American Growing Fight against Sterilizations of Women").

59. Louv, "Sterilization of American Indian Women," 53. The population control ethic evinced by this doctor was in keeping with the HEW philosophy. Angela Davis reported on an HEW pamphlet, designed for distribution to Native Americans, which depicted two families: one with ten children and one horse, the other with one child and ten horses, "[a]s if the ten horses owned by the one child family had been magically conjured up by birth control and sterilization surgery" (Davis, *Women, Race, and Class,* 218).

60. Louv, "Sterilization of American Indian Women," 51; Carmen, "Native American Growing Fight against Sterilizations of Women."

61. On-air interview, *Woman: Concerns of American Indian Women.*

62. Shapiro, *Population Control Politics,* 91.

63. Louv, "Sterilization of Indian Women," 100; emphasis added.

64. Ibid. During this period, feminists and activists at rape crisis centers mounted a highly visible challenge to identical attacks on rape victims accused of changing their stories to cover up their "indiscretions."

65. Ibid., 57.

66. By 1974, more than twice this number came forward with stories of sterilization abuse, but for various reasons did not sign on as coplaintiffs.

67. Robert E. McGarrah Jr., "Voluntary Female Sterilization: Abuses, Risks, and Guidelines," *Hastings Center Report* 4, no. 3 (June 1974): 5.

68. Del Castillo, "Sterilization," 68.

69. Velez, "Se me acabó la canción," 78. Half of the twenty-five interns and residents surveyed at L.A. County Hospital disclosed that they had engaged in pushing sterilization on the women they treated (McGarrah, "Voluntary Female Sterilization," 6). As for informed consent elsewhere in the country, McGarrah noted that Baltimore City Hospitals used a detailed, six-paragraph waiver for its radiology patients, but only a six-sentence consent form for women about to be sterilized.

70. Velez, "Se me acabó la canción," 78. In 1974, McGarrah reported on twelve women, almost all between the ages of eighteen and twenty, sterilized in Baltimore City Hospitals. They were given consent forms to sign just minutes before cesareans. "In none of these instances was there evidence of a previously expressed interest in sterilization, although it is routine hospital procedure to ask a woman if she wants sterilization when, weeks or months prior to delivery, she registers at the clinic" (McGarrah, "Voluntary Female Sterilization," 5).

71. Hernandez, "Chicanas and the Issue of Involuntary Sterilization," 6–7. Acosta was not informed of the procedure until two months later when she returned requesting birth control pills. Her husband had signed the consent form after being told it was a release for a caesarean section (Dreifus, "Sterilizing the Poor," 14).

72. Dreifus, "Sterilizing the Poor," 14.

73. Hernandez, "Chicanos and the Issue of Involuntary Sterilization," 31.

74. Ibid., 8.

75. Del Castillo, "Sterilization," 69.

76. Emphasis added; Velez, "Se me acabó la canción," 86, 87.

77. CESA papers, cited in Shapiro, *Population Control Politics*, 114.

78. Ibid.

79. Petchesky, "Reproduction, Ethics, and Public Policy," 35; and Patricia Donovan, "Sterilizing the Poor and Incompetent," *Hastings Center Report* 6, no. 5 (October 1976): 7. See also *Douglas v. Holloman*, USDC, SN.Y., Civil Action no. 76 Civ. 6.

80. Shapiro, *Population Control Politics*, 187.

81. Petchesky, "Reproduction, Ethics, and Public Policy," 36.

82. It is important to remember that policies of individual NOW chapters may have differed from the policies of the national office (as was the case with the Sierra Club's position on California's proposition 187 in 1994). Likewise, during World War II, some Planned Parenthood staff called for a greater focus on housing and health care issues, a view that was not shared by the home office, which actually suggested that birth control could compensate for cuts in health and welfare programs (Gordon, *Woman's Body, Woman's Right*, 350).

83. Judy Norsigian, "The Women's Health Movement in the United States," in *Man-Made Medicine*, ed. Moss, 90. Similarly, a pamphlet produced by the Association for Voluntary Surgical Contraception and marketed to those counseling low-income women on sterilization laments, "Poor women often encounter bureaucratic or institutional barriers in their attempts to obtain sterilization surgery, particularly when they seek postpartum services.... most women who encountered such barriers felt regret because they were not able to have the operation while they were in the hospital for delivery" (Association for Voluntary Surgical Contraception, *Counseling Low-Income Women about Voluntary Sterilization* [New York: Association for Voluntary Surgical Contraception, n.d.]).

84. Goldberg, *Racist Culture*, 5.

85. Alison M. Jaggar, *Feminist Politics and Human Nature* (Sussex, England: Harvester Press, 1983), 43, 194–95.

86. As previously noted, however, the regulations rarely protected this latter group, for even health care providers under HEW jurisdiction were chronically out of compliance.

87. Goldberg, *Racist Culture*, 5.

88. Chris Weedon, *Feminist Practice and Poststructuralist Theory* (Oxford, England: Basil Blackwell, 1987), 143.

89. Mills, *The Racial Contract*, 56.

New Technologies, Old Politics

1. Depo-Provera, delivered via injection, inhibits ovulation by altering the mucus and uterine lining (Committee to Defend Reproductive Rights, "Is Depo Provera Safe?" *CDRR News* [San Francisco], spring–summer 1993, 4). Norplant I (six capsules) and Norplant II (two rods), containing a synthetic hormone (progestin levonorgestrel), are inserted in the upper arm where they can remain for up to five

years. The progestin seeps through the capsule walls and into the bloodstream. Norplant thickens cervix mucus, hindering sperm entry; alters the womb's lining; and prevents monthly discharge of the mature egg from the ovaries ("Updating a Revolution," *Washington Post*, 7 January 1991, 3; and "The National Latina Health Organization Takes a Look at Norplant," *Morena* [March–April 1992]: 3).

2. It is impossible to discuss reproductive technologies without situating them in a global context. There has been substantial work done in this field, most notably by scholars such as Betsy Hartmann and countless writers, theorists, and activists, including members of FINRRAGE (Feminist International Network of Resistance to Reproductive and Genetic Engineering), who have exposed and explicated the population control–imperialism–human experimentation configuration. If the following account of Norplant and Depo-Provera abroad is brief, it is due to a desire to build on their work rather than to replicate it.

3. Carol Levine, "Depo-Provera and Contraceptive Risk: A Case Study of Values in Conflict," *Hastings Center Report* 9, no. 4 (August 1979): 11.

4. Committee to Defend Reproductive Rights, "Is Depo Provera Safe?" 4; Hartmann, *Reproductive Rights and Wrongs*, 202; Levine, "Depo-Provera and Contraceptive Risk," 9.

5. "Depo Provera—Get the Facts before You Get the Shot," informational flyer produced by Women against Imperialism, San Francisco. Hartmann notes that many of these side effects are dismissed as minor by doctors and administrators. She cites Gena Corea, "Depression is a minor side effect which merely destroys the entire quality of a woman's life" (Hartmann, *Reproductive Rights and Wrongs*, 202). Due to the fact that Depo-Provera is an injectable administered every three months (though the effects linger much longer), a woman suffering side effects is literally trapped in her own body—there is no way, save the passage of time, for her to rid her system of Depo-Provera once she's had a shot.

6. Committee to Defend Reproductive Rights, "Is Depo Provera Safe?" 4.

7. Luz Alvarez Martinez, "Women of Color and Reproductive Health" session participant, *Dangerous Intersections*, 61; Women against Imperialism, "Get the Facts before You Get the Shot," *WAI News* 1 (spring–summer 1994).

8. Martinez, "Women of Color and Reproductive Health" session participant, *Dangerous Intersections*, 61.

9. Levine, "Depo-Provera and Contraceptive Risk," 8.

10. Hartmann, *Reproductive Rights and Wrongs*, 204.

11. Raymond, *Women as Wombs*, 117.

12. Hartmann, *Reproductive Rights and Wrongs*, 204. Race was a critical determinant of who was targeted outside the United States as well. According to Hartmann, apartheid South Africa singled out Black and "mixed race" women to receive Depo-Provera, often the only contraceptive available, just as family planning was the only free health care provided to Black women (206). Similarly, in France, only 4 percent of birth control–using French-born women used it, compared to 15 percent of Algerian women and 20 percent of sub-Saharan African women living in France—this despite the fact that the African women requested contraception other than Depo-Provera more than twice as often as the native French women (Committee to Defend Reproductive Rights, "Is Depo Provera Safe?" 7).

13. Suzanne Shend, "What I Saw on Guantanamo Bay," *Gay Community News*, April 1993, 1. Shend noted that Haitians alone were tested for HIV prior to their ar-

rival in this country. "Non-Haitians, even if they are intercepted in the same waters or even in the same boats, are not subjected to HIV screening." The discourse that enveloped Haitians in the eighties and nineties, including their early designation as one of AIDS's "4Hs," along with homosexuals, heroine addicts, and hemophiliacs, laid the groundwork for coerced acceptance of Depo-Provera. The detainees at Guantanamo, Shend reported, "were forcibly tested, and then told of their status over a loud speaker in an airplane hangar."

In 1999, the *New York Times* ran a page-one story on AIDS research in a Cornell Medical College clinic in Port-au-Prince. "Haitians are ideal research subjects because they are not receiving the kind of care now standard in the world's developed countries." The *Times* reported that, despite the clinic's annual federal grant of $786,000, federal oversight is minimal, and the standard of care significantly lower than at Cornell's affiliated hospital in New York City. "If the research were done in the United States, experts agree, the physicians would be obligated to prescribe the anti-retrovirals and deliver the most effective counselling against unprotected sex" (Nina Bernstein, "For Subjects in Haiti Study, Free AIDS Care Has a Price," *New York Times*, 6 June 1999, 1; Bernstein, "Oversight Agencies Give Program Scant Review," *New York Times*, 6 June 1999, 10).

14. Levine, "Depo-Provera and Contraceptive Risk," 11.

15. Ibid., 8. Fifteen years later, the injectable was second only to the pill in rate of use among Kenyan women, and as Hartmann reported at the time, was being aggressively promoted across the continent, despite the fact that only 1 percent of birth control users globally use it (Hartmann, *Reproductive Rights and Wrongs*, 200–201).

16. Jennifer McGuire, "Health Educator Says Depo-Provera and Norplant Harmful," *Indian Country Today*, 22–29 July 1996, C1.

17. William Booth, "Updating a Revolution," *Washington Post*, 7 January 1991, 3.

18. Raymond, *Women as Wombs*, 15.

19. UBINIG, "'The Price of Norplant is Tk.2000! You Cannot Remove It': Clients Are Refused Removal in Norplant Trial Bangladesh," *Issues in Reproductive and Genetic Engineering* 4, no. 1 (1991): 45 (in 1991 Tk.2000 was equivalent to U.S.$55); Margot Zimmerman et. al., "Assessing the Acceptability of NORPLANT Implants in Four Countries," *Studies in Family Planning* 21, no. 2 (March–April 1990): 99.

20. One woman in Bangladesh eventually lied to doctors to compel them to extract the implants, telling them that her children had drowned and her husband wanted another baby (UBINIG, "Research Report on Norplant, The Five Year Needle: An Investigation of the Norplant Trial in Bangladesh from the User's Perspective," *Issues in Reproductive and Genetic Engineering* 2, no. 3 [1990]: 225). For more information on international and UN-sponsored Norplant campaigns, see UBINIG, "The Price of Norplant Is Tk.2000"; and Ana Regina Gomes Dos Reis, "Norplant in Brazil: Implantation Strategy in the Guise of Scientific Research," *Issues in Reproductive and Genetic Engineering* 3, no. 2 (1990).

21. Hartmann, *Reproductive Rights and Wrongs*, 77.

22. Judy Mann, "New Contraceptive Advances Freedom, Responsibility," *Washington Post*, 12 December 1990, B33.

23. UBINIG, "The Price of Norplant Is Tk.2000," 46. A study of clinics in Bangladesh that dispense the implants found that 100 percent of the Norplant-receiving patient population was poor or lower middle-class and that 80 percent

could not read. Staff claims that a leaflet listing the pros and cons of Norplant was read to women prior to the surgery was not confirmed by a single woman. Likewise, the clinics' assertions that they conducted pre-implantation tests for hypertension, asthma, jaundice, and other conditions contraindicated by Norplant were discredited by women who reported only blood pressure tests, weight checks, some urine tests (probably to screen for pregnancy), and, very rarely, pelvic exams. Sixty percent of the women were breast-feeding at the time the capsules were inserted (UBINIG, "Research Report on Norplant, The Five Year Needle," 224, 225).

24. "The National Latina Health Organization Takes a Look at Norplant," 3. The NLHO warns that it should never be used by women with breast cancer, circulation or heart problems, blood clots, jaundice, liver disease, or by women who smoke, are breast-feeding, or who may be pregnant. While Norplant is less effective in women weighing more than one hundred and fifty pounds—who are more apt to develop ectopic pregnancies as a result of Norplant acceptance than other women—the hormonal contraceptive was readily available and routinely prescribed to them (Susan Gerber, Carolyn Westoff, Maria Lopez, and Laurie Gordon, "Use of Norplant Implants in a New York City Clinic Population," *Contraception* 49, no. 6 [June 1994]: 561; and McGuire, "Health Educator Says Depo-Provera and Norplant Harmful," C1).

25. UBINIG, "Research Report on Norplant," 220.

26. "Abstract Arguments, Real Births," *New York Times,* 20 March 1991, 28.

27. Linda Roach Monroe, "New Expensive Birth Control Gets Slow Acceptance," *Los Angeles Times,* 16 May 1991.

28. "The National Latina Health Organization Takes a Look at Norplant," 3. Norplant capsules, like splinters, can travel subdermally, making extrication extremely difficult. A nurse practitioner in Sacramento told a *New York Times* reporter, "Sometimes I've had to poke around for an hour, at which point the tissue is so swollen you can't find anything, and you have to ask the person to come back a month later after it's healed, and try again." Some women have had to undergo surgery under general anesthesia to have Norplant removed from their bodies (Lewin, "'Dream' Contraceptive's Nightmare," *New York Times,* 8 July 1994, 7). In September 1993, Chicago lawyer Jewel Klein filed suit against Wyeth-Ayerst on behalf of women injured during Norplant removal. The following year it was certified as a class action suit with four hundred complainants. By the end of 1995, fifty thousand women, including Klein's clients, sought to sue the company. The class action suit sought damages and an injunction barring Wyeth-Ayerst from selling Norplant to doctors who have not been trained in implantation and extrication of the system (ibid.; Geoffrey Cowley with Susan Miller, "The Norplant Backlash," *Newsweek,* 27 November 1995, 52). In 1999, American Home Products, Wyeth-Ayerst's parent company, settled, paying out $1,500 to each of thirty-six thousand women who suffered headaches, nausea, depression, and irregular menstrual bleeding. The settlement did not cover injuries sustained during Norplant removal. The company did not admit to any wrongdoing.

29. Annie Chang et. al., "Local Reactions at the Insertion Site of the Norplant Contraceptive System," *Journal of Long-Term Effects of Medical Implants* 3, no. 4 (1993): 305; Uel E. Crosby et. al., "A Preliminary Report on Norplant Insertions in a Large Urban Family Planning Program," *Contraception* 48 (October 1993): 359; Cowley and Miller, "The Norplant Backlash," *Newsweek,* 27 November 1995, 52.

30. Lewin, "'Dream' Contraceptive's Nightmare," 7. In 1991, the *Wall Street Journal* reported that 40 to 65 percent of women in Norplant clinical studies requested its removal before the full five years elapsed (Charlotte Allen, "Norplant—Birth Control or Coercion?" *Wall Street Journal*, Western ed., 13 September 1991, 12).

31. Margaret L. Frank, L. Bateman, and Alfred N. Poindexter, "The Attitudes of Clinic Staff as Factors in Women's Selection of Norplant Implants for Their Contraception," *Women and Health* 21, no. 4 (1994): 80, 83.

32. Peter Baker, "Virginia Assembly Approves Norplant Money Despite Critics," *Washington Post*, 12 February 1993, D1, D3.

33. April Taylor, "Securing Reproductive Rights for Women of Color" workshop participant, *Dangerous Intersections*, 74; Andrea Smith, Friday plenary session participant, *Dangerous Intersections*, 6.

34. McGuire, "Health Educator Says Depo-Provera and Norplant Harmful," C1, C3.

35. Martinez, "Women of Color and Reproductive Health" session participant, *Dangerous Intersections*, 61. This echoes the experience of a woman in Bangladesh who had been using Norplant for two years when she requested removal. The doctors told her and her husband that they would remove the Norplant only if the couple would pay the full cost, roughly $400. Unable to pay the ransom, the woman was forced to continue with the implants. Another woman, bleeding severely, made multiple entreaties to clinic staff to take the Norplant out of her arm. She was repeatedly refused and told by the doctor, "There is no medicine for stopping bleeding, so we cannot give you any medicine. And removal of the method? Let us know when you die, we will take it out" (UBINIG, "The Price of Norplant is Tk.2000," 46).

36. Judy Norsigian, "The Women's Health Movement in the United States," in *Man-Made Medicine*, ed. Moss, 87.

37. Smith, Friday plenary session participant, *Dangerous Intersections*, 6.

38. Mann, "New Contraceptive Advances Freedom, Responsibility," B3.

39. Mark Stein, "Judge Stirs Debate with Ordering of Birth Control," *Los Angeles Times*, 10 January 1991, 31.

40. "Judge Orders Birth Control Implant in Defendant," *Washington Post*, 5 January 1991, 1; Mark Stein, "Judge to Let Birth Control Order Stand," *Los Angeles Times*, 11 January 1991, 3. Broadman later maintained that he had used gestures to explain Norplant to Johnson at sentencing.

41. "Judge Orders Birth Control Implant in Defendant," 10; "Judge to Reconsider Probation Requiring Birth Control," *Los Angeles Times*, 9 January 1991, 16.

42. Helen R. Neuborn, "In Norplant Case, Good Intentions Make Bad Law," *Los Angeles Times*, 3 March 1991, M5.

43. In fact, one could argue Broadman placed her children in greater jeopardy by releasing from custody a woman who admitted to beating them.

44. Emphasis added; "Norplant," *60 Minutes*, airdate 10 November 1991, transcript *CBS News* 24, no. 8 (1991): 8. Broadman's concern did not extend to prenatal care, as he was ready to sentence Johnson, pregnant at the time, to prison if she refused Norplant.

45. In 1990, Arthur Caplan, a bioethicist at the University of Minnesota, warned, "It is only a matter of time before a judge in some community orders a Norplant implant for some woman who is engaged in what the judge believes could be dangerous or risky behavior for a fetus. I guarantee you it will happen" (Malcolm Gladwell, "Implant for Birth Control Stirs Debate," *Philadelphia Inquirer*, 31 October 1990, 1–2).

46. Del Castillo, "Sterilization," 68.

47. Stein, "Judge Stirs Debate with Ordering of Birth Control," 31.

48. Tamar Lewin, "5-Year Contraceptive Implant Seems Headed for Wide Use," *New York Times,* 29 November 1991, 26.

49. "Norplant," *60 Minutes,* 9.

50. Stein, "Judge to Let Birth Control Order Stand," 3. In fact, Broadman's jurisdiction, Tulare County, is extremely poor. It has the highest youth poverty rate and the highest rate of teen pregnancy in California (Males, *The Scapegoat Generation,* 62). It is not unreasonable to suggest that Broadman, eager to prevent future births to poor women of color (the county has a large Chicano population), seized the child abuse charges as an opportunity to intervene.

51. Del Castillo, "Sterilization," 67–68; Blank, *Fertility Control,* 69.

52. "Teenager Sentenced to Birth Control," *Gay Community News,* 25 November–8 December 1990, 3.

53. Broadman removed himself from the case after being shot at in court by a man opposed to the Johnson ruling ("Norplant," *60 Minutes,* 10). Broadman was later brought before the Commission on Judicial Performance on willful misconduct charges for several incidents, including talking to the media about the Johnson case while it was still under appeal. The commission opted for the much less serious finding of "conduct prejudicial to the administration of justice" (Ramon G. McCleod, "No Misconduct in Norplant Case," *San Francisco Chronicle,* 27 August 1996, 15). Johnson's appeal, filed by the ACLU, was declared moot by a state appeals court after she was arrested for a drug-related probation violation. Unlike appeals court decisions, trial court rulings cannot set legal precedent in California ("Birth Curb Order Is Declared Moot," *New York Times,* 5 April 1992, 3).

54. Allen, "Norplant—Birth Control or Coercion?" 12.

55. Alexander Cockburn, "Beat the Devil," *Nation* 259 no. 3 (18 July 1994): 80. The 1994 bill finally introduced by Jerke dictated that a prison's executive director decide the "appropriate services to be provided and methods for providing such services," as long as those services did not include abortion. In 1996, the California Legislature moved to have men twice convicted of child molestation "[p]eriodically injected after release from prison" with Depo-Provera unless they opted for vasectomy. State legislators, pointing to the fact that Depo-Provera lowers testosterone levels, shrinks testicles, and results in impotence, maintained this action would protect children (B. Drummond Ayres Jr., "California Child Molesters Face Chemical Castration," *New York Times,* 27 August 1996, 1, 8; Blank, *Fertility Control,* 89). For more on the history of Depo-Provera and chemical castration, see "'Chemical Castration': Another Use for DepoProvera," *Hastings Center Report* 9, no. 4 (August 1979): 10.

56. Allen, "Norplant—Birth Control or Coercion?" 12.

57. Sanger, "Sterilization: A Model Program for Human Health and Welfare," 5 June 1951, 1, MSP-CDS.

58. Mann, "New Contraceptive Advances Freedom, Responsibility," B3; *Philadelphia Inquirer,* 12 December 1990, 18. The *Inquirer* editorial was blasted by the paper's readers, its own staff, the Philadelphia Association of Black Journalists, and journalists across the country. The *Inquirer* was picketed and eventually published a second editorial apologizing for the first ("An Apology," 23 December 1990, 4-C).

59. "Norplant: A Tool against Women," *off our backs* 21, no. 3 (March 1991): 13; Allen, "Norplant—Birth Control or Coercion?" 12. It is unclear whether the follow-up visits were to monitor women's continued health while on the hormone or to ensure they were holding up their end of the bargain and not having the implants removed "prematurely."

60. Ellen Goodman, "The Politics of Norplant," *Boston Globe*, 19 February 1991; "Norplant," *60 Minutes*, 9.

61. This figure was grossly misleading, given the fact that 75 percent of recipients are off welfare within two years and only 8 percent remain on welfare for more than eight years (Martinez, "Women of Color and Reproductive Health" session participant, *Dangerous Intersections*, 62); "Norplant: A Tool against Women," 13.

Like Sanger before him and William Shockley after him, William Vogt, postwar president of PPFA, also advocated "bonuses" for individuals who submitted to sterilization. "Since such a bonus would appeal primarily to the world's shiftless, it would probably have a favorable selective influence.... From the point of view of society, it would certainly be preferable to pay permanently indigent individuals, many of whom would be physically and psychologically marginal, $50 or $100 rather than support their hordes of offspring that both by genetic and social inheritance would tend to perpetuate fecklessness." Vogt, a conservationist and ornithologist, maintained that ecologists' primary concern should be population control (James Ridgeway, "Behind the Bell Curve," *Village Voice*, 15 November 1994, 16).

62. Helen R. Neuborne, "In the Norplant Case, Good Intentions Make Bad Law," *Los Angeles Times*, 3 March 1991; "Norplant: A Tool against Women," 13. This may represent a discrepancy in analyses of Norplant policy in decentralized state offices vis-à-vis each other and the national office. Alternately, it may signify a NOW position that stresses the issue of the venue, whereby Norplant may be mandated by the statehouse, but not the courthouse.

63. Barbara Katz Rothman, *Recreating Motherhood: Ideology and Technology in a Patriarchal Society* (New York: W. W. Norton, 1989), 62.

64. Lewin, "5-Year Contraceptive Implant Seems Headed for Wide Use," 26; "Birth Control Implant Gains among Poor under Medicaid," *New York Times*, 17 February 1992, 1; Allen, "Norplant—Birth Control or Coercion?" 12 (Texas); Stephanie Denmark, "Birth Control Tyranny," *New York Times*, 19 October 1991 (Louisiana); Mehler, "In Genes We Trust," 12 (Connecticut); Alexander Cockburn, "Beat the Devil," *Nation*, 18 July 1994, 79 (Florida).

65. "On this perverted logic," wrote Alexander Cockburn, "five children starve so that one might not be born" (Cockburn, "Beat the Devil," 79, 80).

66. Martinez, "Women of Color and Reporductive Health" session participant, *Dangerous Intersections*, 62.

67. Theresia Degener, "Female Self-Determination between Feminist Claims and 'Voluntary' Eugenics, between 'Rights' and Ethics," *Issues in Reproductive and Genetic Engineering* 3 no. 2 (1990): 88. Degener was speaking primarily of disability and reproduction, but her argument is equally apropos of women welfare recipients, especially in the wake of Clinton's 1996 Welfare Reform Bill.

68. Sawhill in Malcolm Gladwell, "Implant for Birth Control Stirs Debate," *Philadelphia Inquirer*, 31 October 1990, 2-A; Norplant consortium in Paul W. Valentine and Amy Goldstein, "Baltimore to Try Norplant to Reduce Teen Pregnancy," *Washington*

Post, 4 December 1992, 1; Norplant in schools in "Baltimore's Lead in Contraception," 14 December, 1993, 16; Steven A. Holmes, "Norplant Is Getting Few Takers at School," *New York Times,* 3 May 1994, 16.

69. A few articles mentioned charges of genocide, raised by some in the African-American community, but failed to report on the basis of these claims in any meaningful way. Genocide might seem an extreme charge, yet given the press coverage (not to mention the history of provider-controlled contraception and population control in this country), it is understandable. For example, on 4 December 1992, the *New York Times* ran a page-one story under the headline "Baltimore School Clinics to Offer Birth Control by Surgical Implant," directly above another, "Population Growth Outstrips Earlier U.S. Census Estimates," reporting on Census Bureau findings that the white population of the United States would stop growing by 2029 while the Black population was expected to double in the next sixty years (Tamar Lewin, "Baltimore School Clinics to Offer Birth Control by Surgical Implant," *New York Times,* 4 December 1992, 1; Robert Pear, "Population Growth Outstrips Earlier U.S. Census Estimates," *New York Times,* 4 December 1992, 1).

70. Researchers at the University of Texas found that 71 percent of teenagers on Norplant experienced irregular menstrual bleeding as compared to 53 percent of adult women ("Teenagers in Study Favor Norplant Use," *Washington Post Health,* 17 August 1993). An article published that same month noted that 33 percent of teenage girls on Norplant cited emotional disturbances as a side effect (Abbey B. Berenson and Constance M. Wiemann, *Pediatrics* 92, no. 2 [August 1993]: 257). At Parkland Memorial Hospital in Texas, 63 percent of the 431 immediate postpartum Norplant recipients were under nineteen years old; 23.5 percent were younger than sixteen (Crosby et. al., "A Preliminary Report on Norplant Insertions," 359, 361). Disregarding the already visible side effects and the need for further research, each of the above authors concluded that Norplant was ideal for young women and girls. Their recommendation recalls the American College of Obstetricians and Gynecologists's 1969 move to drop suggested age-parity guidelines for sterilization.

71. "Is Depo Provera Safe?" 4; Judith Scully, "Women of Color and Reproductive Health" session participant, *Dangerous Intersections,* 64; Hartmann, *Reproductive Rights and Wrongs,* 212; Martinez, "Women of Color and Reproductive Health" session participant, *Dangerous Intersections,* 65.

72. "Bold Attack on Teenage Pregnancy," editorial, *Los Angeles Times,* 7 December 1992, B6.

73. *Vital Statistics of the United States, 1940–1991,* vol. 1, cited in Males, *The Scapegoat Generation,* 62; Hartmann, *Reproductive Rights and Wrongs,* 144. Also relevant here is a National Public Radio report (broadcast 21 November 1994), which cited an increase in the number of African-American students graduating from high school—up from 67 percent in 1964 to 75 percent in 1994.

74. Cited in Marc Lacey, "Teen-Age Birth Rate in U.S. Falls Again," *New York Times,* 27 October 1999, A16.

75. Males, The *Scapegoat Generation,* 220, 88.

76. Ibid., 88.

77. Karen De Witt, "Teen-Agers Split on Birth Control Plan," *New York Times,* 5 December 1992, 7.

78. Males, *The Scapegoat Generation,* 82.

79. Paul W. Valentine, "In Baltimore, a Tumultuous Hearing on Norplant," *Washington Post*, 12 December 1993, D5; Males, *The Scapegoat Generation*, 9.

80. Cited in Males, *The Scapegoat Generation*, 219.

81. Francis X. Clines, "Computer Project Seeks to Avert Youth Violence," *New York Times*, 24 October 1999.

Disability and Eugenics

1. Disability rights advocates are an obvious exception to this.

2. Blank, *Fertility Control*, 83.

3. Robert A. Jones, "This Beats a Sombrero Any Day," *Los Angeles Times*, 19 May 1991, 3.

4. Cited in Lisa Blumberg, "Eugenics vs. Reproductive Choice," *Sojourner*, January 1995, 17.

5. Cited in ibid.

6. Ibid.

7. Ibid.

8. *Problems in Eugenics*, First International Eugenics Conference, 466.

9. Emphasis added; Dr. Victor C. Vaughan, cited in Morton Aldrich et. al., *Eugenics: Twelve University Lectures* (New York: Dodd, Mead, 1914), 74–75. Michigan passed its first compulsory sterilization law in 1913. It allowed for the eugenic, punitive, and/or "therapeutic" sterilization of "Idiots, imbeciles, the feebleminded, and epileptics." It was later declared unconstitutional because it applied only to the institutionalized and not to the public at large. It was amended several times, the result being an ever-widening circle of sterilization candidates, including, by 1929, "moral degenerates" and "sexual perverts showing hereditary degeneracy" (Landman, *Human Sterilization*, 305, 70).

10. Mehler, "The History of the American Eugenics Society," 95–96.

11. P. W. Whiting, "Selection, the Only Way of Eugenics," *Birth Control Review* 9, no. 6 (June 1925): 166. Whiting also cited Roswell Johnson's claim that venereal disease might even be eugenic because "it sterilizes the stupid and immoral." Johnson, an active member of the American Eugenics Society, coauthored *Applied Eugenics* with Paul Popenoe (Mehler, "The History of the American Eugenics Society," 383).

12. Sanger, "Birth Control and Racial Betterment," *Birth Control Review* 3, no. 2 (February 1919): 12.

13. Sanger, "Margaret Sanger's Own Corner," *Birth Control Review* 8, no. 5 (May 1924).

14. The ABCL, the editorial explained, did believe that sterilization "has its place in the aim of the American Birth Control League. . . . we recognize the value of sterilization in cases where Birth Control is likely to fail, and we are deeply interested in the work that is being done in this field." The occasion for these remarks was a preface to "Sterilization: A Symposium," which appeared in *Birth Control Review* in April 1928. Contributors included Laughlin, who advocated mandatory sterilization, and Ellis, who wrote that while sterilization was "often desirable" it nevertheless "ought not to be inflicted as a sort of punishment, but accepted freely by the subject, or those responsible. When this is done, there is no need for special legislation" ("Sterilization: A Symposium," *Birth Control Review* 12, no. 3 [April 1928]: 73, 90).

15. Sanger, *The Pivot of Civilization*, 86, 101–2. Four years later, *Birth Control Review* published an opposing viewpoint in its 1928 symposium on sterilization. Warren S. Thompson, who would later serve on the board of directors of the American Eugenics Society, favored segregation over sterilization. "The feeble-minded girl in the small community, who becomes prey of most of the young men living there, is quite generally regarded by eugenists *[sic]* as a menace to the community largely because of the children she is likely to bear. Manifestly, if sterilized, this menace is removed and the eugenic conscience appears clear. To the sociologist this view is too simple. The half-dozen children—more or less—that such a woman might leave behind her seem less to be feared than the demoralizing effects of her presence in the neighborhood." Thompson therefore recommended segregation: "She cannot spread disease throughout the community, she cannot set low moral standards which more normal people may be inclined to follow, she cannot be the means of developing careless sex habits in her neighbors. . . . Thus in this case it is evident that sterilization is insufficient to curtail the *power for evil* of such a person and it is unnecessary if the proper segregation is provided for, as this latter deals with both the biological and social aspects of the case at the same time" (emphasis added; Thompson, "Sterilization: A Social View," *Birth Control Review* 13, no. 3 [April 1928]: 76). Neither segregation nor sterilization protected women from the abuse or violence Thompson described. Neither he nor Sanger registered any concern over the well-being of the girls and women they conjured (or condemnation for the young men who might victimize them). Rather, it was their role as eugenic threat, demoralizer and infecter of men, and their drain on the state that captivated them.

16. Chesler, *Woman of Valor*, 195.

17. Ruth Hubbard, "Eugenics and Prenatal Testing," *International Journal of Health Services* 16, no. 2 (1986): 234. A similar, but more exhaustive, campaign was going on in Germany at the time that Kennedy's article was published.

18. Pauling, cited in Duster, *Back Door to Eugenics*, 46. Likewise, calls for genital tattooing of HIV-positive people in the 1980s clearly targeted the already marginalized and imperiled, not AIDS.

19. Michael Bayles, "The Legal Precedents," *Hastings Center Report*, June 1978, 38.

20. Petchesky, "Reproduction, Ethics, and Public Policy," 38. I would add that statutes such as the one upheld in North Carolina take as axiomatic that childlessness is universally preferable to parenting a disabled child.

21. Levine, "Depo-Provera and Contraceptive Risk," 8; Hartmann, *Reproductive Rights and Wrongs*, 204.

22. Macklin and Gaylin, *Mental Retardation and Sterilization: A Problem of Competency and Paternalism* (New York: Plenum Press, 1981), cited in Blank, *Fertility Control*, 61.

23. Petchesky, "Reproduction, Ethics, and Public Policy," 37. Given the increased danger of sexual abuse faced by the developmentally disabled, I would add educational programs dealing with power and consent to Petchesky's list.

Quinacrine, the Next Wave

1. Quinacrine pellets are inserted into the uterus where they dissolve, resulting in inflammation that in turn produces scar tissue at the end of the fallopian tubes, blocking the egg's entry into the uterus (Hartmann, *Reproductive Rights and Wrongs*, 256).

2. Norsigian, "The Women's Health Movement in the United States," in *Man-Made Medicine*, ed. Moss, 88; Hartmann, *Reproductive Rights and Wrongs*, 286, 257. The Vietnamese government, bowing to concerns voiced by women's advocates and health organizations, discontinued quinacrine sterilizations pending further evaluations ("Women's Views Influence Contraceptive Use," *Network*, no. 1, September 1995, 16). The Association of Voluntary Surgical Contraception reports that quinacrine users in Vietnam were paid three dollars a visit (Charles S. Carignan, Deborah Rogow, and Amy E. Pollack, "The Quinacrine Method of Nonsurgical Sterilization: Report of an Experts Meeting," AVSC working paper no. 6, July 1994, 8).

3. Carignan, Rogow, and Pollack, "The Quinacrine Method of Nonsurgical Sterilization." Bleeding, chronic pelvic pain, dizziness, headaches, ectopic pregnancies, uterine perforation during insertion, and irritation and burning have all been reported by women on quinacrine.

4. *Dangerous Intersections*, appendix.

5. Hartmann, *Reproductive Rights and Wrongs*, 99; Mehler, "The History of the American Eugenics Society," 340, 318.

6. Guy Irving Burch, "Immigration Control," *Birth Control Review* 13, no. 6 (June 1929): 163, 164. The remainder of the article centered on Burch's fear of a repeal of the 1924 Immigration Act: "[I]t is up to the great majority of the American people who are not represented in the 1890 foreign born basis to see that Congress under hyphenated [-American] pressure does not play politics with the bloodstream of this country." Carole McCann has maintained that Sanger and her adherents opposed immigration restriction, though "their rhetoric required assimilation of an old-stock perspective" (McCann, *Birth Control Politics in the United States*, 132). If her classification of Sangerists' antirestriction beliefs is accurate, then both the article's publication in *Birth Control Review* and Burch's ascendancy to the ABCL leadership remain somewhat inexplicable. According to Gordon, Burch explained his support for birth control as an outgrowth of his commitment to "prevent the American people from being replaced by alien or Negro stock" (Burch, cited in Gordon, *Woman's Body, Woman's Right*, 279). In 1939, Burch campaigned to bar the entry of Jewish orphans/refugees into the United States (Hartmann, *Reproductive Rights and Wrongs*, 99).

7. Hartmann has noted that Expo organizers apologized for the oversight that enabled Collins and Epstein's presence in Washington, D.C. (Hartmann, Saturday plenary session participant, *Dangerous Intersections*, 23).

8. Asoka Bandarage, "A New and Improved Population Control Policy?" *Sojourner* 20, no. 1 (September 1994): 17.

9. RU486 is contraindicated by anemia, and therefore, as Jael Silliman notes, is "not an attractive option" for many women in developing countries and elsewhere. (Silliman, "Cairo and Beyond: International Perspectives on Population and the Environment" workshop participant, *Dangerous Intersections*, 78). Anemic or not, women may suffer side effects including vomiting, prolonged bleeding, severe pain, and, in some cases, thrombosis and cardiac arrhythmia (FINRRAGE Position Paper [on] RU486, Feminist International Network of Resistance to Reproductive and Genetic Engineering, Hamburg, Federal Republic of Germany, October 1991). Many thanks to Rachel Rosenbloom for making this material available to me. Hartmann reports, "Hemorrhages requiring transfusions occur much more frequently with RU486 than with surgical abortion" (Hartmann, *Reproductive Rights and Wrongs*,

264). To date, these are the only known side effects. Until the long-term effects of RU486 and quinacrine are known, truly informed consent remains an impossibility.

10. Karen Houppert, "How Prolife Forces Strangle Research," *Village Voice*, 1 October 1996, 25. Houppert falls into this trap by assigning opposition to Norplant and similar "contraceptive advances" solely to anti-choice groups and registering her own support of the new technology without examining its dangers.

Conclusion

1. Charles A. Boston, "A Protest against Laws Authorizing the Sterilization of Criminals and Imbeciles," *Journal of the Institute of Criminal Law and Criminology* (September 1913).

2. Erica Goode, "Roe v. Wade Resulted in Unborn Criminals, Economists Theorize," *New York Times*, 20 August 1999.

3. Mosse, *Nationalism and Sexuality*, 133, 134.

4. *Proceedings of the First National Conference on Race Betterment*, 478.

5. The heading is a quote from Ruth Hubbard, "Irreplaceable Ewe," *Nation*, 24 March 1997, 4.

6. While Proposition 187 was still in court, politicians in Florida, Illinois, and New York expressed a desire to see similar legislation enacted in their own states, and there were early indicators that Proposition 187 would pave the way for a federal overhaul of immigration policy, regardless of the outcome in the California courts. Indeed, federal measures have superseded many of the provisions of Proposition 187. Under the one-strike-and-you're-out clause of the Anti-Terrorism Act signed by President Clinton in April 1996, longtime legal residents convicted of "serious crimes" are subject to mandatory detention and deportation (regardless of how long ago they were convicted, sentenced, and released), without the rights they previously had to apply for a waiver of deportation or to appeal in federal court ("Immigrant's Peril—'One Strike and You're Out,'" *San Francisco Chronicle*, 3 September 1996, 1).

7. Cited in Vern Bullough, "Homosexuality and the Medical Model," *Journal of Homosexuality* 1, no.1 (1974): 106.

8. Minkowitz, "Trial by Science," 28, 29. The failure of biologists to reconcile their theories with bisexuality certainly predates this trial. In 1941, for example, Dr. L. A. Kerr classified bisexuality, along with hermaphroditism, as "biological stuttering" ("Bi-Sexuality in Mankind," *Sexology* 8, no.4 [March 1941]: 221).

9. Emphasis added; Hamer and Copeland, *The Science of Desire*, 53.

10. Kate Bornstein, *Gender Outlaw* (New York: Routledge, 1994), 3.

11. Patricia Donavan, "Sterilizing the Poor and Incompetent," *Hastings Center Report* 6, no. 5 (October 1976): 9.

12. Étienne Balibar, "Racism and Nationalism," in *Race, Nation, Class*, 60.

Index

ABCF. *See* American Birth Control Federation

ABCL. *See* American Birth Control League

Ableism, 129

Abortion, 167, 171, 184, 204, 260n21, 261n42; mandatory, 113–14; refusing, 197; self-induced, 165; sterilization and, 174

"Abortion Hope after 'Gay Genes Finding'" *(Daily Examiner)*, 63

Abourzek, James, 172, 173

Abuse: sexual, 272n23; of sterilization, 170, 176, 200, 255n66, 262n66

ACLU. *See* American Civil Liberties Union

ACLU News, 260n29

ACOG. *See* American College of Obstetricians and Gynecologists

Acosta, Guadalupe, 174

ACPS. *See* American Coalition of Patriotic Societies

Adair, B. A. Owens: on sterilization, 133

Additive causation theories, xxi, 60, 68, 82, 88, 103, 104, 107, 123, 211

Advocate: on E hormone, 113; Murphy in, 115, 118

AES. *See* American Eugenics Society

AFDC. *See* Aid to Families with Dependent Children

Affirmative action, 4, 214

African-Americans. *See* Blacks

AGA. *See* American Genetic Association

Aggression: genetically determined, 214

AIDS, 64, 66, 88, 121, 192, 211, 237n3, 241n1, 272n18; bisexuality and, 65; homophobia and, 120; homosexuality and, 119; responses to, xx, 120

Aid to Families with Dependent Children (AFDC), 170, 190, 191, 193; Norplant and, xxvii, 185; sterilization and, 166, 168

Alabama Department of Pensions and Security, 170

Alcoholism: homosexuality and, 239n15

Allen, Clifford: on chemical castration, 111

Allen, Lauren, xx

Alpines, 27

Alpine Slavs, 26–27, 39

Alvarez Martinez, Luz, 180, 184

Amendment 2, xxiii, 68, 68, 78; challenging, 61, 121

American Association for the Advancement of Science, xx, 113, 247nn38, 40

American Birth Control Federation (ABCF), 152, 153

American Birth Control League (ABCL), 141, 144, 152, 199, 203, 253n27, 273n6; *BCR* and, 143; guinea pigs for, 149; on sterilization, 271n14

American Breeders Association, 11, 222n11, 227n55

American Breeders Journal, xvi

American Breeders Magazine, 5, 39, 227n55; Ward in, 7, 222n11

American Child Health Association, 151

American Civil Liberties Union (ACLU), 167, 190, 194

American Coalition of Patriotic Societies (ACPS), 47, 48, 235n9

American College of Obstetricians and Gynecologists (ACOG), 168, 176, 270n70

American Enterprise Institute, 22, 188

American Eugenics Society (AES), 11, 39, 46, 144, 198, 203, 228n75, 229n90, 230n103, 232n16, 243n33, 253n27, 255n58; charter members of, 231n5; child labor law and, 236n8; contests and, 33; Johnson and, 25; Popenoe and, 251n13; Wiggam and, 142

American Friends Service Committee, 149

American Genetic Association (AGA), 16, 20–21, 227n55

American Home Products, 266n28

American Immigration Control Foundation, 55, 235n16

American Journal of Psychiatry: Radzinski in, 47–48

American Medical Association, 139, 176

American Negro as a Dependent, Defective, and Delinquent, The (McCord), 162

American Psychiatric Association (APA), 89, 114

American Psychological Association (APA), 88, 89, 106

American race, xvii, 211

Amsterdam News: on hormone treatment, 108

"Analysis of America's Modern Melting Pot, An" (Laughlin), 30

Anderson, Benedict: quote of, 1

Andrews, Laurie: on sexual orientation, xxiv

Androgyny, 116

Androsterone, 113

And the Band Played On (Shilts), 83

Antianarchists, 5, 20, 22

Anti-choice proponents, 137, 204

Anticommunists, 5; immigration and, 20

Antidiscrimination protections, xxiii, 61

Antigay initiatives, xxi, xxiv

Anti-immigrant groups, 48, 203, 208, 210, 218n11, 225n17, 228n66

Anti-immigration, xv, xvii, 3–4, 6, 15, 24, 45, 54, 55, 56

Anti-Jewish statutes, 43

Antimiscegenation laws, xii, xv, 39, 43, 133

Anti-Semitism, 99

Antisterilization movement, 153

Anti-Terrorism Act, 274n6

APA. *See* American Psychiatric Association; American Psychological Association

Apomorphine, 112

Applied Eugenics (Whiting and Popenoe), 271n11

Applied Eugenics in Present-Day Germany (film), 46

Armes, Irene Headley, 156

Army Mental Tests, 25, 34, 228n80

Asetoyer, Charon, 183, 184

Association for Voluntary Sterilization, 176

Association for Voluntary Surgical Contraception, 263n83

Asylums, 30, 79, 135, 148, 240n31; foreign-born in, xvi; Jews in, 99

Atascadero State Hospital, 114

Atlantic Monthly, 49; Ewald and Cochran in, 60; Fairweather in, 42–43; on homosexuality germ, 59

Aversion therapy, 102, 106, 112, 114, 116

"Award in Planned Parenthood of the Albert and Mary Lasker Foundation," 154, 155

Bailey, J. Michael, 88, 115, 248n52; on
CGN, 89–90; on lesbianism and
bisexuality, 59; twin studies by, xx,
67, 89, 118, 120
Balibar, Étienne: mythical national type
and, 8; national purification and,
213–14; on racial and cultural purity,
xv; on racism and eugenics, 6–7, 16
Baltimore City Hospitals, 261n50,
262n69
Baltimore school system: birth control
in, 193
Bandarage, Asoka: on reproductive
freedom, 204
Barahal, Hyamn S., 247n30
Bardin, Wayne: on Norplant, 182
Barlow, Thomas W., 251n15
Barry, Warren E.: on undocumented
immigrants, 54
Bateson, William, 143
Bay Area Reporter: on Hamer, 63
Bayless, Judge H. Jeffrey: on
Amendment 2, 121
BCR. See Birth Control Review
Bell, Alexander Graham, 231n5
Bell Curve, The (Herrnstein and
Murray), xi, xii, xviii, xix, 48, 129, 214,
224n10
Benkert, Karoly Maria, 70, 211
Bery, Charles: on chemical castration, 111
Besharov, Douglas: on class prejudice,
188
Bethlehem Steel, 23
Beyondism, 230–31n4
Bill 64 (1937), 255n63
Binet, Alfred, 11, 223–24n8
Binet-Simon intelligence tests, 138–39
Bingham, Theodore: on criminal
element, 14
Biological Aspects of Immigration, xvi,
1, 29
Biological determinism, xiv, 4, 40, 122,
209, 214, 219n30; eugenics and, 60–
61; politics and, 123
Biological discourses, 102–3
Biological immutability, 122, 212
Biologism, 9, 71, 123; antigay initiatives
and, 66; public policy and, xiv–xv

Biology, 110, 113; consolidation of, 101
Birth control, 158, 167, 176, 249–50n8,
264n12, 271n14; advice on, 149;
concept of, 128; doctrine of, 143;
eugenics and, xxv, 131, 136–37, 157,
159, 253n35; failed, 191; fit and, 136;
immigration control and, 203;
imposing, 186; Jews and, 100; Native
Americans and, 180; neo-Malthusian,
157; poverty and, 178, 188; propaganda on, 143; racism and, xxvi;
sterilization and, 147, 262n59;
temporary forms of, 200; using,
141–42
Birth control movement, 137, 141, 150, 213
Birth control pill, 151, 152, 256n72,
262n71
Birth Control Politics in the United
States, 1916–1945 (McCann), xxv–
xxvi, 137
Birth Control Review (BCR), 5, 147, 198,
199, 273n6; ABCL and, 143; on birth
control, 142–43, 150; Burch in, 203;
contributors to, 144; differential birth
rates in, 146; Duvall in, 146; eugenics
and, xxv, 148; Lidbetter in, 51; Sanger
in, 136, 157; on sterilization, 272n15;
Wiggam in, 141–42
Birth rates, differential, 140, 141, 146
Bisexuals, xiv, 65, 105, 242n6, 274n8;
antidiscrimination protections for,
61; attacks on, xiii; interventions on,
xxi; scientific assessments of, 211;
twin, 59
Blacks: contraception and, 256n79; HIV
incidence among, 217n2; inferiority
of, 26, 36, 38, 46, 214; insanity and,
232n16; mental-racial hierarchy and,
232n16; migration of, 232n18;
Norplant and, 183, 185; poverty for,
189–90; sterilization and, xxvi–xxvii,
162, 163, 165; unfit, 162
Blair, Ralph, 112
Blank, Robert, 195; technofixes and,
xxv, 129, 249n3
Blumberg, Lisa, 196, 197
Boring, E. G., 228n77
Born criminals, 14, 225n23

Bornstein, Kate, 212

Boston, Charles A., 206

Boston City Hospital, 261n50

Boston Medical and Surgical Journal, 76

Box, John C., 23–24, 34, 35, 231n10

Boyarin, Jonathan, xxiv, 220–21n44

Brandeis, Louis, 250n11

"Breaking Up a Homosexual Fixation by the Conditional Reaction Technique" (Max), 106

Breast cancer gene, 214

Brigham, Carl C., 25; on Alpine Slavs, 39; on European Jews, 26–27; on immigrants, 26, 28

Brightman, Lee, 172

British Medical Journal, xxi, 111, 112

British Medical Society, 65

Broadman, Howard, 189, 190, 267nn43, 44, 268n50; gender bias of, 187; Johnson and, 185, 186, 187, 188, 268n53; Norplant and, 267n40

Broca, Paul, 38

Brown, Shirley, 167–68, 174

Brown v. Board of Education (1954), 214

Bryant, Anita, 114, 247n44

Buck, Carrie, 81, 135, 167, 199, 201

Buck v. Bell (1927), 134, 186, 250n11, 251n12; compulsory sterilization and, 32, 81, 135, 201

Bullough, Vern, 240n31

Burch, Guy Irving, 203, 273n6

Bureau of Alcohol, Tobacco, and Firearms, 194

Butch and femme, 95, 96

Butler, Justice Pierce: *Buck* and, 250n11

Calavita, Kitty, 221n2

Calderone, Mary S., 257n94

California Coalition for Immigration Reform, 55, 210

California Court of Appeals: birth control and, 186

Camp, Talcott, 187–88

Campbell, C. G., 40, 43–44

Caplan, Arthur: on Norplant, 267n45

CARASA. *See* Committee for Abortion Rights and against Sterilization Abuse

Carnegie Institution, 30

Carolina Association for Retarded Children v. North Carolina (1976), 200

Carpenter, Edward, 71, 115

Cartwright, Samuel A., 162, 258n8

Castration, xxi–xxii, 61, 76, 80, 100, 107; alternatives to, 220n33; bilateral, 84; of body and mind, 86; chemical, 111; court-mandated, 78; curing with, 86; execution and, 77–78; hormone treatment and, 111; sex crimes and, 77; sterilization and, 185; vasectomies and, 240n34

Cattell, Raymond: Beyondism and, 230–31n4

Center for Constitutional Rights, 175

Center for Research on Population and Security, 202

Centers for Disease Control, 119, 219n29

Cerebral disease: perversion and, 75

CESA. *See* Committee to End Sterilization Abuse

CGN. *See* Childhood Gender Nonconformity

Chase, Allan, 217n4

Chauncey, George, 70, 74, 95

Chelf, Frank: on Trevor, 48

Chemical treatments, 60–61, 102

Chesler, Ellen, xxv, 138, 145, 199, 254n58; eugenics and, 157; on Sanger, xxvi, 137

Chicago Birth Control Conference: Ross at, 255n60

Chicago College of Physicians and Surgeons, xxi, 74, 162

Chicago Municipal Airport, 25

Chicago Tribune, 63

Chicanas: sterilization of, 175

Chideckel, Maurice, 94, 96–97, 98–99

Childhood Gender Nonconformity (CGN), xxii, 89–90, 101, 114, 209

Chinese Exclusion Act (1882), xvi, 6, 223n7, 228n70; laborers and, 13, 218n12

Choice, 137, 177, 190

Christian Coalition, 205

Christian Identity, 3

Christian Right: eugenics and, 205
Christian Scientists: lesbian and gay, 66
Churchill, Winston, 231n5
Cincinnati Sanitarium, 133
Citizens Responsible for Immigration, 228n66
Civil rights, 48, 112, 121; anatomy and, xxiii; eugenics and, 260n22; gay and lesbian, xxii, 61, 62, 66, 107, 214, 244n40; genetics and, 4; natural minorities and, 67; Pioneer Fund and, 46; science and, 83
Classism, 4, 74, 130, 177, 178, 187–88, 209
Cleveland, Grover, 10
Clinton, Bill, 269n67, 274n6
CLIP. *See* Colorado Legal Initiatives Project
Clitoridectomies, 61, 76, 91, 93, 99, 136
Clone Age: Adventures in the New World of Reproductive Technology, The (Andrews), xxiv
Coalition for the United States Population Stabilization, 49–50
Cochran, Gregory, 59, 60
Cockburn, Alexander, 269n65
Cofer, L. E., 12–13, 225n17
Collins, Donald, 203, 273n7
Colorado Legal Initiatives Project (CLIP), 61–62, 212
Colt's Patent Firearms Manufacturing Company, 23
Commission of Relations with Japan, 11
Commission on Judicial Performance, 268n53
Commission on Wartime Relocation and Internment of Civilians, 160
Committee for Abortion Rights and against Sterilization Abuse (CARASA), 175–76
Committee for the Study of Sex Variants, 29
Committee on Eugenic and Dysgenic Effects of Birth Control, 243n33
Committee on Immigration (AGA), 16, 20–21, 33
Committee on Popular Education, 142
Committee on Research Problems in Eugenics (AES), 253n27

Committee on the Eugenic and Dysgenic Effects of Birth Control (AES), 255n58
Committee on Women, Population, and the Environment, 183
Committee to End Sterilization Abuse (CESA), 175, 176
Communism, 5, 20
Concerned Women for America, 205
Congregational Mission Hospital, 152
Congress of Industrial Organizations: immigrant workers and, 226n30
Conscientious objectors, 26
Consent, 162, 174; informed, 171, 173, 184
Conservatism, 203; eugenics and, 129; nationalism and, 55
Construction of Homosexuality, The (Greenberg), 73
Contraception, 139, 145, 164, 254n44, 270n69; Blacks and, 256n79; chemical, xxvii; economic-based, 131; failure of, 161, 259n17; information on, 138, 143; long-term, 189; patient-controlled, 183; physician-dependent, 128; poverty and, 191; voluntary, 199; welfare and, 189
Contraception (journal): on Norplant, 182
Cook v. State (1972), 187
Coolidge, Calvin, 31, 52, 236n7
Copeland, Peter: on heredity, xi–xii
Cornell Medical College: on AIDS research, 265n13
Cox, Nial Ruth, 166–67, 174, 260n29
Cradle competition, 140
Craniology, 94, 223n8
"Crash Made Him Gay, Jury Made Him Rich" *(Oakland Tribune)*, 59
Crime, 74, 79, 94; eugenics and, xxii, 14, 40; Norplant and, 188–89; sexual inclination and, 220n33
Crime, Degeneracy, and Immigration (Orebaugh), 39
Criminal anthropology, 94
Criminal children, 210
Culture: eugenics and, 14
Culture of poverty, 52, 128, 158, 172
"Cure for Homosexuals?" (Reidner), 104

Curtis, Jesse, 175
Curve (magazine), 88
"Cycle of dependency" theories, 52

Daniel, F. E., 78, 133, 208; on castration, xxi–xxii, 77; on criminality and sexual inclination, 220n33; on hysteroepilepsy, 91
Darwin, Leonard, 143, 144
Davenport, Charles, 11, 34, 141, 144, 222n11, 253n27; AES and, 231n5; Galton Society and, 228n67
Davis, Angela, xxv, 137, 168, 262n59
Davis, R. E., 80, 81
Davis, Secretary James J., 36
Defectives, 139, 198, 200, 226n37; Black, 162; deportation of, 13; Jewish philanthropy and, 52; sterilization of, 131
Degener, Theresia, 191, 269n67
Degeneracy, 100, 259n10; hereditary, 71, 105; racial, 40; sterilization and, 134
Degler, Carl, 217n4
De Lapouge, Count, 234n47
Del Castillo, Adelaida, 175, 260n25
Delinquente nato, 77
Deneuve (magazine), 88
Department of Eugenics, 30
Department of Health Education and Welfare (HEW), 151, 170, 171, 213, 263n86; regulations by, 176–77, 178, 179, 200
Depo-Provera, xiv, xxvii, 169, 179, 180, 192, 200, 202, 205, 260n36, 263n1, 264n2; birth control and, 264n12; as curative, 131; mandatory, 268n55; marginalized women and, 180, 201; Norplant and, 181; side effects of, 184, 264n5; sterilization and, 181, 194; support for, 128–29, 183; welfare and, 128
Deportation, xix, 13, 17, 18, 22, 42, 44, 226n46, 274n6
Deportation Bill (1925), 226n46
Derelicts, mental and physical, 17
Detroit Lancet: Kiernan in, 94
Deukmejian, George, 185

Deviance, 120; gender, xxii, 59, 90; inborn, xxii; sexual, 75–76, 90, 101, 114
Diagnostic and Statistical Manual of Mental Disorders (DSM-IV) (APA), 89
Diaz, Mary, 174
Dickinson, Robert Latou, 29, 94, 149, 243n33, 245n1, 254n58, 255n58
Difference, 26, 113, 120
Dipsomania, 71
Disabled, 147, 197; costliness of, 147; dehumanization of, 114, 195–96; developmentally, 201, 272n23; physically, 200; sterilization and, 155
Discrimination, racial, 134
Diseases of Society, The (Lydston), 76, 91
Division of Child Welfare, 35
DNA scans, xxiii
Donohue, John, 207, 225n23
Dörner, G., 248n58; rat study by, 95, 116–17; on teratophysiology, 117–18
Douglas, Justice William O., 250n12
Draper, Wycliffe: Pioneer Fund and, 45
Draper Project, 46
Drapetomania, 258n8
DSM-III, 114
DSM-IV. See Diagnostic and Statistical Manual of Mental Disorders
Du Bois, W. E. B.: on reforms, 153
Duggan, Lisa, 122, 219n30
Duke, David: on welfare and Norplant, 191
Duster, Troy, 217n4, 251n12
Duvall, John C., 146
Dysesthesia, 258n8
Dysgenesis, xii, 15, 17, 76, 140, 153, 193

East, Edward: Sanger and, 144–45
Eastland, James, 235n8
Economics: eugenics and, 146, 187
Edwards, D. J., 26
Eidelsberg, Joseph, 108
Electric shock therapy, 61, 102, 106–7, 112, 123, 209
Eliot, Charles, 231n5

Ellis, Havelock, xxii, 106, 118, 140, 143, 144, 240n20, 241n7, 243n29; on castration, 86; eugenics and, 86, 87, 253n35; on "History XXXVII," 93; on homosexuality, 85–86; on inversion, 240n23; on legal and social redress, 122; on sterilization, 271n14; theories of, 82, 84, 94, 102; on tribadism, 97–98

Ellis, Thomas F., 235n9

Ellis Island, 11, 12, 56, 225n14

E Magazine, 49

Emergence (lesbian and gay organization), 66

Emetine, 112

Endocrimes, 110

Endocrinology, 109

Environmentalism, 18; economic, 139; eugenics and, 50; FAIR and, 49

Ephedrine, 112

Epilepsy, 13, 29, 79, 90, 135, 136, 140, 182, 199, 225n17, 230n94; homosexuality and, 105; sterilization and, 56, 257n88, 271n9

Epstein, Sally, 203, 273n7

Equal Rights Amendment, 107

ERA. See Eugenics Record Association

Erblehre und Rassenhygiene im volischen Staat (Eugenical News), 42

Erotic mania, 76

Ervin, Sam, 114

Espionage Act (1917), 228n72

Essence (magazine): advertising in, 183

Estriol, 246n27

Ethnic laws: immigration law and, 53

Etiocholano, 113

"Eugenic," "genetic" and, 195

Eugenical News, xvi, 5, 38, 39, 40–41, 44, 223n5, 253n27; anti-Jewish statutes and, 43; on Filipino immigration, 37; Frick in, 42; Laughlin and, 35, 144; on mass sterilization programs, 42; Ploetz in, 43; Third Reich and, 41

Eugenical Sterilization in the United States (Laughlin), 29

Eugenicists, 31, 48, 138, 157, 204, 233n36; agenda of, xvi; antimiscegenation sympathies of, 39; biological and

political future and, 22; birth control and, 131, 150; immigration and, xvi, 70, 210; legislation and, 4, 20; legitimacy for, 143; political right and, xxv, 249n6; poverty and, 15; socioeconomic considerations by, 164; sterilization and, 160; xenophobia and red-baiting and, 23

Eugenic Reform (Darwin), 144

Eugenics: appeal of, xii, xiii, 214; characteristics of, 140; close reading of, 213–14, 215; commitment to, 86; do-it-yourself, 219n29; history of, 203; humane, 78, 258n6; ideology of, 33, 124, 203, 207, 227n65; Lamarckian, 249n6; legislative goals of, xvii; Mendelian, 249n6; national, 8; negative, 8; nimbleness of, xxvii, xxviii; as pan-class solidarity, 14–15; race-based, 160; restrictionist, 210; social implications of, xvii, 62; strategy of, 124; treatises on, 196. See also Nazi eugenics

Eugenics Board of North Carolina: sterilization and, 165

Eugenics boards: sterilization and, 78, 79, 134, 165

Eugenics Committee of the United States of America, 24, 33, 141

Eugenics doctrine, 14, 40, 139, 144, 198, 209

Eugenics enterprises, 128, 209

Eugenics movement, xxvi, 6, 118, 208, 209, 250n8; history of, xxv, 7, 206–7

Eugenics Publishing Company, 94

Eugenics Record Association (ERA), 24, 25, 30, 32, 33, 37, 40, 41, 43, 44, 229n90, 233n41

Eugenics Record Office, 144, 227n65, 253n27

Eugenics Research Association, xvi, 142

Eugenics Society of Northern California, 38

"Eugenic Value of Birth Control Propaganda, The" (Sanger), 140

Eugenik, 249n6

Evolutionary theories, 102, 138

Ewald, Paul, 59, 60

Exclusionists, 9, 25, 47, 210
Executive Order 9066, 161
Experimentation, 93, 100, 111, 246n28
Extermination policies, 172
Ezzel, Harold, 49

FAIR. *See* Federation for American
    Immigration Reform
Fairchild, Henry Pratt, 203, 231n5
Fairweather, Nicholas: on Nazi
    platform, 42–43
Family planning, 171, 191, 204, 264n12
Family Planning Clinic, 169
Farrall Instrument Company, 247n34
Fascism, 129, 203
FDA. *See* Food and Drug
    Administration
Federal Emergency Relief
    Administration: birthrate and, 148
Federation for American Immigration
    Reform (FAIR), 221n1; Pioneer Fund
    and, 3, 48–49, 235n16; Proposition 187
    and, 202, 208; rhetoric of, 210; Zero
    Population Growth and, 49
Feebleminded, xxvi, 5, 8, 11, 17, 25, 29,
    30, 37, 80, 139, 140, 156, 157, 194,
    225n17, 259n10; assaults on, 195;
    Blacks, 162, 163; deportation for, 18;
    immigration and, 40; screening for,
    12; sterilization of, 56, 79, 153, 154, 165,
    199, 201, 271n9
Feighan, Michael A., 47
Felsen, Jim, 173
Feminism, 177, 204
Feminist Expo, 203
Feminist International Network of
    Resistance to Reproductive and
    Genetic Engineering (FINRRAGE),
    264n2
Femme and butch, 95, 96
Fertility, 54, 146, 211; national, 32;
    woman-controlled regulation of, 164
Fetal rights, 186
Fight the Right Project (National Gay
    and Lesbian Task Force), 220n40
FINRRAGE. *See* Feminist International
    Network of Resistance to Reproduc-
    tive and Genetic Engineering

Fishberg, Maurice, 52, 229n84
Fit, 136; unfit and, 140, 141
"Fitter Families Contests," 33, 142
Fluoride: homosexuality and, 237n1
Foam-powder-sponge method, 150, 161,
    255n59
Food and Drug Administration (FDA):
    Depo-Provera and, 179, 180, 181, 200;
    Norplant and, 184, 191, 192
Forel, August, 86, 95
Fortenberry, J. Dennis, 194
Foucault, Michel, xxii, 71
Fourteenth Amendment: *Skinner V.*
    Oklahoma and, 250n12
Fox, Ruth, 108
Free choice, 176, 177, 190
Free love advocates, 144, 145
Freud, Sigmund: on homosexuality, 103
Futuyma, Doug, 68, 219n30

Gall, Franz Joseph, 243n30
Galton, Francis, 222n11, 229n84;
    Christianity and, 230n4; eugenics
    and, xii, 242n6
Galton Society, 24, 41, 228n67, 253n27
Gamble, Clarence, 127, 148, 157, 158,
    255n59, 258n6; birth control
    advocacy by, 152–54; on German
    sterilization policies, 258n5; guinea
    pigs for, 149; legacy of, 130–31;
    National Committee on Maternal
    Health and, 254–55n58; on Negro
    Project, 153; Planned Parenthood
    and, 152; poor and, xxvi; Puerto
    Rico plan of, 151, 152; Sanger and,
    149, 150, 152, 151, 155, 161, 257nn98,
    99; sterilization and, 161–62, 165
GAO. *See* General Accounting Office
Gay and Lesbian Medical Association,
    220n42
Gay cancer, 119
Gay children, 85–86, 219n29
Gay gene, xii, xxiv, 60, 66, 68, 117,
    248nn52, 60; search for, 63, 88, 118,
    207
Gaylin, Willard, 201
Gay-man-as-true-woman theory,
    245n62

Gay-related immune deficiency (GRID), 119
Gay rights movement, 103, 119
Gay rights ordinance, 219n29, 247n44
Gays, xiv, 60, 67, 242n6; biological difference for, 113; Black, xxii; castration of, 100, 107; characterizations of, 83, 99; genetic and hormonal failings of, 114; hazards for, 87; hormonal experimentation in, 93; medical misinformation on, 103; ostracism for, xiii, 87, 112–13; political goals of, 122; protecting, 61, 87; registering, 123; scientific assessments of, 211; sterilization laws and, 240n39; straight men and, 242n11; testosterone levels in, 116; treatment for, xxi, 102–3
Gay sons/fathers/brothers, 63, 64
Gay Sunshine: on lobotomies and homosexuality, 114
Gender: sterilization and, 159
Gender bias, 187
Gender conformity, xxii, 90
Gender equity, 137
Gender Identity Disorder, 89
Gene pool, xix, 3–4, 199
General Accounting Office (GAO), 171, 172, 173
General Electric, 23
General Motors, 23
"Genetic," "eugenic" and, 195
Genetic testing, xxiii–xxiv, 131, 219n29
Genetic theories, 67, 242n5; antigay bias of, 68
Gengle, Dean, 115
Genitals: masculinization of, xxiii; sense of smell and, 97
Gentlemen's Agreement, 13, 223n7, 228n70
Germ plasm, xvi, xvii, 22
Germ theory, 59, 60, 207
Gesell, Gerhard, 170
Gifted children, 229n84
Gilman, Sander, 99, 217n4, 219n30
Gingrich, Newt, 239n15

Glass, H. B.: on mandatory abortions, 113–14
Goddard, Henry Herbert, 11, 25, 222n11, 224n11; Ellis Island and, 225n14; IQ test and, 56
Goethe, C. M., 37, 41, 55, 233n41
Goldberg, David Theo, 128, 137, 177, 178
Goldstein, Richard, 120, 219n30
Gonadotropins, 110
Good Housekeeping: Coolidge in, 52–53
Gordon, Linda, xxv, 137, 144, 148, 273n6
Gould, Stephen Jay, 217n4
Graefenberg Ring, 161
Graham, Loren R.: on eugenics, 129, 249n6
Grand Jury of King's County: literacy tests and, 10
Grant, Madison, 25, 33, 34, 42, 50, 55, 141, 142; AES and, 231n5; anti-miscegenation laws and, 39; Coolidge and, 236n7; criticism by, 17, 38; Galton Society and, 228n67; on Jewish characteristics, xvi; Jewish immigration and, 18; national suicide and, 210; racial demography and, 218n17; sentimental impulses and, 51, 52; on "yellow and black peril," 39
Great Depression, 149; birth control and, 148; eugenics and, 232n16
Great Migration, xiv, xvii, 4, 232n18
Green, Richard, 62, 113, 118
Greenberg, David, 73, 103, 217n4, 239n15, 243n30
GRID. See Gay-related immune deficiency
Gromacki, Tom, 127
Gulick, Sidney, 11, 15, 16, 223n7
Gunther, Hans F. K., 141, 253n28
"Gynecology of Homosexuality, The" (Dickinson), 94

Habitual criminals, 8, 56, 80
Haitians, 264–65n12, 265n13; sterilization of, 181

Hamer, Dean, xx, 117, 118, 219n29, 220n42, 239n20, 248n60; gay gene and, 60, 248n52; on gay sexual direction, 103; on genetic basis, xxiii–xxiv, 59, 62; on heredity, xi–xii; homophobia and, 121; research by, 63, 64, 66, 88, 120, 238n7; subpoena for, 61

Hamilton, Luther, 165

Hansen, Elias, 81

Harden, Linda Byrd: on Norplant, 183

Harding, Garrett, 48

Harding, Warren G., 10, 24, 52

Hardwick, Senator Thomas: on white people, 19

Harpers, 14, 49

Harris, Eugene, 164–65

Hartmann, Betsy, xxv, 150, 217n4, 264n2, 265n15, 273n7; *BCR* and, 143; Depo-Provera and, 180–81, 264nn5, 12; on investment incentives, 255n66; on pill studies, 256n72

Hastings Center Report: on Depo-Provera, 179

HBF. *See* Human Betterment Federation

Head Start, 211

Hearst, William Randolph, 231n5

Heart and Soul (magazine): advertising in, 183

Helms, Jesse, 235n9

Hereditary Health Laws (1933), 30, 41, 161

Hereditary pauperism, 146

Heredity, xi–xii, 106, 214; defective, 198; masturbation and, 92; theories of, xxii, 102, 103, 114–15

Heredity and Twelve Social Problems (Princeton League of Women Voters), 36

Heredity-based models, xxii

Hermaphroditism, 75, 105, 274n8

Hernandez, Antonia, 168, 217n4

Hernandez, Nancy, 186

Herrnstein, Richard J., xi, 129, 224n10, 253n28; on immigration, xviii, xix; Pioneer Fund and, 48

Heterosexual-as-norm/homosexual-as-disease paradigm, 111–12

Heterosexuality, 70, 115, 213; eugenics and, 207

HEW. *See* Department of Health Education and Welfare

Hill, Robert, 194

Hirschfeld, Magnus, xxii, xxiii, 61, 87, 93, 99, 115, 241n7; biological models and, 85; on causation theories, 82; on legal and social redress, 122; on male inverts, 243n29; medical models and, 84; on third sex, 71

Historical revisionism, 7, 128

Hitler, Adolf, 41, 85, 155, 258n6; eugenics and, 30, 234n49

HIV, 120, 217n2, 264–65n12

Holmes, Oliver Wendell, 81, 199; on reform, 72–73; on sterilization, 135, 239n11

Holmes, S. J., 37

Holocaust, 42, 252n16

Homophobia, xiv, xxii, 61, 87, 115, 119, 124, 209, 211; AIDS and, 120; antidote for, 60; eugenics and, xxvii; increase in, xx; liberation from, 67; persistence of, 121, 123; political right and, 62–63; racism and, 244n40

Homosexuality: aberrance of, 103; acquired, 104; additive approach to, 107; animal, 95; biological, xii, xiii, xxii, 61–62, 67, 68, 102, 103, 105, 121, 212; brain center for, 84, 88; causes of, 59, 88; characterizations of, 70–71, 77, 102, 105; congenital, 61, 84, 102, 103, 104, 114–15; curing, 60, 70, 84, 86, 104, 108, 112, 114, 118, 211, 241n9; decriminalization of, xxii, 77, 84; documentation and diagnosis of, 97; environmental, 102, 122; eugenics and, 74, 77, 102, 104, 212; explanations for, 69, 70; genetic, xiii, xiv, 59, 60, 62–64, 68, 83, 113, 120, 219n29; hereditary, 61, 71–75, 102, 122–23; inate, 105, 114–15; inborn, 61, 123; morality-based approach to, 73; neuroses of, 111; origins of, xx, 92, 211; pathologization of, 70; preventing, 114, 117, 124, 248n58; psychiatric, xxii, 102, 103; race- and class-based configurations of, 90;

Homosexuality *(continued)*: scientific approach to, xiv, 82, 107, 212; situational, 198; sterilization and, 80; tolerance for, 86; treatises on, xxii

Homosexuals: differences with, 96, 120; parents of, 105; procreative duty for, 96; sexual rights for, 87; warm climates and, 98

Hoover, Herbert, 25, 150–51

Hormonal curatives, 110–11

Hormone treatment, 102, 108, 110–11, 112, 118, 247n30

Houppert, Karen: on Norplant, 274n10

House Committee on Immigration and Naturalization, xvi, 24, 29, 33; Box and, 23; Laughlin and, 1, 31, 35

House Committee on the Judiciary, xviii, 47

House Un-American Activities Committee (HUAC), 46

"How Much of Us Is in the Genes?" *(New York Times Book Review)*, xi

H.R. 10384, 13–14, 17

HUAC. *See* House Un-American Activities Committee

Hubbard, Ruth, 196, 200, 217n4

Human Betterment Association, 257n94

Human Betterment Federation (HBF), 135, 155, 156

Human Genome Project, xxiii, 131, 214

Human rights, 61, 121, 129, 175

Hurtt, Rob, 55

Huss, Magnus, 239n15

Hypnosis, 106, 112

Hypothalamus, 59, 88, 116, 211, 247n38; gay, xx, 245n62; of male-to-female transsexuals, 59; sexuality and, 64–65

Hysterectomies, xxvi, xxvii, 61, 76, 136, 173, 179, 251n15; ban on, 170; nonconsensual, 159; tubal ligation and, 261n50

Hysteria, 136

Hysteroepilepsy, 91

Identity: collective, 178; ethnic, 9; gender, 115; national, 9, 208; racial, 9

Idiocy, 75, 79, 135, 136, 139, 271n9

Ignatiev, Noel, 249n7

IHS. *See* Indian Health Service

Illegitimate children: sterilization and, 165

Imagined Communities (Anderson), 1

Imbeciles, 18, 79, 135, 271n9

Immigrants, 74, 81, 121; agriculture and, 15; Anglo-Saxon, 7; Asian, xix, 14; attacks on, xvi–xvii, 134, 214; Chinese, 13, 219n23, 223n7, 228n70; debates and hearings on, xvi; disease and, 24, 210; eugenics and, xiv, xvi, xviii, 6, 13, 28, 29–30, 42, 56, 132; Filipino, 37; illegal, 55, 237n8; intellectual underdevelopment of, 37; Italian, 5; Japanese, 219n23, 223n7, 228n70; Jewish, 18; Mexican, 29, 35, 36–37, 54, 55, 233n41; national defense and, 210; policy on, 4, 53, 219n24, 274n6; restrictions on, xii, xv, xvii, 23, 35, 133, 273n6; Russian, 27; Sikh, 227n50; Southern/Eastern European, 7; unassimilable, 54–55; undocumented, 3, 49, 54

Immigration Act (1917), 4, 22, 24, 25, 52, 56, 81–82, 209; economic terms of, 19; feeblemindedness and, 17; Jewish immigration and, 18; racist intent of, 19–20; rhetoric of, xv; World War I and, 17

Immigration Act (1924), xv, 35, 52, 56, 209, 273n6

Immigration Act (1965), xviii–xix, 45, 48

Immigration and Naturalization Service, 3, 241n44

Immigration Commission, 13

"Immigration Control" (Burch), 203

Immigration laws, xix, 16; ethnic law and, 53; homosexuality and, 241n44; Pioneer Fund and, 47

Immigration Quota Acts (1921–24), 41

Immigration Restriction Act (1924), 28

Immigration Restriction League (IRL), 10, 24, 36, 223n5

Immorality, 165, 259n10

INAH 3. *See* Interstital nuclei of the anterior hypothalamus

Incest: sterilization and, 80

Indian Health Service (IHS), 131; birth control and, 180; Depo-Provera/Norplant and, 183, 184; investigation of, 173; racism at, 171; sterilization and, 172, 209
Indian Women United for Social Justice, 172, 175
Industrial Workers of the World, 144, 226n30
Inequity, 124
Inferiority: Black, 26, 36, 38, 46, 214; intellectual, 34, 214; psychopathic, 18
Insanity, 29, 30, 80, 135, 140, 199, 210, 225n17; Blacks and, 232n16; masturbation and, 92; orificial surgery and, 91–92; sexuality and, 90; sterilization for, 154, 199; women and, 251n15
Institute of Experimental Endocrinology, 117
Institute of Sex Research, 61
Intelligence: measuring, 37, 228–29n80; research on, race-based, xiii
Intermediary races, 71
Intermediate sex, 71
International Congress of Criminal Anthropology, Lombroso at, 77
International Eugenics Conference: sterilization and, 198
International Eugenics Congress, 231n5
International Federation of Eugenic Organizations, 41, 253n27
International Planned Parenthood Federation (IPPF), 181, 255n66, 258n99
"Inter-Racial Council," 23
Interracial relationships, 97, 145
Intersexes, 71
Interstital nuclei of the anterior hypothalamus (INAH 3), 65, 66, 248n60
Inversion, xxii, 70, 71, 86, 87, 90, 104, 105, 109, 240n23, 243n29
IPPF. See International Planned Parenthood Federation
IQ tests, xi, xii, xv, 5, 11, 25, 26, 138, 207, 211, 218n6, 224n10, 228n80; immigration and, xviii
IRL. See Immigration Restriction League

Issues in Reproductive and Genetic Engineering (Degener), 191
IUDs, 161, 169, 183, 191, 202

Jackson, Don, 238n13
Jaggar, Alison, 177
Janiger, Oscar, 113
Japanese-American internees: sterilization of, 160
Jennings, Herbert, 230n103
Jensen, Arthur, 235nn13, 14
Jerke, Bill: on "voluntary" sterilization, 188
Jews: as Alpine Slavs, 26–27; Ashkenazi, 224n10; in asylums, 99; birth control and, 100; discourse on, 99–100; effeminacy of, 100–101; eugenic assessment of, 224n10; homosexuality and, 99, 100; intellectual superiority of, 224n10; massacre of, 17
Jews, The (Fishberg), 229n84
Jim Crow, 42, 97, 162, 208
Johnson, Albert, 23, 25, 31, 33
Johnson, Darlene, 210; birth control and, 185, 186; Broadman and, 187, 188; gender bias and, 187; sentence for, 185, 190, 267n44, 268n53
Johnson, Jed, 160
Johnson, Roswell: on venereal disease, 271n11
Johnson-Reed Act (1924), 24, 31, 46, 210; Laughlin and, 134; passage of, 32, 41
Journal of Heredity, xvi, 5, 12, 16, 23, 135, 222n11, 227n55; on eugenic propaganda, 14; sterilization and, 28; Woods in, 22
Journal of Long-Term Effects of Medical Implants: on Norplant, 182
Journal of Orificial Surgery, xxi, 76, 91, 92, 99
Journal of the American Psychiatric Association: Hubbard in, 200

Kadetsky, Elizabeth: on Pioneer Fund, 235n16
Katz, Jonathan Ned, 217–18n4
Kavinoky, Nadina, 149, 150
Kearny, Frank, 186

Kelves, Daniel, 13, 218n4, 222n14, 223n5
Kemp, Jack, 23
Kempton, J. H., 234n49
Kennedy, Donald, 181
Kennedy, Foster, 200
Kerr, L. A.: on bisexuality and hermaphroditism, 274n8
Kiernan, James: on phrenology, 94
Kilhefner, Don: on E hormone, 113
Kinsey Report, 112
Kirsch, John: on homosexuality, 116
Klein, Jewel, 266n28
Kleptomania, 71
Knox, Howard, 12, 18
Ku Klux Klan: Pioneer Fund and, 46
Kurtenbach, Dick, 190
Kushner, Tony, 123

Labor: immigration and, 20; threat to, 15–16
Lake, George, 104, 116
Lambda Legal Defense and Education Fund, 88, 220n39
Lamm, Richard, 221n1
Lancet (journal), 65
Lang, Theo, 245n62
Larsen, Nils P., 161
Lasker, Mary: Negro Project and, 154
Latham, Kathleen, 110
Laughlin, Harry H., xvi, xvii, 33, 36, 42, 45, 46, 127, 140, 210, 222n11, 250n6, 251n16; AES and, 231n5; *BCR* and, 144; epilepsy for, 230n94; on eugenic and dysgenic breeders, 136; Eugenics Record Office and, 227n65; Galton Society and, 228n67; Immigration Act and, 25; lobbying by, 134; on Mexicans and relief agencies, 35; on sterilization, 29–30, 32, 136, 207–8, 271n14; testimony of, 1, 25, 31
"Lecture on Sexual Perversions, Satyriases, and Nymphomania, A" (Lydston), 74–75
Leland Fikes Foundation, 202
Lesbianism, xiv, 63, 91, 94, 242n6; anatomy of, 95–96; Black, xxii, 96, 101; characterizations of, 83; criticism of, xiii, 96–97; development of, 88–

89; excessive testosterone and, 115–16; gender-transgressing, 94; genetic and hormonal failings and, 114; hormonal experimentation in, 93; hypermasculinity of, 101; political goals of, 122; protecting, 61, 87; research on, 88, 103; role of, 95; scientific assessments of, 211; treatment for, xxi, 61, 90, 101, 102–3, 107, 112–13; in twins, 59; vice and, 92
"Letter to an American Mother" (Freud), 103
LeVay, Simon: on gay men's brain tissue, 65–66; homophobia and, 121; hypothalamus and, xx, 59, 64, 211, 245n62; on INAH 3, 66; research by, 88, 120; sex genes and, 65; on Xq28, 248n60
Levin, Joseph, 169, 170
Levitt, Steven: unborn criminals and, 207, 225n23
Lewin, Tamar: on Norplant, 187
Lewontin, Richard C., 56, 219n30, 238n7
Leys-Stepan, Nancy, 100
LGBT community, 113, 211, 212; AIDS crisis and, 119; CGN and, 101; model rejection by, 121; visibility for, 112
Liberalism, 177, 193; body-by-body approach of, 213; eugenics and, xiv, xxvii, 208; "free choice" doctrine of, 190; individualism and, 178; nationalism and, 55; racism and, 131; self-determination and, 203; sterilization and, 209
Lichtenstein, P. M.: on clitoris and lesbians, 96
Lidbetter, E. J.: on chronic pauper stocks, 51
Literacy tests, 10, 11
Little, C. C., 144, 231n5
Lobotomies, xxii, 61, 102
Lodge, Henry Cabot, 223n5
Lombroso, Cesare, 77, 207, 240n31; born criminals and, 14, 225n23
London *Daily Examiner*: on Hamer, 63
Long Island Rest Home: Murray at, 107

Los Angeles County Medical Center, 173, 261n50, 262n69; sterilizations at, 131, 168, 174

Los Angeles Times, 113; on birth control, 182; on disabled, 195–96

Lydston, G. Frank, xxi, 240n20; cerebellum and, 91; on homosexuality and eugenics, 77; on perverts, 74–75, 76, 162; sterilization statutes and, 78

Macklin, Ruth, 201

Madrigal v. Quilligan, 173, 174, 175

Malaria: treatment for, 202

Male hormone therapy, 109

Males, Mike: quote from, 192–93

Mankind Quarterly (journal), 235n8

Marginalization, 60, 67, 180, 201

Margolese, M. Sydney, 113, 118

Marmor, Judd: on temporal lobe pathology, 62

Marr, Wilhelm, 99

Marriages: common-law, 163; consanguinary, 105; inter-, 39; miscegenetic, 40

Masochism, 99

Massachusetts Audubon Society, 50

Massachusetts Institute of Technology, 22

Masturbation, 66, 91, 93, 98, 136; castration for, 77; eugenic consequences of, 72; heredity and, 92; homosexuality and, 72, 92; insanity and, 92; Jews and, 99; perversion and, 75

Max, Louis, 106, 112

Mayne, Xaviar: on intersexes, 71

McCann, Carole R., xxv, 253n27; on birth control movement, 137; on Davenport, 141; eugenics and, 138, 157; on Sanger, xxvi, 137, 138, 139, 145, 273n6

McCarran-Walter Act (1952), xviii, 46, 218–19n19

McCarthy, Eugene, 221n1

McCarthy, Joseph, 46

McCord, Charles, 162, 259nn10, 15

McGarrah, Robert E., Jr.: on consent forms, 262n69; on tubal ligation and hysterectomy, 261n50

Medicaid, 130, 167, 170, 171; Norplant and, 201; sterilization and, 168

Medical Committee for Human Rights, 175

Medicalization, xxi, 62, 64, 68, 74, 84, 99, 112, 145, 211

Medicojuridicial complex, 209

Mehler, Barry, 24, 218n4, 228n72, 253n27, 271n11; on Galton, 242n6; on Grant, 39; on Pioneer Fund, 48; on Wiggam, 142

"Memorial on Immigration Quotas: To the President, the Senate, and the House of Representatives," 34

Mengele, Joseph: Mankind Quarterly and, 235n8

Menopause: perversion and, 75

Mental illness, 14

Mental incompetence, 12–13, 156, 199–200

Metaphors, 100–101

Mexicans, 231n10; attacks on, 36, 175; labor by, 36–37; relief agencies and, 35; scrutiny of, 32; sterilization of, 175

Michigan State Board of Health, 198

Mill, John Stuart, 258n4

Miller, Adam, 48

Mills, Charles, xiv, 19, 137

Minkowitz, Donna, xxiii, 212, 219n30

Minnesota College Republicans, 127, 205

Minors: sterilization of, 170, 176

Miscegenation, 4, 5, 40

Misogyny, 124, 130

Mississippi appendectomy, xxvi

Mitchell, Alice, 94

Mixed race group, 38, 39

Moll, Albert, 97, 105, 118, 244n55, 245n62; "association-therapy" and, 106; intellectual evolution of, 104; on inversion, 90; on tribadism, 98

Mongolism, 114

Montgomery Community Action Committee, 169

Moore, Arch A.: on Trevor, 48

Moral degenerates, 73, 80; sterilization of, 56, 79, 80, 271n9

Morality, 73, 165, 259n10
Moral turpitude, xxi, 13, 30, 68, 82, 241n44
Moreau, Paul, 73, 98
Morgan, John T.: on Chinese immigration, 6
Moron communities, 198
Morton, Thomas G., 251n15
Mosaic-2000, 194
Mosse, George, xv, 218n4; on eugenic travesties, 100; on nationalism, 55; on racism, 9, 207
Moynihan Report, 260n28
Muller, D.: on curing gays, 65
Murphy, Norman C., 115, 116, 118
Murray, Charles, xi, 129, 193, 224n10; on immigration, xviii, xix; Pioneer Fund and, 48; on poor, 51
Murray, Pauli, 246n16; civil rights and, 107; on homosexuality, 108, 246n21; hormone therapy and, 108, 109
Mussolini, Benito, 155

NAACP, 183
Nacke, Paul, 88
NAFTA, 55
Nashville Board of Health, 165
National Asian Women's Health Organization, 180
National Black Feminist Organization, 176
National Black Women's Health Project, 180, 184
National Cancer Institute, 61
National Center for Health Statistics, 192
National character: biological foundations of, 42
National Committee on Maternal Health, 254–55n58
National Council of Negro Women, 256n79
National Gay and Lesbian Task Force, 220nn40, 42
National Institutes for Mental Health, 89
National Institutes of Health, xx, 70, 63, 66, 212

Nationalism, 22, 213, 218n16; eugenics and, xiv, xv, 4, 6–7, 8, 56, 100, 124, 207, 210; immigration and, xv; masculinity and, 248n58; racism and, 210; war and, 20
National Latina Health Organization (NLHO), 180, 182, 266n24
National Organization for Women (NOW), 178, 180, 190, 213; Norplant and, 269n62; Proposition 187 and, 263n82; self-determination and, 177
National Origins Act (1924), xvii, 5, 24, 139
National Public Radio, 3
National purification, 213–14
National Sharecroppers Week, 109
National stock: degeneration of, 39
National Welfare Rights Organization (NWRO), 170, 176, 177
National Women's Health Network, 180, 192
Native Americans: birth control and, 180; sterilization and, 262n58
Native American Women's Health Education Resource Center, 180, 183
Nativism, 4
Naturalization, 30, 35, 43, 218n19, 223n7, 226n48
Nature versus nurture paradigm, 69, 115, 121, 211
Nazi eugenics, 40, 43, 159–60, 161, 200, 221n44, 258n5; U.S. eugenics and, 41–42, 44, 258n6, 220n44
Negative Population Growth (NPG), 49, 50
Negro People's Committee to Aid Spanish Refugees, 107
Negro Project, 152, 153, 154
Negro Question, xxvii
Nehira, Alice Tenabe, 160, 161, 258n2
Nelson, Alan, 49, 55, 210
Neo-Malthusianism, 49, 146, 147, 157, 164, 204, 251n17
"New Contraceptive Advances Freedom, Responsibility" (Washington Post), 189
New Dealism: eugenics and, 148

New England Journal of Medicine, 247n38

New Leader (labor weekly), 160

Newsweek, 83; on genetic postulations, 67; on Hamer, 63, 88

New York Post: Wiggam in, 142

New York Times, 270n69; on AIDS research, 265n13; on Hamer, 63; on Norplant, 182–83, 187, 266n28

New York Times Book Review, xi, 60

New York Times Magazine: Murray in, 51

Nicolosi, Joseph: on "pre-homosexual" boys, 89

Nightline: Hamer and LeVay on, 59–60, 63

NLHO. *See* National Latina Health Organization

"Nordic Movement in Germany, The" (*Eugenical News*), 40–41

Nordics, 27, 28, 32, 33

Norplant, xiii, xxvii, 179, 190, 195, 202, 205, 209, 220n33, 263–64n1, 264n2, 266n24, 270n70; approval of, 182–83, 188, 191–92; criminal justice system and, 188–89; as curative, 131; mandatory, 188, 267n45; Medicaid and, 201; monitoring with, 187; opposition to, 184, 191–94, 267n44; politics of, 185; problems with, 185, 192, 266n28; removal of, 192, 266n28, 267nn30, 35, 269n59; side effects of, 181, 182, 183, 184, 266n23; sterilization and, 185, 194; support for, 128–29, 183, 189; using, 181–82, 191, 204, 265–66n23; welfare and, 128, 189–90, 191, 193, 194, 266n28, 270n70

North Carolina: sterilization in, 161, 163, 164, 165

North Carolina Eugenics Commission, 154, 260n29

North Carolina Supreme Court: *Carolina Association* and, 200

NOW. *See* National Organization for Women

NOW Legal Defense Fund, 190

NPG. *See* Negative Population Growth

Nuremburg trials: eugenicists and, xviii, 42, 159–60

NWRO. *See* National Welfare Rights Organization

Nymphomania, 76, 91, 99

Oakland Tribune, 59

Office of Economic Opportunity (OEO), 170, 171

Office of Research Integrity (National Institute of Health), 63

Oltman, Rick, 237n8

Onanism, 72, 239n8

Oophorectomy, 251n15

Oral progesterone, 152

Orebaugh, D. A., 39

Oregon Court of Appeals: Cook and, 187

Orificial surgery, xxii, 76, 102, 136, 211; -as-cure-all mentality, 99; for insane women, 91–92

Ortho Pharmaceutical, 149

Osborn, Frederick, 45

Osborn, H. F., 34, 37, 231n5

"Our Immigration Laws from the Viewpoint of Eugenics" (Ward), 8

Overpopulation, 49, 154, 255n60

Page Law (1875), xvi, 13, 218n12

Paquin Middle-Senior High School, 192, 193, 194

Paragraph 175: defenders of, 84

Parents and Friends of Lesbians and Gays (PFLAG), 67

Parkland Memorial Hospital: Norplant at, 270n70

Parsons, Arlene, 231n11

Passing of the Great Race, The (Grant), 39, 51–52, 55, 141

Paternalism, 131, 133, 172, 177, 204; eugenics and, 87, 161–70; racial- and class-based, 167

Pathologization, xiii, xxiii, 62, 70, 107, 123

Path to National Suicide—An Essay on Immigration and Multiculturalism, The (American Immigration Control Foundation), 55

Patrick, Kerry, 190–91, 269n61

Patton, Cindy: on disease reportage, 120

Pauling, Linus: on defective genes, 200
Pauper stocks, 51, 259n15
PBCF. *See* Pennsylvania Birth Control Federation
Pearl, Raymond, 146, 254n44
PEB. *See* Population-Environment Balance
Pederasty, 73, 98
Pennsylvania Birth Control Federation (PBCF), 148, 149, 254n57
Perinatal stress, 248n56
Perkins, Muriel, 95, 96
Perkins, Penny: on sexism, 88
Perversion, 105, 121, 162, 198; acquired, 75, 106; congenital, 75; eugenics and, xxii; transmission of, 75–76
Perversions of the Sex Instinct (Moll), 106
Perverts, 71, 72, 76, 99, 162; physical and moral, 74–75; sterilization of, 56, 79, 80, 271n9
Petchesky, Rosalind, 177, 200
PFLAG. *See* Parents and Friends of Lesbians and Gays
Philadelphia Association of Black Journalists, 268n58
Philadelphia Committee on Lunacy, 251n15
Philadelphia Inquirer, 189, 268n58
Phin, Helen, 129
Phrenology, 94
PHS. *See* U.S. Public Health Service
Physiological treatment, 107
Pierce, Clovis, 167–68
Pillard, Richard C., 88, 115, 248n52; on CGN, 89–90; on lesbianism and bisexuality, 59; research by, xx, 67, 89, 118, 120
"Pill Planted in Body Turns Weak Effeminate Youths into Strong Virile Men" (*World-Telegram*), 108
Pilocarpine, 112
Pincus, Gregory, 151, 152
Pioneer Fund, 44–46, 210, 234n3, 235n9; eugenics and, 49; FAIR and, 3, 48–49, 235n16; immigration legislation and, 47; publications by, 235n8; racial betterment and, xviii, 3

Pittsburgh Courier: on sterilization laws, 153
Pivot of Civilization, The (Yerkes and Terman), 138, 143, 199
Planned Parenthood, 127, 139, 176, 182, 203; Gamble and, 152; housing and health care issues and, 263n82; population control and, 213
Planned Parenthood Federation of America (PPFA), 154, 180, 252n16, 256n84, 257n94
Planned Parenthood of Kansas: Patrick and, 190
Ploetz, Alfred: Nazi eugenics and, 43
Politics, 14, 20; biological determinism and, 123
Poor, 148; eugenics and, 4, 130; heredity-based attacks on, 254n45; sterilization of, 131; unfit and, 146
Popenoe, Paul, 135, 251n13, 271n11; differential birth rate and, 141; sterilization and, 28
Population Committee, 50
Population control, 129, 152, 158, 210, 213, 236n21, 262n59; coercive, 178; in developing nations, 204; eugenics and, 130, 203; involuntary, 130; neo-eugenic, 191
Population Council, 181, 182
Population-Environment Balance (PEB), 50
Population pressure, 150, 236n21, 252n16
Populism, 23, 208, 227n65
Poverty, 74, 124, 128, 197; alleviating, 149, 188, 189–90; birth control and, 178, 188; Black, 189–90; choice and, 190; contraception and, 191; culture of, 52, 128, 158, 172; cycle of, 129, 192; dysgenesis and, 15; eugenics and, 161–70; overpopulation and, 49; teen pregnancy and, 268n50
"Poverty and Norplant: Can Contraception Reduce the Underclass?" (*Philadelphia Inquirer*), 189
PPFA. *See* Planned Parenthood Federation of America
Pratt, E. H., 77, 91
Prenatal stress: homosexuality and, 117

Prenatal testing, 128, 196, 197–98
Press, Black, 153
Prevention (magazine): on fluoride and homosexuality, 237n1
Princeton League of Women Voters, 36
Privacy rights, 186
Procreation, 5, 86, 164; punishment and, 185–89
Proposition 187 (1994), xiii, 55, 210, 237n8, 263n82, 274n6; FAIR and, 202, 208; passage of, xix, 3–4
Prostitution, 14, 135, 260n21
"Protest against Laws Authorizing the Sterilization of Criminals and Imbeciles, A" (Boston): quote from, 206
Prudential Insurance Company of America, 259n9
Pseudobiology, 54
Pseudohermaphroditism, 108
Psychiatry, 101, 107
Psychobiological hypothesis, 112
Psychosurgery, 102
Psychotherapy, 102–3; hormone inundation and, 111
Public Aid Commission: sterilization and, 259n21
Public Art Works, 237n8
Public Citizen, 171, 261n50
Puerto Rican Emergency Relief Administration, 151
Puerto Rican Socialist Party, 175
Puerto Rico, 38; birth control in, 255n63
"Purpose of Eugenics, The" (Duvall), 146

Queerness, 60
Queers, 248n1; discarding, xxiii; experimentation on, 100; image of, 116; medicalization of, xxi, 62; nature versus nurture paradigm and, 69, 211; pathology and, xiii, 123; social control over, xxii
Quick fixes, 115, 136
Quinacrine, 202, 208, 209, 272–73n1; side effects of, 273n3
Quota Act, 36
Quotas, 5, 36, 37, 47, 218–19n19, 232n21

Race: eugenics and, 41; sterilization and, 159
Race demonization, 54
Race mixture, 38, 39, 40, 47–48
Race suicide, 34, 141, 255n60
Racial betterment, xviii, xxii, 3, 42, 45, 147; eugenics and, 56; impediments to, 208
Racial Contract, The (Mills), 19
Racial deterioration, 40, 141
Racial Elements of European History, The (Gunther), 141
Racial hygiene, xxvi, 43, 133
Racial Hygiene (Rice), 227n61
"Racial Limitation of Bolshevism, The" (Woods), 22
Racial maternalism, 157
Racism, 38, 110, 124, 128, 129, 134, 136, 145, 166, 178, 204, 209, 210, 218n16, 237n3; bolstering, 4, 90; environmental, 197; eugenics and, xv, 6, 7, 45–46, 67, 130, 140, 210, 218n17, 222n11, 227n50, 258n6; exclusive, xv, 6; genetics and, 4; homophobia and, 244n40; immigration and, xv, xviii, 4; inclusive, xv; liberalism and, 131; nationalism and, 210; as scavenger ideology, 207; stereotyping and, 9; sterilization and, 80, 167, 172; top-down, 235n14; as visually centered ideology, 9. *See also* Scientific racism
Radzinski, John M., 47–48, 236n7
Rape: sterilization and, 80
Rassenhygiene, 40, 43–44, 249n6
Rats, 95; studying, 116–17
Raymond, Janice, 128
Reagan, Ronald, 235n9
Reason: eugenics, rationality and, 53
"Reason IX—The Preservation of Civilization" *(BCR)*, 150
Reed, Adolph, xxvii, 19, 20, 193; Immigration Act and, 24; on liberals, 129
Reform: eugenics and, 208; immigration and, xviii
Refugees, xvii, 16
Reidner, Kermit: on homosexuality cure, 104
Reilly, Philip, 154, 165, 218n4

Relf, Katie, 169
Relf, Mary Alice, 168, 169, 170
Relf, Minnie Lee, 168, 169, 170
Relf v. Weinberger, 170
Religious tolerance, 122
Reproduction: regulation of, 28;
    technologies, 128, 159
Reproductive rights, xxvi, 145, 184, 204
Rice, Thurman, 227n61
Richardson, Charles, 109
Ricketts, Wendell, 68, 95, 110
Right to access, 177
*Rising Tide of Color against White World
    Supremacy, The* (Stoddard), 25, 55,
    144
Robertson, Pat, 23
Rock, John, 151
Rockefeller, John D., Jr., 227n65, 231n5
Rodman, James, 116
Rodrique, Jessie, 153
Roeder, F.: on curing gays, 65
Roe v. Wade (1973), 207, 225n23, 242n5
Rogers, Don, 3
Roosevelt, Eleanor, 107
Roosevelt, Franklin D., 34, 151
Rose, Florence, 153
Ross, Edward A., 34, 226n30; race
    suicide and, 255n60
Rothman, Barbara Katz, 190, 221n48
Rothman, Stanley, 218n6
Rowland, Alen, 172
RU 486, 184, 205; side effects of, 273–
    74n9
Rukeyser, Muriel, xxviii
Rushton, J. Philippe, 217n2

Sacramento Church Federation: on
    Puerto Ricans, 38
Sadism, 99
Saleeby, C. W., 222n11, 224n10
Salk Institute, 64, 212
Sanchez, Marie, 172, 173
San Francisco Chronicle: on illegal
    immigrants, 55
San Francisco Examiner, 60, 63
San Francisco Spokesman, 153
Sanger, Margaret, 125, 153, 254n51,
    257nn88, 98, 99, 269n61, 272n15;

Address of Welcome quote, 125; anti-
    Semitism of, 254n43; *BCR* and, 5, 143;
    birth control and, 49, 128, 139, 143,
    144, 184, 213, 249–50n8; Catholics
    and, 256n82; choice and, 137; on
    contraceptive service, 259n17;
    criticism of, 130, 141, 142–43; on
    disabled, 147, 200; eugenics and, xxv,
    129–30, 137, 138, 140, 146, 157–58, 203,
    213; on feebleminded, 140; on fit, 136;
    Gamble and, 149, 150, 151, 152, 155, 161;
    HBF and, 156; HEW regulations and,
    177; on immigration restrictions,
    273n6; IQ data and, 207; legacy of,
    127–28, 130–31, 204; pension plan
    and, 190; political identification of,
    157; population control and, 130;
    poster of, 127; racism of, xxvi, 130,
    145; sterilization and, 154, 155, 156,
    158, 199; "wrongful life" suits and,
    257n93
Saphists, 73
Satyriasis, 76, 91
Save Our State (S.O.S.), 3
Save the Children campaign, 247n44
Sawhill, Isabel, 191
Scapegoat Generation (Males), 192–93
Scapegoating, xix, 14, 215, 236n27
Schaeffer, William Donald, 193
Schmidt, Gunter, 84, 117–18
Scholastic Aptitude Test, 26
Schoofs, Mark, 219n29
Science, xx, 84; bogus, 210–14; civil rights
    and, 83; as consensus builder, 129;
    eugenics and, xiv; internationalism,
    xxiv–xxv; judiciary and, 79; legal and
    social redress and, 122; in news stories,
    217n3; -as-savior paradigm, 122
Science (journal): LeVay in, 64
Science of Desire, The (Hamer and
    Copeland), xi, xiii, 60, 62, 103
Scientific-Humanitarian Committee, 61
Scientific racism, xi, xvi, xix, xxiv, 9, 40,
    136, 145, 157, 204, 213, 231n5; circula-
    tion and validation of, 162–63;
    eugenics and, 56, 161–70, 222n11; gene
    pool and, 3–4; metaphors and, 100–
    101; understanding, xxi

Scott, James Foster: on Jews and venereal disease, 99; on manifestations of sex, 98; on masturbation, 72; on onanism, 239n8; on sexual instinct, 72

Scott, Ralph, 234n3

Segregation, 198, 199, 208, 272n15

"Selection, the Only Way of Eugenics" (Whiting), 198

Self-determination, 137, 177, 203

Self-restraint, 87, 102

Senate Subcommittee on Labor and Public Welfare, 231n11

Sentimental impulses, 51, 52

Sex: discourse on, 71; impulses, 91; manifestations of, 98; nonprocreative, xxii, 72, 83, 92

Sex and Germs: The Politics of AIDS (Patton), 120

Sex crimes, xxi–xxii, 77, 81, 116, 240n43

Sexism, 88, 172

Sexologists, 91

Sexology (magazine), xxi, 86, 98, 110, 246n4; contributors for, 104; on homosexuality and whistling, 243n29; on sexual inversion, 109

Sexual acts, mapping out, 97–98

Sexual Instinct: Its Use and Dangers, as Affecting Heredity and Morals: Essentials to the Welfare of the Individual and the Future of the Race, The (Scott), 72

Sexual Inversion (Ellis), 85, 93, 102, 240n20

Sexuality, xxi, 104, 212; biologically determined, 110, 123; Black, 163; characterization of, 83; conventional, 139; hypothalamus and, 64–65; insanity and, 90; of native children, 244n52; regulating, 5; tropical climates and, 98

Sexual orientation, 242n8; biologic component of, 121; centers, 89, 114; changes in, 111, 116, 124; determining, xx; genetic component of, 121; measuring, 212; treatments for, xxiii

Sexual Question, The (Forel), 95

Sex Wars: Sexual Dissent and Political Culture (Duggan), 122

Shaw, Margery, 196

Sheehan, Katherine, 182

Shend, Suzanne: on Haitians and HIV, 264–65n12

Shilts, Randy, 83, 241n1

Shockley, William, 257n88, 258n6; Pioneer Fund and, 235n14; on sterilization, 269n61

Sickle cell anemia, 196

Sierra Club, 49, 50, 263n82

Silliman, Jael: on RU 486, 273n9

Sims, J. Marion, 258n7

Sin: ideology of, 123

Sissy Boy Syndrome, The (Green), 62

Sixth International Neo-Malthusian and Birth Control Conference, 146, 198; Sanger at, 125, 139

60 Minutes, 186, 187

Skinner v. Oklahoma (1942), 250–51n12

Smith, Andrea, 184

Smith, Julia Holmes, 92, 93

Smith, Ruth Proskauer, 156

Social and Rehabilitation Services, 170

Social Darwinism, xii, 162, 248n1

Socialism: nationalism and, 55

Socialization, 67, 114

Socioeconomic factors: eugenics and, 164

Sociological classification, 99

Sodomy, 73, 80, 81; laws, 78, 123

Somerville, Siobhan, xxi, 39, 71, 90, 97

S.O.S. See Save Our State

South Carolina Medical Association, 168

Southern Poverty Law Center, 169

Southern White Citizens Councils, 46

Standardized tests, 10, 11

Standard Oil Company of New Jersey, 23

State Hospital for Negroes, 163

State Hospital for the Insane, 251n15

Station for Experimental Evolution, 253n27

Statistical Abstract of the United States, 27

Steinach, E., 84

Steinberg, Stephen: on liberalism and racism, 131

Stepan, Nancy Leys, 218n4
Stereotypes, 9, 85–86, 89, 110
Sterilization, xiii, xiv, xv, 111, 151, 153, 164, 169, 183, 191, 199, 203, 204, 254n57, 269n61; abuse of, 170, 176, 200, 255n66, 262n66; chemical, 202; class-based, 155, 160; compulsory, xxv, xxvi, 32, 49, 79, 81, 127, 130, 133, 134, 148, 157, 163, 166, 173, 174, 175–76, 195, 200, 201, 207–8, 212–13, 259n21, 271n14; consent for, 76, 82, 173, 174, 178, 262n69; demographics of, 168; disability-based, 155, 160; eugenic, xxvi, xxvii, 42, 78, 79, 81, 133, 134, 136, 147, 153, 158, 160, 161–70, 171, 194; German policies on, 258n5; illegal, 134; Japanese-American, 160; limiting, 155; long-term, 180; mass, 172; nontherapeutic, 171; postpartum, 161, 172; race-based, xxvi, 160; side effects of, 260n29; states' right to, 135, 239n11; support for, 128–29, 130, 155, 163; surgical, 180; temporary, 179; therapeutic, 81, 156, 171, 271n9; voluntary, 168, 188
Sterilization-for-probation deals, 88, 209, 214
Sterilization for the Human Betterment (Popenoe), 251n13
Sterilization hearings: legal counsel for, 135
Sterilization laws, 41, 78, 80, 133, 153, 206; American, 233n44; compulsory, 79, 134, 135, 271n9; early, 195, 233n35; eugenics and, 56; gays and, 240n39; German, 233n44; model, 29
Stevens, Frances, 88
Stilbesterol, 110, 246n27
Stith, Rosetta, 193
Stocking, George, 9–10
Stoddard, Lothrop, 42, 55, 144, 228n67
Stone, Harlen, 250n11, 250–51n12
Stonewall uprising, xiii, 112
"Story of a Subsidized Family, or How to Populate the Earth with the Unfit, The" (BCR), 147
Straight gene, xxiv, 70
Stryker, Susan, 246n4

Studies in the Psychology of Sex (Ellis), 85
Study of American Intelligence, A (Brigham), 26, 28
Stumpf, Walter E., 65
Sub-Committee on Selective Immigration, 33
Subpersons, 137
Suggested Experimental Federal Law, 198, 231n5
Sunnen, Joseph, 255n70
Supreme Court of Utah: Walton and, 80–81
Swaab, Dick, 57, 59, 64
Synderman, Mark, 218n6
Syphilis, 79, 162, 246n28; Blacks and, 259n9; sterilization and, 56, 199

Taft, William Howard, 10, 250n11
Tanton, John, 54
Tapia, Victoria, 187
Tay-Sachs disease, 196
Technofixes, xxv, xxvii, 128, 129, 178, 197, 209, 249n3; encrypted use of, 194; eugenics and, 131; reproductive, 159
Teen pregnancy, 192, 194, 209; poverty and, 268n50; social costs of, 193
Temporal lobe pathology, xxiii, 62
Temporary Assistance to Needy Families, xxvii, 191, 221n2
"Ten Good Reasons for Birth Control" (BCR), 150
Terman, Lewis, 28, 29, 138, 229nn84, 90; IQ data and, 207
Terry, Jennifer, 90–91
Testosterone, 112; excessive, 115–16; injections of, 110
Thind, Bhagat Singh, 227n60
Third International Congress of Eugenics, 41, 253n27
Third Reich: Eugenical News on, 41; eugenics and, xviii, xxiv; gays and, 245n62; race laws of, 40; sterilization during, 30
Third sex, xxii, 71
Thirtieth Annual Meeting Luncheon (PPFA), 154, 155
Thompson, Warren S., 272n15

"Three Disappointing Cases" (Smith), 93
Three Percent Restrictive Act (1921), 5, 24, 52
Tide of Immigration, The (Warne), 10
Tijerina, Katherine Harris, 172
Time magazine, 148
"To the Men of America" (Trumball), xi
Transgendered people, xiii, xiv; interventions on, xxi; scientific assessments of, 211
Trevor, John, 47, 236n7
Trevor, John, Sr., 47, 48
Tribadism, 73, 97–98
Trumball, Rose, xi
Tubal ligations, xxvi, xxvii, 167, 168–69, 172, 173, 179, 183, 209; hysterectomy and, 261n50; invasiveness of, 240n34; nonconsensual, 159
Tuberculosis, 84, 136, 162, 225n17, 256n72, 259n9
Tuskegee experiments, 111, 246n28
20/20: Hamer and LeVay on, 59–60
Twins, xx, 59, 67, 120; CGN and, 90; studies on, 88–89, 118, 238n7, 242nn6, 7

Ulrichs, Karl Heinrich, 243n29; on homosexuality, 61; on legal and social redress, 122; on urning, 71
Unborn, protecting, 186
Unborn criminals: theory of, 207
Unconceived: rights of, 186
Unfit, 158, 160, 194, 208, 225n17; Blacks, 162; breeding by, 147; fit and, 140, 141; poor and, 146; sterilization of, 133
UNFPA. See United Nations Fund for Population Activities
Union Law School, 74
United Market stores: boycotting, 237n8
United Nations Fourth World Conference on Women, 203
United Nations Fund for Population Activities (UNFPA), 181, 182
United Native Americans, 172

United States v. Ajkoy Kumar Mazumdar (1913), 226n48
United States v. Balsara (1910), 226n48
United States v. Bhagat Singh Thind, 226n48
University of California Medical Center, 114
University of Oklahoma Health Sciences Center, 194
"Unprofitable Children: Are These Bodies Fit Temples for Immortal Souls?" (BCR), 147
Upjohn: Depo-Provera and, 179, 180, 181
Uranists, 104–5, 106, 244n55
Urban Institute, 191
Urbanization, 74, 121
Uri, Connie, 171–72, 218n4, 261–62n54
Urning, 71
U.S. Agency for International Development, 182
USA Today, 60, 63
U.S. Public Health Service (PHS), 12, 225n17; Tuskegee experiments and, 246n28
Utah State Prison: Walton at, 80

Vaid, Urvashi, 220n42
Vasectomies, xxvii, 61, 76, 220n33, 240n34, 268n55
Venereal disease: Jews and, 99
Venerologists, 117
Vice, 75, 76, 92
Village Voice, xxiii, 212
Villerme, Louis-Rene, 244n37
Virginia Law (1924): test case for, 135
Vogt, William, 269n61
Voluntary motherhood, 127
Volunteer Sterilization Bonus Program, 257n88
Von Krafft-Ebing, Richard, 71–72, 93
Von Verschner, Ottmar: Mankind Quarterly and, 235n8

Wadsworth, Glen, 98
Walker, Mrs. Virgil, 167–68, 170
Walter, Francis, xviii, 46
Walton, Esau, 80–81

Ward, Robert DeCourcey, 33, 34, 36, 210, 222nn10, 11, 225n14; American race and, xvii, 211; Chinese/Japanese immigration and, 223n7; on deportation, 226n46; on eugenics, 7, 8, 15, 16; literacy test and, 10; on mentally defective aliens, 226n37; mythical national type and, 8; percentage limitation principle and, 228n70; race and nationality and, 218n16; on refugees, 16; sentimental impulses and, 51, 52

Warden Davis v. Walton (1929), 80–81

WARN. *See* Women of All Red Nations

Warne, Frank Julian, 10, 12, 223n6

War on Poverty, 170

Washington County Hospital: sterilization at, 166

Washington Post, 60; on contraceptives, 189; on Hamer, 63; on Norplant, 185

Wattenberg, Ben J., 22–23

Weirick, C. A., 91, 92, 243nn19, 24

Welfare, 50, 159, 171, 194, 204, 255n59, 263n82, 269n61; birth control and, 187; contraception and, 189; Depo-Provera and, 128; Norplant and, 128, 189–90, 191, 193; reliance on, 130; sterilization and, xxvi, 161, 165, 166, 167, 185, 257n88, 260n33; termination of, xviii

Welfare Reform Act (1996), 193, 269n67

Western Hemisphere Quota Act, 37

What Is Eugenics (Darwin), 144

Whiteness: membership requirements for, 19

White supremacy, xv, 3, 5, 124, 132, 160, 210, 213; eugenics and, 207, 258n6

Whiting, P. W., 198, 199, 271n11

Whitman Walker Clinic, 66

Whitney, Leon F.: on Depression and eugenics, 232n16

"Whose Country Is This" (Coolidge), 52–53

Wiggam, Albert E., 141–42

Wilde, Oscar, 85

Wilder, Tracy, 187–88

Wilson, E. O., 62, 116

Wilson, M. G., 26

Wilson, Woodrow, 20, 225n22

Winsco, James P.: on tolerance, 86

Witelson, Sandra, 64, 121

Wobblies, 144, 226n30

Woman of Valor: Margaret Sanger and the Birth Control Movement in America (Chesler), xxv, 138

Women against Imperialism, 264n5

Women of All Red Nations (WARN), 183

Women's Economic Agenda Project, 180

Women's Hospital (Los Angeles County Medical Center), 261n50

Wood, H. Curtis, Jr., 156, 157, 168, 260n33

Woods, Frederick Adams, 22

Woodside, Moya, 161, 163, 165, 258n6, 259n17; on general consent, 162; on sterilization, 164

Workfare, 52

World Bank, 182

World Health Organization, 181, 192

World Population Emergency Campaign, 257n94

World-Telegram: on hormone treatment, 108

Wrongful birth/life litigation, 196, 197

Wyeth-Ayerst: suit against, 266n28

X chromosome, 64, 117

Xenophobia, xv, xix, xxiv, 5, 16, 22, 43, 99, 209, 210, 222n11; eugenics and, 4, 7, 23, 56; immigration and, 49, 50

Yerkes, Robert M., 25, 26, 28, 138, 228n77; data from, 27, 207; on military psychology, 42

Young Women's Health (magazine): advertising in, 183

Zero Population Growth, 49, 213

Nancy Ordover holds a Ph.D. in ethnic studies from the University of California at Berkeley. She lives and works in New York City.